The Application of EEG in Neurorehabilitation

The Application of EEG in Neurorehabilitation

Guest Editor

Ryouhei Ishii

MDPI

Basel • Beijing • Wuhan • Barcelona • Belgrade • Novi Sad • Cluj • Manchester

Guest Editor
Ryouhei Ishii
Graduate School of
Rehabilitation Science
Osaka Metropolitan
University
Osaka
Japan

Editorial Office
MDPI AG
Grosspeteranlage 5
4052 Basel, Switzerland

This is a reprint of the Special Issue, published open access by the journal *Brain Sciences* (ISSN 2076-3425), freely accessible at: https://www.mdpi.com/journal/brainsci/special_issues/1PVIBV3LSI.

For citation purposes, cite each article independently as indicated on the article page online and as indicated below:

Lastname, A.A.; Lastname, B.B. Article Title. *Journal Name* **Year**, *Volume Number*, Page Range.

ISBN 978-3-7258-5939-9 (Hbk)
ISBN 978-3-7258-5940-5 (PDF)
https://doi.org/10.3390/books978-3-7258-5940-5

Contents

About the Editor

Ryouhei Ishii

Ryouhei Ishii is a professor and academic researcher in occupational therapy at Osaka Metropolitan University. His research is focused on the clinical application of neurophysiology, particularly using electroencephalography (EEG) to investigate pain, motor function and recovery mechanisms in neurological and psychiatric disorders. His work includes projects on EEG-based biomarkers for stroke rehabilitation, brain wave analysis for mental health interventions and the development of healthcare technologies to support patient recovery. He has a strong interest in applying advanced data analysis techniques to brain signals and is actively involved in academic publishing and research ethics.

Preface

As the Guest Editor, it is with great pleasure that we introduce this Reprint on the "Application of EEG in Neurorehabilitation". The field of neurorehabilitation stands at a fascinating and critical intersection of clinical science, neuroscience and engineering. Its fundamental goal—to restore function and improve quality of life for individuals with neurological impairments—remains a profound challenge. In recent years, electroencephalography (EEG) has evolved from a primarily diagnostic tool into a dynamic and versatile instrument capable of driving therapeutic innovation. The motivation behind this collection was to capture the breadth and depth of this evolution, showcasing how EEG is being leveraged not just to observe brain activity but to actively understand, guide and enhance the process of recovery. This Reprint consolidates pioneering research that highlights the synergy between advanced EEG methodologies and their practical application across a spectrum of neurological conditions.

The scope of the articles presented in this publication is a testament to the field's vibrant and multidisciplinary nature. The works traverse a wide range of technological frontiers. Several contributions demonstrate the powerful integration of machine learning and EEG, developing sophisticated algorithms for subject-independent brain–computer interfaces (BCIs) and optimizing the timing of non-invasive brain stimulation in stroke patients. These studies push the boundaries of how we interpret complex neural data to create responsive, personalized interventions. Furthermore, the research explores the combination of EEG with other powerful modalities, including transcranial magnetic stimulation (TMS) and non-invasive brain stimulation techniques like tACS and tDCS. This integration is pivotal for developing closed-loop systems that can modulate brain plasticity with increasing precision. Other works address essential methodological questions, examining how different signal processing techniques can enhance the accuracy and reliability of BCI systems, a critical step for their translation into clinical practice.

The ultimate aim of these technological advancements is to address pressing clinical needs. The articles collected here are firmly grounded in real-world applications, targeting conditions such as stroke, cerebral palsy and age-related cognitive and motor decline. The research delves into crucial functional domains, from investigating the neural underpinnings of stepping accuracy to prevent falls to exploring brain connectivity patterns during working memory tasks in immersive virtual reality environments. One particularly insightful study examines how the therapist–patient relationship and verbal encouragement influence neural networks in children with cerebral palsy, reminding us that technology must serve to enhance the human element of care. Collectively, these studies aim to identify robust, neurophysiological biomarkers of motor recovery, cognitive function and even the effects of reward processing, laying the groundwork for more objective and data-driven rehabilitation protocols.

This Reprint is addressed to a diverse audience, including clinicians on the front lines of neurorehabilitation such as physiatrists, neurologists and occupational therapists, as well as neuroscientists, biomedical engineers and data scientists who are building the next generation of therapeutic tools. It is our hope that this collection will not only serve as a valuable reference on the current state of the art but also inspire new avenues of inquiry and foster collaboration across disciplines. The path toward restoring brain function is complex, but as the work in this Reprint demonstrates, EEG-based approaches are illuminating that path with unprecedented clarity. We extend our sincerest gratitude to the authors for their outstanding contributions and to the readers for their interest in this exciting and impactful field of study.

Ryouhei Ishii
Guest Editor

brain sciences

MDPI

Editorial

Editorial: The Application of EEG in Neurorehabilitation

Ryouhei Ishii [1,2,3]

[1] Department of Occupational Therapy, Osaka Metropolitan University Graduate School of Rehabilitation Science, 3-7-30 Habikino, Habikino 583-0872, Osaka, Japan; ishii@psy.med.osaka-u.ac.jp; Tel.: +81-72-950-2111; Fax: +81-72-950-2131
[2] Department of Rehabilitation, Osaka Kawasaki Rehabilitation University, 158 Mizuma, Kaizuka 597-0104, Osaka, Japan
[3] Department of Psychiatry, Osaka University Graduate School of Medicine, 2-2 Yamadaoka, Suita 565-0871, Osaka, Japan

1. Introduction

The field of neurorehabilitation stands at a pivotal juncture, driven by a confluence of advancements in neuroscience, engineering, and data science. At the heart of this transformation is the imperative to move beyond one-size-fits-all therapeutic models towards personalized, evidence-based interventions that are precisely tailored to the individual's unique neural landscape. Electroencephalography (EEG), a technology with a long and storied history in clinical neurology, has re-emerged as a powerful and versatile tool in this new era. Its non-invasive nature, exceptional temporal resolution, portability, and relative affordability have positioned it as an ideal modality for exploring the dynamic processes of neural recovery and guiding rehabilitation strategies. This Special Issue of *Brain Sciences*, titled "The Application of EEG in Neurorehabilitation", brings together a collection of cutting-edge research that collectively illuminates the expanding role of EEG in assessing brain function, optimizing therapeutic interventions, and pioneering the next generation of assistive technologies [1].

For decades, the primary application of EEG in neurology was diagnostic, focused on identifying epileptiform activity or characterizing sleep stages. However, the last two decades have witnessed a paradigm shift. The integration of sophisticated signal processing algorithms, machine learning, and mobile brain/body imaging (MoBI) has unlocked the potential of EEG to serve not just as a diagnostic tool but as a dynamic window into the functioning brain [2]. This is particularly crucial in neurorehabilitation, where the ultimate goal is to understand and promote neural plasticity—the brain's intrinsic ability to reorganize itself following injury from events such as stroke, traumatic brain injury (TBI), or in the context of neurodevelopmental conditions such as cerebral palsy [3].

Despite this progress, significant gaps in our knowledge and technical challenges remain. One of the most persistent hurdles is inter-subject variability; neural signatures that are robust in one individual may be weak or absent in another, complicating the development of universally effective brain–computer interfaces (BCIs) or biomarkers [4]. Furthermore, translating findings from controlled laboratory settings into complex, real-world clinical environments remains a formidable challenge [5]. There is a pressing need for more robust, interpretable models that can handle the noisy, high-dimensional data characteristic of EEG recordings during movement and interaction. Additionally, while we have become proficient at identifying neural correlates of motor function, the neurophysiological underpinnings of cognitive and motivational factors—which are critical for patient engagement and successful rehabilitation—are less understood.

This Special Issue directly confronts these challenges, presenting a multifaceted view of how researchers are pushing the boundaries of what is possible with EEG. The nine articles featured herein span a wide spectrum of applications, from foundational investigations into neural mechanisms to the development of advanced computational models and the exploration of novel therapeutic paradigms.

2. Advancing Brain–Computer Interfaces and Computational Models

A central theme of this collection is the relentless pursuit of more accurate and reliable BCI systems, which hold immense promise for restoring function to individuals with severe motor impairments. Three papers in this Special Issue tackle key challenges in BCI development from a computational perspective. Mao et al. (Contribution 1) address the critical issue of subject-independent EEG recognition in motor imagery tasks. Their proposed Mirror Contrastive Learning with Sliding Window Transformer (MCL-SWT) model is a significant step forward, leveraging neurophysiological principles (contralateral event-related desynchronization) to inform the architecture of their deep learning model. By improving performance on benchmark datasets, their work paves the way for BCIs that require less user-specific calibration, a major barrier to clinical adoption.

Complementing this, Suresh et al. (Contribution 2) explore the synergy between EEG-based machine learning and non-invasive brain stimulation (NIBS). Their study demonstrates that transcranial direct current stimulation (tDCS) not only facilitates motor recovery but also enhances the decodability of movement-related EEG signals in chronic stroke patients. This finding is profound, suggesting that NIBS can be used to "prime" the brain, making neural states more distinct and therefore easier to classify. This creates a potential positive feedback loop where stimulation improves both neural function and the BCI's ability to assist that function. The work by Adolf et al. (Contribution 3) further underscores the complexity of the BCI pipeline, conducting a systematic investigation into how different preprocessing choices—such as artifact rejection, filtering, and transfer learning—impact classification accuracy. Their results highlight that there is no single "best" pipeline; rather, processing techniques must be carefully selected and tailored to specific network architectures and even individual subjects, reinforcing the need for personalized approaches.

3. Uncovering the Neural Signatures of Motor Control and Recovery

To build effective rehabilitation strategies, we must first understand the fundamental neural mechanisms of motor control and how they are altered by injury. Okuyama et al. (Contribution 4) provide a blueprint for this by investigating the cortical networks underlying stepping accuracy in healthy individuals using mobile EEG. Their work, integrating eLORETA-ICA with microstate analysis, reveals a distributed network where the anterior cingulate cortex (ACC) plays a central role in performance monitoring. This provides a normative model against which patient populations can be compared, offering potential targets for therapeutic intervention aimed at improving gait and preventing falls.

Lacerda et al. (Contribution 5) apply a similar neurophysiological lens directly to stroke survivors. Their longitudinal study powerfully demonstrates the clinical utility of EEG- and TMS-derived biomarkers. They identify the theta/alpha ratio (TAR) in the lesioned hemisphere as a robust predictor of motor outcomes and a potential signature of adaptive compensatory processes. This research exemplifies the goal of translational neuroscience: to identify reliable biomarkers that can predict recovery trajectories and help clinicians stratify patients for targeted therapies, moving us closer to a precision medicine model for stroke rehabilitation.

4. Exploring the Psycho-Social and Contextual Dimensions of Rehabilitation

Successful rehabilitation is not merely a matter of retraining motor circuits; it is deeply influenced by motivation, social interaction, and cognitive engagement. Two particularly innovative articles in this Special Issue use EEG to explore this often-overlooked dimension. Yamauchi et al. (Contribution 6) investigate how the brains of children with cerebral palsy respond differently to verbal encouragement from their mothers versus their physical therapists. Their findings suggest that the source and tone of voice elicit distinct patterns of neural activity, with a mother's voice potentially engaging more internal, self-referential processing. This work highlights the critical importance of the therapeutic alliance and communication in pediatric rehabilitation.

Similarly, Ozel (Contribution 7) uses a virtual reality (VR) environment to dissect how the brain processes social versus non-social cues during a working memory task. By creatively applying Leading Eigenvector Dynamics Analysis (LEiDA) to EEG data, the study reveals that social avatar cues trigger a unique brain state characterized by heightened connectivity in self-referential and memory networks. This suggests that embedding social elements into VR-based rehabilitation tasks could significantly enhance cognitive engagement and, by extension, therapeutic outcomes.

5. Foundational Research for Future Therapies

Finally, this Special Issue includes two contributions that lay the groundwork for future therapeutic development. Morales Fajardo et al. (Contribution 8) conduct a focused investigation into the effects of high-definition transcranial alternating current stimulation (HD-tACS) on beta oscillations in healthy older adults. By demonstrating frequency-specific and spatially focal modulation of brain activity, their work provides crucial mechanistic insights that are necessary for designing effective tACS protocols to enhance motor control in aging and neurological populations. In a comprehensive review, Koloski et al. (Contribution 9) make a compelling case for using neurophysiological markers, such as cortico-striatal beta oscillations related to reward processing, to inform preclinical TBI research. They argue that these markers are highly translatable between animal models and humans and could be used to develop and test targeted neuromodulation therapies for cognitive impairments, bridging the gap between basic science and clinical application.

6. Future Directions and Concluding Remarks

The papers collected in this Special Issue paint a vibrant picture of a field in rapid evolution. Looking ahead, several key research trajectories emerge. The first is the continued push towards closed-loop systems. The ultimate goal is to create integrated systems that can simultaneously record EEG signals, decode user intent or brain state in real time, deliver a therapeutic intervention (e.g., functional electrical stimulation, robotic assistance, or NIBS), and then use subsequent EEG feedback to adapt the intervention on the fly. The work on ML classification and NIBS modulation provides the foundational components for such systems.

A second major direction is the expansion of rehabilitation in naturalistic environments. Technologies such as mobile EEG and VR are critical for moving assessment and therapy out of the constrained laboratory and into settings that more closely mimic the challenges of daily life [2]. Future research should focus on developing robust algorithms that can function amidst the noise and complexity of real-world environments.

Third, the theme of personalization will only grow in importance. As demonstrated by Lacerda et al. and Adolf et al., group-level findings must be refined to account for individual differences in brain injury, neurophysiology, and cognitive profiles. This will require larger datasets, more sophisticated individual-level modeling, and the integration of multimodal data, including genomics, structural imaging, and behavioral assessments [3].

The research presented here collectively demonstrates that EEG is far more than a passive measurement tool; it is an active agent in the rehabilitation process. It provides the essential feedback for BCIs, offers biomarkers to guide treatment selection, and serves as a direct target for neuromodulatory therapies. As we continue to refine our ability to interpret the brain's complex electrical symphony, EEG will undoubtedly become an even more indispensable component of the neurorehabilitation toolkit, helping us to better understand, restore, and enhance human brain function.

Conflicts of Interest: The author declares no conflicts of interest.

List of Contributions

1. Mao, Q.; Zhu, H.; Yan, W.; Zhao, Y.; Hei, X.; Luo, J. MCL-SWT: Mirror Contrastive Learning with Sliding Window Transformer for Subject-Independent EEG Recognition. *Brain Sci.* **2025**, *15*, 460. https://doi.org/10.3390/brainsci15050460
2. Suresh, R.E.; Zobaer, M.S.; Triano, M.J.; Saway, B.F.; Grewal, P.; Rowland, N.C. Exploring Machine Learning Classification of Movement Phases in Hemiparetic Stroke Patients: A Controlled EEG-tDCS Study. *Brain Sci.* **2025**, *15*, 28. https://doi.org/10.3390/brainsci15010028
3. Adolf, A.; Köllőd, C.M.; Márton, G.; Fadel, W.; Ulbert, I. The Effect of Processing Techniques on the Classification Accuracy of Brain-Computer Interface Systems. *Brain Sci.* **2024**, *14*, 1272. https://doi.org/10.3390/brainsci14121272
4. Okuyama, K.; Maeda, K.; Yamauchi, R.; Harada, D.; Kodama, T. Neural Oscillatory Mechanisms Underlying Step Accuracy: Integrating Microstate Segmentation with eLORETA-Independent Component Analysis. *Brain Sci.* **2025**, *15*, 356. https://doi.org/10.3390/brainsci15040356
5. Lacerda, G.J.M.; Silva, F.M.Q.; Pacheco-Barrios, K.; Battistella, L.R.; Fregni, F. Adaptive Compensatory Neurophysiological Biomarkers of Motor Recovery Post-Stroke: Electroencephalography and Transcranial Magnetic Stimulation Insights from the DEFINE Cohort Study. *Brain Sci.* **2024**, *14*, 1257. https://doi.org/10.3390/brainsci14121257
6. Yamauchi, R.; Ito, H.; Kitai, K.; Okuyama, K.; Katayama, O.; Morita, K.; Murata, S.; Kodama, T. Effects of Different Individuals and Verbal Tones on Neural Networks in the Brain of Children with Cerebral Palsy. *Brain Sci.* **2025**, *15*, 397. https://doi.org/10.3390/brainsci15040397
7. Ozel, P. Dynamic Neural Network States During Social and Non-Social Cueing in Virtual Reality Working Memory Tasks: A Leading Eigenvector Dynamics Analysis Approach. *Brain Sci.* **2025**, *15*, 4. https://doi.org/10.3390/brainsci15010004
8. Morales Fajardo, K.; Yan, X.; Lungoci, G.; Casado Sánchez, M.; Mitsis, G.D.; Boudrias, M.-H. The Modulatory Effects of Transcranial Alternating Current Stimulation on Brain Oscillatory Patterns in the Beta Band in Healthy Older Adults. *Brain Sci.* **2024**, *14*, 1284. https://doi.org/10.3390/brainsci14121284
9. Koloski, M.F.; Menon, R.; Krasnyanskiy, V. Neurophysiological Markers of Reward Processing Can Inform Preclinical Neurorehabilitation Approaches for Cognitive Impairments Following Brain Injury. *Brain Sci.* **2025**, *15*, 471. https://doi.org/10.3390/brainsci15050471

References

1. Tsiamalou, A.; Dardiotis, E.; Paterakis, K.; Fotakopoulos, G.; Liampas, I.; Sgantzos, M.; Siokas, V.; Brotis, A.G. EEG in Neurorehabilitation: A Bibliometric Analysis and Content Review. *Neurol. Int.* **2022**, *14*, 1046–1061. [CrossRef] [PubMed]
2. Cioffi, E.; Hutber, A.; Molloy, R.; Murden, S.; Yurkewich, A.; Kirton, A.; Lin, J.P.; Gimeno, H.; McClelland, V.M. EEG-based sensorimotor neurofeedback for motor neurorehabilitation in children and adults: A scoping review. *Clin. Neurophysiol.* **2024**, *167*, 143–166. [CrossRef]
3. Jin, W.; Zhu, X.; Qian, L.; Wu, C.; Yang, F.; Zhan, D.; Kang, Z.; Luo, K.; Meng, D.; Xu, G. Electroencephalogram-based adaptive closed-loop brain-computer interface in neurorehabilitation: A review. *Front. Comput. Neurosci.* **2024**, *18*, 1431815. [CrossRef]

4. Vavoulis, A.; Figueiredo, P.; Vourvopoulos, A. A Review of Online Classification Performance in Motor Imagery-Based Brain–Computer Interfaces for Stroke Neurorehabilitation. *Signals* **2023**, *4*, 73–86. [CrossRef]
5. Wang, P.; Cao, X.; Zhou, Y.; Gong, P.; Yousefnezhad, M.; Shao, W.; Zhang, D. A comprehensive review on motion trajectory reconstruction for EEG-based brain-computer interface. *Front. Neurosci.* **2023**, *17*, 1086472. [CrossRef] [PubMed]

brain sciences

MDPI

Article

MCL-SWT: Mirror Contrastive Learning with Sliding Window Transformer for Subject-Independent EEG Recognition †

Qi Mao [1], Hongke Zhu [2], Wenyao Yan [1], Yu Zhao [1], Xinhong Hei [2] and Jing Luo [2,*]

[1] School of Data Science and Engineering, Xi'an Innovation College of Yanan University, Xi'an 710100, China; maoqi_xiancxxy@163.com (Q.M.)

[2] Human-Machine Integration Intelligent Robot Shaanxi University Engineering Research Center, School of Computer Science and Engineering, Xi'an University of Technology, Xi'an 710048, China

* Correspondence: luojing@xaut.edu.cn

† This paper is an extended version of our paper published in Luo, J.; Mao, Q.; Shi, W.; Shi, Z.; Wang, X.; Lu, X.; Hei, X. Mirror contrastive loss based sliding window transformer for subject-independent motor imagery based EEG signal recognition. In Proceedings of the Human Brain and Artificial Intelligence, 4th International Workshop, HBAI 2024, Held in Conjunction with IJCAI 2024, Jeju Island, Republic of Korea, 3 August 2024.

Abstract: Background: In brain–computer interfaces (BCIs), transformer-based models have found extensive application in motor imagery (MI)-based EEG signal recognition. However, for subject-independent EEG recognition, these models face challenges: low sensitivity to spatial dynamics of neural activity and difficulty balancing high temporal resolution features with manageable computational complexity. The overarching objective is to address these critical issues. **Methods**: We introduce Mirror Contrastive Learning with Sliding Window Transformer (MCL-SWT). Inspired by left/right hand motor imagery inducing event-related desynchronization (ERD) in the contralateral sensorimotor cortex, we develop a mirror contrastive loss function. It segregates feature spaces of EEG signals from contralateral ERD locations while curtailing variability in signals sharing similar ERD locations. The Sliding Window Transformer computes self-attention scores over high temporal resolution features, enabling efficient capture of global temporal dependencies. **Results**: Evaluated on benchmark datasets for subject-independent MI EEG recognition, MCL-SWT achieves classification accuracies of 66.48% and 75.62%, outperforming State-of-the-Art models by 2.82% and 2.17%, respectively. Ablation studies validate the efficacy of both the mirror contrastive loss and sliding window mechanism. **Conclusions**: These findings underscore MCL-SWT's potential as a robust, interpretable framework for subject-independent EEG recognition. By addressing existing challenges, MCL-SWT could significantly advance BCI technology development.

Keywords: BCI; EEG; motor imagery; contrastive learning; sliding window transformer

1. Introduction

Brain–computer interface (BCI) is a human–computer interaction technology that enables information exchange and control between the brain and external devices. It has received extensive attention in the fields of brain science and artificial intelligence [1,2]. BCI systems operate by translating brain activity into information or commands that devices can understand or execute, thereby facilitating the connection between an individual's mind and the external world [3]. According to the different ways of capturing signals, BCI technologies can be classified into invasive BCI and non-invasive BCI [4]. An ethical issue often encountered with BCI as an assistive technology is that while BCI can improve the quality of life for people with disabilities and their loved ones, BCI devices may increase the stigma of disability associated with the individual, which may influence potential users not to use BCI [5]. Non-invasive BCI refers to BCI technology that does not require surgery or direct contact with brain tissue, such as Electroencephalography (EEG). Motor imagery (MI) is a crucial paradigm in the development of BCI. The MI paradigm in BCI involves the mental imagery of limb movements or other body parts, allowing users to control external devices or perform specific tasks through mental imagery, which has been applied in post-stroke motor function rehabilitation [6,7]. The MI EEG captures EEG signals during MI tasks, which is convenient, simple, flexible, non-invasive, and minimally demanding on the environment [8]. Achieving accurate decoding of EEG signals is crucial for the development of stable MI-BCI-based rehabilitation systems.

With the advancement of machine learning technologies, many machine learning algorithms have been applied to EEG signal decoding [9]. For example, linear discriminant analysis (LDA) [10] and support vector machine (SVM) [11], but these methods require the manual design of feature extraction methods. As an improved variant of CSP [12], the Filter Bank Common Spatial Pattern (FBCSP) [13] overcomes the limitation of relying on a single specific frequency band and further enhances performance through a multi-band analysis strategy. End-to-end feature extraction and classification methods based on CNNs have shown excellent performance in the field of MI-BCI [6,14,15]. Schirrmeister et al. explored the impact of different CNN model designs on the classification of EEG signals from MI [16]. Although these methods perform well in subject-dependent classification tasks, their performance in subject-independent scenarios still requires improvement. Notably, most current MI-based BCI applications rely on subject-dependent settings, which poses significant challenges in practical implementations: on the one hand, each new subject requires the collection of a substantial amount of individual data for model calibration. On the other hand, classification models designed for specific subjects often perform poorly on other subjects [17,18]. This undoubtedly increases system complexity while reducing practicality and real-time performance. Therefore, the research of subject-independent BCI systems focuses on solving the problem of generalization between subjects, based on which BCI systems can effectively adapt to multiple subjects, expanding their scope of application and enhancing their adaptability.

A decrease in frequency energy in relevant areas of the cerebral cortex is known as event-related desynchronization (ERD), and an increase in frequency energy is known as event-related synchronization (ERS) [19]. ERD/ERS in the contralateral sensorimotor cortex was analyzed using standardized electroencephalographic protocols. The baseline mu-rhythm power (8–13 Hz frequency range) was established during the pre-movement interval from 500 ms to 100 ms before movement onset. Post-movement cortical dynamics were evaluated through non-overlapping 300 ms epochs spanning 0–1500 ms after movement initiation. ERD/ERS magnitude was quantified as the percentage power decrease relative to baseline, with statistical significance assessed through cluster-based permutation testing (5000 iterations) to address multiple comparison challenges. ERD and ERS exhibit spatial variability across subjects, and Independent Component Analysis (ICA) [20] is commonly employed to mitigate this spatial variability [21]. Kim et al. [21] extracted ERD/ERS using EEG-independent electron sources obtained by ICA and used them for ERSP-based movement classification. Zhao et al. [22] proposed an end-to-end deep domain adaptation method to improve the classification performance of a single subject using valid information from multiple subjects (source domains) in order to address the significant differences between subjects. Xu et al. [23] proposed an online pre-alignment strategy based on Riemannian Procrustes Analysis (RPA) [24] for aligning the EEG distributions of different subjects prior to the training and inference process,thereby reducing cross-dataset variability. Traditional BCI classification utilizes discriminative features representing ERD/ERS to classify MI [25]. Most methods use a transfer learning strategy to address inter-subject differences. MI of the left or right hand induces ERD in the brain's contralateral sensory areas and simultaneously induces ERS in the ipsilateral sensory areas [19]. Therefore, monitoring the locations of ERD/ERS phenomenon in the brain's sensorimotor areas provides an important foundation for classifying MI EEG signals. The spatial variance in ERD/ERS phenomenon across different subjects presents significant research challenges [25]. Consequently, improving the model's ability in recognizing ERD and ERS is crucial for enhancing the performance of subject-independent MI EEG signal recognition.

While deep learning techniques have made significant advancements in MI based EEG signal recognition, these models are often viewed as black boxes. Their decision-making processes are difficult to interpret and comprehend, leaving us uncertain whether they are grounded in the identification of ERD/ERS patterns within the sensorimotor regions of the brain, which is a key factor in classifying MI. In addition, the success of the transformer model in natural language processing and computer vision has drawn increasing attention from BCI researchers. The attention mechanism has become a focal point in research on MI EEG decoding, demonstrating a strong ability to extract discriminative features from EEG signals [8,26,27]. Luo et al. proposed a shallow Transformer model to capture discriminative segments in multi-subject EEG signals through a multi-head self-attention mechanism to improve the classification accuracy of MI tasks [8]. Altaheri et al. designed an attention-based temporal convolutional network (ATCNet) model, which combines a multi-head self-attention mechanism and a spatio-temporal convolutional network to efficiently extract the spatio-temporal features of MI-EEG and improve the classification performance with fewer parameters [28]. But the computational complexity of the global multi-head self-attention mechanism in transformer models increases quadratically with the length of the input sequence. At present, most transformer models applied in MI-EEG recognition utilize short input sequences, which restricts the temporal resolution of the features obtained [27].

To address the issues mentioned above, we propose a mirror contrastive learning-based sliding window transformer (MCL-SWT) to enhance subject-independent MI-based EEG signal recognition. This manuscript is a conference extended paper, and the original paper was published in [29].

The main contributions of this paper are as follows:

(1) According to research in the field of neurology, mental imagery of left or right-hand movements induces ERD in the contralateral sensorimotor regions of the brain. Based on this finding, this study introduces mirror contrastive learning (MCL), which enhances the accuracy of identifying the spatial distribution of ERD/ERS by contrasting the original EEG signals with their mirror EEG signals obtained by exchanging the channels of the left and right hemispheres.

(2) For subject-independent EEG recognition, a temporal sliding window transformer (SWT) is proposed to achieve high temporal resolution in feature extraction while maintaining manageable computational complexity. Specifically, the self-attention scores are computed within temporal windows, and as these windows slide along the EEG signal's temporal dimension, the information from different temporal windows can interact with each other.

(3) The experimental results on subject-independent based MI EEG signal recognition demonstrate the effectiveness of MCL-SWT method. Parameter sensitivity experiments demonstrated the robustness of the MCL-SWT model, and feature visualization and ablation studies further validated the effectiveness of the MCL method.

2. Method

This section provides the detailed description of the MCL-SWT model, and the framework of the MCL-SWT model is illustrated in Figure 1.

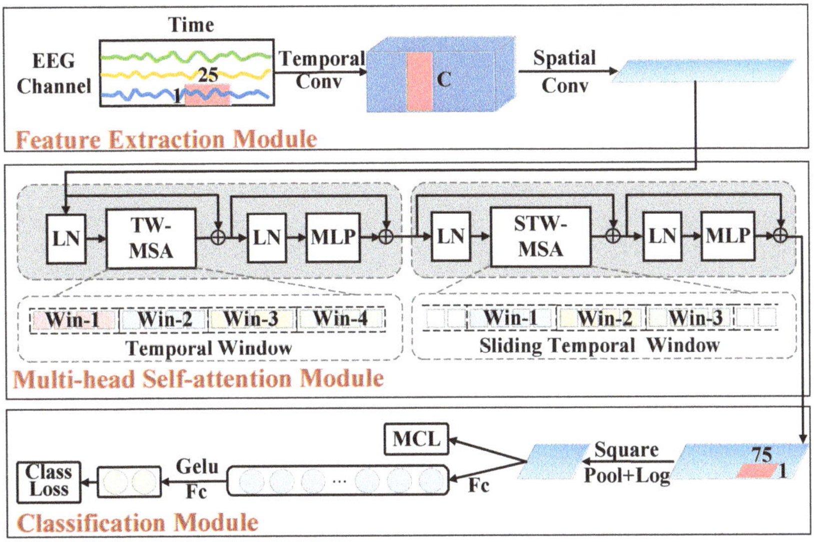

Figure 1. Illustration of the overall architecture of the MCL-SWT model.

2.1. Notations and Definitions

The EEG data collected in a session is defined as $\{(X_i, Y_i)\}_{i=1}^{n}$, where n represents the number of EEG trials. $X \in R^{T\times C}$ denotes the EEG trial, T is the number of sampling points and C is the number of EEG channels. The raw EEG signal X serves as the input of the MCL-SWT model, with a batch size of B, resulting in an input data dimension of $B \times 1 \times T \times C$.

2.2. Sliding Window Transformer Model

For subject-independent EEG recognition, a temporal sliding window transformer (SWT) is proposed to achieve high temporal resolution in feature extraction while maintaining manageable computational complexity. The overall architecture of the SWT model consists of three parts: a feature extraction module, a multi-head self-attention module, and a classification module, as shown in Figure 1.

2.2.1. Feature Extraction Module

Initially, drawing inspiration from the bandpass filters used in the filter bank common spatial pattern algorithm [13], we used a temporal convolution C_t with a kernel size of $T \times 1$ to process temporal information. Larger convolution kernels help to achieve a wider range of signal transformations at this stage. Secondly, in order to simulate the spatial filter in the common spatial pattern algorithm, a spatial convolutional layer C_s with a convolutional kernel size of $1 \times C$ is used to fuse the individual channels of the EEG. Then, a batch normalization layer BN is added, which is a technique used to normalize the model parameters with the aim of increasing the training speed of the neural network model, improving the convergence of the model, and reducing the occurrence of overfitting [30]. This CNN-based feature extraction module is able to learn shallow features with spatio-temporal information and is able to reduce the dimensionality of the shallow spatio-temporal features, thus reducing the amount of computation required to perform the self-attention computation afterwards, which can be described as:

$$F = \mathrm{BN}(C_s(C_t(X))) \tag{1}$$

2.2.2. Multi-Head Self-Attention Module

A multi-head self-attention module based on temporal sliding window is introduced to capture the global temporal dependencies of EEG features. This module consists of two blocks: a multi-head self-attention block based on a temporal window and another based on a sliding temporal window, with the primary distinction being the segmentation of the temporal window, as illustrated in Figure 2. The different segmentations of the temporal window enable effective information exchange between sequences in neighboring windows.

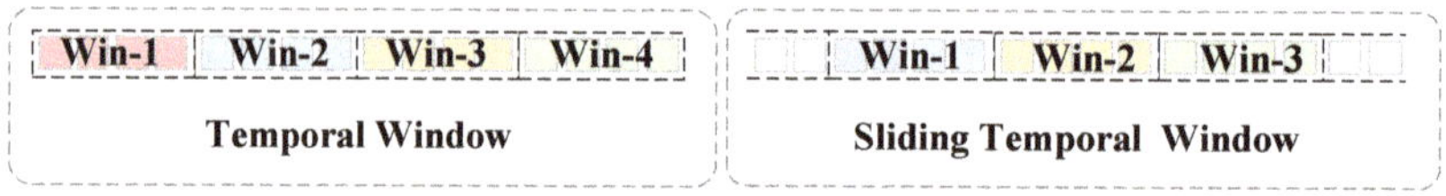

Figure 2. Illustration of temporal window and sliding temporal window.

Each multi-head self-attention block incorporates a sequence of components: a layer normalization (LN), a temporal window-based multi-head self-attention layer (TW-MSA), a layer normalization, and a multi-layer perceptron layer (MLP), along with two additional residual network structures to enhance feature learning and stability. Each layer is detailed as follows:

(1) LN: The LN layer performs normalization along the feature dimensions of each sample, ensuring that the output features are normalized as described in [31]. This normalization enhances the network's ability to learn and generalize, particularly when handling data with complex feature distributions.

(2) TW-MSA: The TW-MSA layer is designed to capture local temporal dependencies by computing self-attention scores within a specified temporal window. As illustrated in Figure 1, the attention score computation is restricted to each local window, which improves the model's sensitivity to local information while enhancing computational efficiency. Based on empirical evidence, the window size for TW-MSA is set to 8. Additionally, a residual connection is incorporated alongside the LN and TW-MSA layers to promote stable gradient flow and model optimization. The specific calculation steps for TW-MSA are as follows:

$$Q_i = W_i^Q \cdot \mathrm{LN}(F) \tag{2}$$

$$K_i = W_i^K \cdot \mathrm{LN}(F)$$

$$V_i = W_i^V \cdot \mathrm{LN}(F)$$

$$h_i = \mathrm{Attention}(Q_i, K_i, V_i) = \mathrm{Softmax}\left(\frac{Q_i K_i^T}{\sqrt{d_k}}\right) V_i \tag{3}$$

$$O = \mathrm{concat}(h_1, \cdots, h_H) W^O + F \tag{4}$$

where $W_i^Q \in R^{d_{\mathrm{model}} \times d_q}$, $W_i^K \in R^{d_{\mathrm{model}} \times d_k}$, $W_i^V \in R^{d_{\mathrm{model}} \times d_v}$ and $W^O \in R^{(H \cdot d_v) \times d_{\mathrm{model}}}$ are the corresponding weight matrices, h_i represents the i-th attention head, H denotes the number of attention heads, d_{model} is the dimension of the input embedding, and d_q, d_k, d_v represent the dimensions of the query, key, and value, respectively.

(3) MLP: After the second LN, an MLP is added, which consists of a positionally fully connected layer (FC), a GELU activation function (G) [32] and another positionally

fully connected layer. Additionally, a residual connection is incorporated alongside the LN and MLP layers to promote stable gradient flow and model optimization. The MLP layer is used to perform further nonlinear transformations and feature extraction on the attentional outputs to enhance the representation and fitting ability of the model, and also to increase the depth of the model's network so that the model is able to learn deeper levels of abstract features.

$$A = O + \mathrm{FC}(\mathrm{G}(\mathrm{FC}(\mathrm{LN}(O)))) \tag{5}$$

To enhance the correlation between local windows and capture the global temporal dependencies of EEG signals, another multi-head self-attention block based on sliding temporal window is added after the first multi-head self-attention block. This block facilitates information interaction by cyclically shifting local windows. As illustrated in Figure 2, the local window is shifted to the right in steps of half the window size, $M/2$, and self-attention scores are then computed within the newly formed temporal windows. By integrating two multi-head self-attention block, the model reduces the computational cost of attention weight calculations while significantly improving its ability to model long-range temporal dependencies. This combination effectively enhances both the efficiency and performance of the model.

The computational complexity of the global multi-head self-attention mechanism is quadratic in the length of the input vectors, as it is necessary to compute the attentional weights between each input position:

$$\Omega(MSA) = 4LD^2 + 2L^2D \tag{6}$$

where L is the input vector length and D is the input dimension. Whereas in multi-head self attention model with sliding window (WMSA), the window size is fixed to M, then the computational complexity is linearly related to the input vector length.

$$\Omega(WMSA) = 8LD^2 + 4MLD \tag{7}$$

Therefore, when the input vector length is much larger than the window size, multi-head self attention model with sliding window has an obvious advantage in reducing the computational complexity.

2.2.3. Classification Module

A square non-linear function is first added to the classification module for activation. Then, an average pooling layer is applied to reduce the temporal feature dimensions, after which a log activation function is added. Then, the EEG features with global temporal dependencies are input into the fully connected layer for classification, and finally a SoftMax function is applied to compute the predicted probabilities. The category with the highest predicted probability value is selected as the classification result of the EEG signals. As the output of the multi-head self-attention module is A, the output of the classification module is:

$$\mathrm{P} = \mathrm{Softmax}(\mathrm{FC}(\log(\mathrm{AvgPool}(A \circ A)))) \tag{8}$$

Finally, the predicted probabilities for the mirror EEG signals (described below) corresponding to left MI and right MI are swapped, and then added to the predicted probabilities of the original EEG trial to obtain the final predicted probabilities, as described in [8]:

$$[p_l, p_r] = [p_l^o + p_r^o] + [p_r^m + p_l^m] \tag{9}$$

2.3. Mirror Contrastive Learning

In EEG signals, when people perform left or right hand MI, ERD occurs in the contralateral sensorimotor region of the brain, while ERS occurs in the ipsilateral region. Therefore, monitoring the locations of ERD/ERS phenomenon in the sensory regions of the brain is an important basis for classifying MI EEG signals. However, existing deep learning models function primarily as black boxes, making it challenging to identify ERD/ERS phenomenon, which presents significant challenges for research. To address this issue and enhance the localization ability of MI recognition models, we propose mirror contrastive learning. The MCL increases the distance in feature space between EEG signals containing ERD from different sides of the sensorimotor cortex (negative samples pair), while reducing the distance between EEG signals containing ERD from the same side of the sensorimotor cortex (positive samples pair). In this way, the model learns to effectively locate the ERD phenomenon.

2.3.1. Mirror EEG Signal

The mirror EEG signal is constructed to increase the number of negative sample pairs in the mirror contrastive learning. This is achieved by swapping the EEG channels between the right and left brain hemisphere regions to create the mirror EEG signals [8,33], as shown in Figure 3. For instance, the data from the original C3 channel is exchanged with the C4 channel, and vice versa. The electrode positions of the mirror EEG signals are the same as those of the original EEG signals in a mirror, hence the term "mirror EEG signals". This approach ensures that the ERD/ERS phenomenon appears on the contralateral side when comparing the mirror EEG signal with the original EEG signal. Additionally, it is crucial to assign the opposite label to the mirror EEG signal relative to the original EEG signal. For example, if the original EEG label is "left hand", the label for the mirror EEG signal will be "right hand".

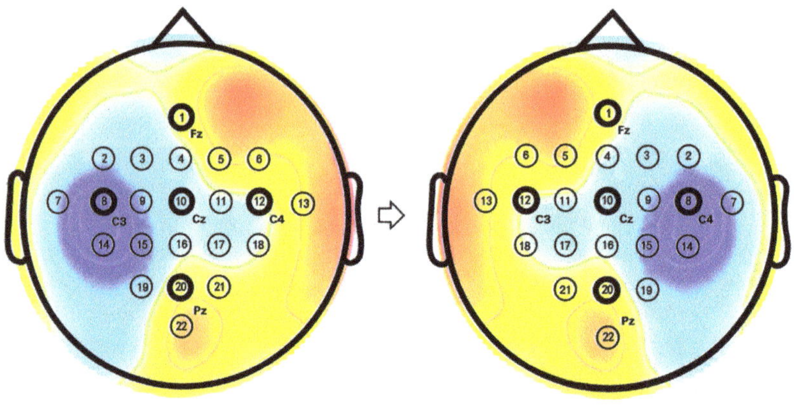

Figure 3. Mirror EEG signals are created to generate negative samples pair for MCL.

Initially, both the mirror and original EEG signals are used in the training process, effectively doubling the number of training samples through this data augmentation technique. Simultaneously, the mirror and original EEG signals are paired as negative samples and applied in the mirror contrastive loss function.

2.3.2. Mirror Contrastive Loss Function

As described in the literature [34], the contrastive loss method guides the model to optimize the feature representation during training by comparing positive sample pairs (similar features) with negative sample pairs (dissimilar features). Specifically, the model is encouraged to bring the features of the positive sample pairs closer together while pushing the features of the negative sample pairs farther apart. However, the successful construction of positive and negative sample pairs is an important factor in contrast loss. By constructing mirror EEG signals, we apply the mirror contrastive loss to enhance the model's sensitivity to ERD/ERS locations by contrasting the original EEG signals with their corresponding mirror EEG signals as shown in Figure 4.

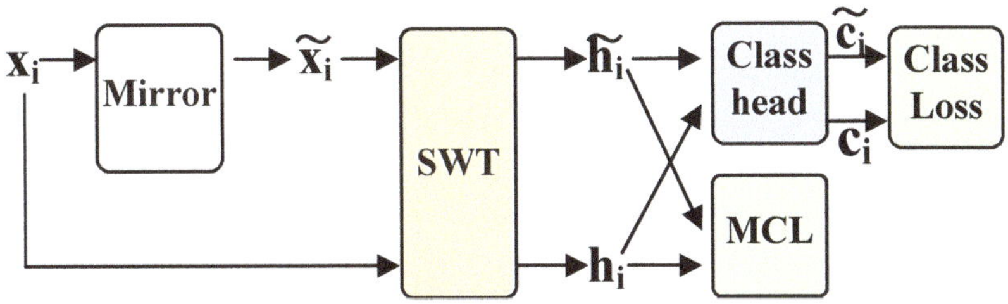

Figure 4. MCL-SWT Model training framework.

As in conventional contrastive learning methods, samples originating from the same MI task category are treated as positive sample pairs, while those originating from different MI task categories are treated as negative sample pairs in this study:

$$\mathcal{L}_o = \sum_{i,j \in OE} \max\left(\left(\alpha + g_{ij}\left(D_{ij} - \beta\right)\right), 0\right) \tag{10}$$

where α controls the degree of separation between positive sample pairs, while β determines the boundary between positive and negative sample pairs. D_{ij} represents the Euclidean distance between the pair of samples i and j. The variable g_{ij} is assigned a value of 1 for positive sample pairs and -1 for negative sample pairs. And OE denote the original EEG signal set. This loss function offers robustness to noise since it permits a certain degree of tolerance and does not overly penalize the model for minor similarities arising from noise.

Additionally, to emphasize the distinct locations of the ERD/ERS phenomenon between the original EEG signal and its mirror counterpart, an additional contrastive loss is applied to the sample pairs formed by the original EEG signal and mirror EEG signal, as follows:

$$\mathcal{L}_m = \sum_{i \in ME, j \in OE} \max\left(\left(\alpha + g_{ij}\left(D_{ij} - \beta\right)\right), 0\right) \tag{11}$$

where ME denote the mirror EEG signal set. Therefore, the mirror contrative loss function is given as:

$$\mathcal{L}_{mc} = w_o \mathcal{L}_o + w_m \mathcal{L}_m \tag{12}$$

where w_o and w_m denote the weights of each loss. By the application of mirror contrastive loss function in the training of model, the model's ability to accurately localize ERD/ERS locations is enhanced by contrasting the original EEG signals with the mirror EEG signals. Specifically, the model is encouraged to push the original EEG signal and its mirrored version apart in the feature space. The mirror contrastive loss is applied to the feature after the log activation function as shown in Figure 1.

2.3.3. Model Training Loss

For MI EEG recognition, the cross-entropy loss was used as the loss function:

$$\mathcal{L}_c = -\frac{1}{N}\sum_{i=1}^{N} y_i \ln(p_i) \tag{13}$$

where N is the total number of MI EEG trials, y_i is the true label of the ith MI EEG trial, and p_i is the probability of the ith MI EEG trial category predicted by the model.

Therefore, the final loss function of the model training is the sum of the mirror contrastive loss and the classification loss:

$$\mathcal{L} = \mathcal{L}_c + \mathcal{L}_{mc} \tag{14}$$

The training framework is given in Figure 4. To provide a clear description, we show the pseudo-code for MCL-SWT in Algorithm 1.

Algorithm 1 The MCL-SWT Method

Input: The training set $\{X, Y\}$, hyperparameters α, β, w_o, w_m, max number of iterations I_{max}.
Output: Model parameters W.
1: initialize W.
2: **for** i=1 to I_{max} **do**
3: **for** l=1 to l_d **do**
4: sample a batch of original EEG signal $\{(\mathbf{X}^O, \mathbf{Y}^O)\}$ from set $\{X, Y\}$;
5: construct a batch of mirror EEG signal $\{(\mathbf{X}^M, \mathbf{Y}^M)\}$ from set $\{(\mathbf{X}^O, \mathbf{Y}^O)\}$;
6: calculate the training loss using (14);
7: backpropagation calculates the error of each layer, calculates the reciprocal of the parameters W, and updates the parameters W;
8: **end for**
9: **end for**
10: **return** W

3. Experiment and Data

In this section, we describe two publicly available EEG datasets, how the data is preprocessed, and how the data is divided. A code implementation of MCL-SWT is available at https://github.com/roniusLuo/MCL_SWT (accessed on 20 April 2025).

3.1. Data

To evaluate our proposed method, we conducted extensive experiments on three public MI datasets, BCI Competition IV datasets 2a, 2b [35], and OpenBMI datasets [36].

(1) BCI Competition IV Dataset 2a: The BCI Competition IV Dataset 2a included the EEG signals of four-category MI recognition tasks (left hand, right hand, feet, tongue) from nine subjects. For each subject, two sessions of EEG signals were collected on different days, and there were total 288 trials (72 trials per class) per session. At the beginning of each training session, approximately 5 min of recording were conducted to assess EOG influence. The recording was divided into three segments: (1) 2 min with eyes open (gazing at a fixation cross on the screen), (2) 1 min with eyes closed, and (3) 1 min with eye movements. Subjects were seated comfortably in an armchair facing a computer screen. Each trial began (t = 0 s) with a fixation cross appearing on a black screen, accompanied by a short auditory warning tone. After 2 s (t = 2 s), a visual cue (arrow pointing left, right, down, or up-corresponding to the four classes) appeared and remained on screen for 1.25 s. No

feedback was provided. Subjects were instructed to perform the motor imagery task until the fixation cross disappeared at t = 6 s.

During the EEG acquisition, according to the prompts on the screen, the subjects performed four different types of MI tasks, and 22 channels of EEG were acquired simultaneously, using electrode positions in the 10–20 international standard lead system. 22 Ag/AgCl electrodes (electrode spacing 3.5 cm) were used to record the EEG signals, and the electrode arrangement is shown in the left side of Figure 3. All signals were recorded monopolarly with the left mastoid serving as reference and the right mastoid as ground. The signals were sampled with 250 Hz and bandpass-filtered between 0.5 Hz and 100 Hz. The sensitivity of the amplifier was set to 100 μV. An additional 50 Hz notch filter was enabled to suppress line noise. In addition to the 22 EEG channels, 3 monopolar EOG channels were recorded and also sampled with 250 Hz. They were bandpass filtered between 0.5 Hz and 100 Hz (with the 50 Hz notch filter enabled), and the sensitivity of the amplifier was set to 1 mV.

(2) BCI Competition IV Dataset 2b: The EEG signals of two-class MI tasks (left hand and right hand) from another nine subjects were collected on BCI Competition IV Dataset 2b. The subjects were right-handed and had normal or corrected-to-normal vision. All volunteers were sitting in an armchair, watching a flat screen monitor placed approximately 1 m away at eye level. For each subject, five sessions are provided, whereby the first two sessions contain training data without feedback (screening), and the last three sessions were recorded with feedback. At the beginning of each session, a recording of approximately 5 min was performed to estimate the EOG influence. The recording was divided into three blocks: (1) two minutes with eyes open (looking at a fixation cross on the screen), (2) one minute with eyes closed, and (3) one minute with eye move- ments. The artifact block was divided into four sections (15 s artifacts with 5 s resting in between) and the subjects were instructed with a text on the monitor to perform either eye blinking, rolling, up-down, or left-right movements. At the beginning and at the end of each task, a low and high warning tone were presented.

Three bipolar recordings (C3, Cz, and C4) were recorded with a sampling frequency of 250 Hz. The recordings had a dynamic range of ±100 μV for the screening and ±50 μV for the feedback sessions. They were bandpass- filtered between 0.5 Hz and 100 Hz, and a notch filter at 50 Hz was enabled. The placement of the three bipolar recordings were slightly different for each subject. The electrode position Fz served as EEG ground.

(3) OpenBMI dataset: The OpenBMI dataset consisted of EEG data from 25 females and 29 males for a total of 54 subjects with an age range of 24 to 35 years. All participants had no history of neurological, psychiatric, or other relevant disorders that could affect the results of the experiment. Of these 54 subjects, 21 were normal motor imagery subjects and 33 were BCI illiteracy subjects, i.e., a group unable to use the BCI proficiently or correctly. For the first 3 s of each trial, a black gaze cross was displayed in the centre of the screen to help prepare the subject for the motor imagery task.The EEG signal was acquired at a sampling rate of 1000 Hz through 62 Ag/AgCl electrodes. Subsequently, an arrow to the right or left was displayed on the screen as a visual cue, and the subject was required to perform the corresponding hand grasping imagery task for 4 s. After each task, the screen would remain blank for 6 s (±1.5 s). The experiment was divided into a training phase and a test phase, each containing 100 trials and an equal number of right- and left-handed motor imagery tasks. The EEG amplifier used in the experiment was a BrainAmp. The channels were nasion-referenced and grounded to electrode AFz. Additionally, an EMG electrode recorded from each flexor digitorum profundus muscle with the olecranon was used for reference.

3.2. Data Preprocessing

(1) BCI Competition IV Dataset 2a, 2b: The experiments in this paper are based on the Braindecode EEG processing toolbox [16]. The experiments used 4.48 s EEG segments, acquired from 0.5 s before the appearance of motor imagery cues (this prompted the subjects to perform the desired motor imagery task) to 3.98 s after cue onset. In order to allow the model to autonomously perform feature learning and detect valid EEG signal segments, only minimal preprocessing was performed on the EEG signal trials. Two types of classification tasks (left- and right-handed) were first performed using the C3, Cz, and C4 EEG channels, and data from these channels were converted from volts to microvolts to improve numerical stability. Subsequently, a third order band-pass filter from 4 to 38 Hz or 0–38 Hz was used for filtering along the time axis to highlight or remove signal components in a specific frequency range, and finally using the channel exponential sliding normalization technique, the initial window size for calculating the mean/variance was calculated with a value of init_block_size set to 1000, and the value of factor_new set to 0.001. In order to maintain numerical stability, eps is set to 0.0001.

(2) OpenBMI dataset: The dataset was divided into different BCI illiteracy paradigms (people who cannot use BCIs correctly), so we excluded data from this group of subjects from the experiment, and only 21 normal MI BCI users (subject 1, subject 2, subject 3, subject 5, subject 6, subject 9, subject 17, subject 18, subject 19, subject 21, subject 22, subject 28, subject 29, subject 32, subject 33, subject 36, subject 37, subject 43, subject 44, subject 45, subject 52) retained EEG signal data for experimental purposes. In this paper, 4.48 s EEG segments with C3, C4, and Cz electrodes were selected for the experiments. The EEG signal data were first filtered along the time axis (axis = 1) using a 3rd order bandpass filter from 8 to 30 Hz to highlight or remove signal components in a specific frequency range, followed by a channel exponential sliding normalization technique to calculate the mean/variance with the initial window size of init_block_size set to 1000 and factor_new set to 0.001, in order to maintain the numerical stability, eps was set to 0.0001, and the EEG signal was resampled using a frequency of 250 Hz.

The aim of this paper is to investigate the classification performance of subject-independent MI EEG signals, and in order to achieve this goal, a cross-validation approach is used to partition the dataset to ensure that the train and test sets are from different groups of subjects. When using the train sessions of all nine subjects in dataset IIa as the train set, the test sessions of all nine subjects in dataset IIb (session 4 and session 5) were used as the test set; conversely, when using the train sessions of all nine subjects in dataset IIb (session 3) as the train set, the test sessions of all nine subjects in dataset IIa were used as the test set. By dividing the data in this way, we are able to ensure that the train data and the test data are from completely different subjects. While OpenBMI dataset has EEG signal from 21 subjects, seven-fold cross validation (leave three subjects out) was applied. Thus, the subjects in the testing set were new subjects because they are completely different from the subjects in the training set.

3.3. Experiment Settings

In the experiments, the window size M was set to 8. In order to divide the input feature vector of length L into integer non-overlapping windows, the experiments used a segment of EEG signal of 4.48 s, from 0.5 s before the onset of the MI cue to 3.98 s after the onset of the MI cue, with a total of 1120 sampling points. We first preprocessed the data, bandpass filtered the EEG signals in the range of 4–38 Hz or 0–38 Hz, followed by channel-wise logarithmic sliding normalization, and finally, input the preprocessed EEG signals into the MCL-SWT model. During model training, Adam optimizer [37] was used as the optimization method and the weight decay parameter was set to 0.05. Due to the new experimental setup with new subjects, two categories of three-channel MI classification tasks (C3, Cz, and C4) from dataset IIb were used in this paper. The temporal convolution kernel T is set to 25, the spatial convolution kernel C is set to 3. The weight coefficients w_o and w_m are empirically in Equation (12) are set to 0.2 and 0.3, and the weight coefficients α and β are set to 0.2 and 1.2, respectively. The definition of epoch is the number of times the entire dataset is completely traversed (forward and backward) when the model is being trained, and 1 epoch is equal to the number of times all the training samples are learnt by the model. For the IIa and IIb datasets, we train 500 epochs, and for the OpenBMI dataset, we train 400 epochs.

4. Results

4.1. Performance Comparison of Subject-Independent MI Recogination

This Section evaluates subject-independent MI recognition performance of the MCL-SWT model. Since the subjects in dataset IIa are completely different from the subjects in dataset IIb, we use this cross-dataset setup the evaluate performance of the subject-independent MI based EEG signal recognition. During the training process, the model typically converges within 500 epochs, so the maximum number of training epochs is set to 500 to ensure model's convergence. While OpenBMI dataset has EEG signal from 21 subjects, seven-fold cross validation (leave three subjects out) was applied. Thus, the subjects in the testing set were new subjects because they are completely different from the subjects in the training set. The maximum number of epochs for the training was set to 500 for dataset 2a and 2b, and 400 for dataset OpenBMI to guarantee convergence of the model. Given the considerable randomness that impacts test accuracy at any given epoch, the evaluation metrics employed in the experiment are as follows: (a) Maximum test accuracy over 500 epochs (Max Accuracy); (b) Average test accuracy from epochs 401 to 500 (Average Accuracy); (c) Test accuracy at the epoch with the lowest training loss (Accuracy). Table 1 presents the results in terms of "Accuracy/Kappa value", with the best results highlighted in bold. The first row indicates the encoding format as "dataset-bandpass filter". For example, "2a-0 Hz" indicates that the model was trained on dataset IIb and tested on dataset IIa, preprocessed using a "0–38 Hz" bandpass filter. Five State-of-the-Art models were compared in the experiments, including Shallow ConvNet, Deep ConvNet [16], EEGNet [38], FBCNet [39], and ATCNet [28]. The SWT and MCL-SWT refer to the SWT model without MCL and with MCL, respectively. In addition, Table 2 presents the results of performance comparison on OpenBMI dataset.

A paired-sample one-sided Student's t-test is conducted to confirm the significance of the precision improvement of Table 1 (the null hypothesis is that the accuracy of the MCL-SWT model is equal to the SOTA models, against the alternative that the accuracy of the MCL-SWT model is greater than the accuracy of original model based on different SOTA models). The p-values are listed in Table 3.

The experimental results in Tables 1–3 lead to the following conclusions: (a) The MCL-SWT model shows an effect in terms of accuracy when compared to several of the compared models. (b) The proposed MCL-SWT model exhibits excellent performance across different datasets and bandpass filters. (c) MCL further improves the recognition accuracy and kappa value of SWT. (d) The proposed SWT model is resistant to overfitting, as evidenced by the small gap between the maximum accuracy/kappa value and the average accuracy/kappa value of SWT. (e) Experimental results show that MCL-SWT has better inter-subject generalization ability. (f) The precision improvement provided by the MCL-SWT is significant.

To illustrate the classification accuracy across different MI task categories in the MCL-SWT model, Figure 5 presents the confusion matrices for SWT and MCL-SWT across nine subjects in dataset IIb. The first and third rows display the confusion matrices for subjects 1 to 9 (sub1–sub9) using the SWT model, while the second and fourth rows show the confusion matrices for the same subjects using the MCL-SWT model. In each diagram, the horizontal axis represents the predicted classes, and the vertical axis represents the true classes. From these results, we can conclude that: (a) MCL-SWT improves classification accuracy for most subjects compared to SWT. (b) The classification accuracy of MCL-SWT is more balanced across each class.

Table 1. Classification performance (Accuracy and Kappa value) of subject-independent MI recognition on dataset 2a and 2b. The best results are highlighted in bold.

		2a-0 Hz	2a-4 Hz	2b-0 Hz	2b-4 Hz
Avg Acc/ Kappa	Shallow	63.72/0.29	60.76/0.25	72.70/0.46	67.06/0.34
	Deep	60.55/0.25	60.80/0.25	71.55/0.44	65.93/0.32
	EEGNet	60.58/0.25	59.52/0.23	71.21/0.44	65.91/0.32
	FBCNet	60.56/0.25	59.73/0.23	70.55/0.43	65.29/0.31
	ATCNet	58.87/0.22	57.45/0.20	68.47/0.36	64.69/0.30
	SWT	66.00/0.33	63.68/0.28	74.50/0.49	69.71/0.40
	MCL-SWT	**66.56/0.33**	**64.74/0.30**	**75.85/0.52**	**72.54/0.45**
Acc/ Kappa	Shallow	63.66/0.28	61.18/0.26	73.45/0.47	68.70/0.36
	Deep	61.27/0.26	61.34/0.26	72.89/0.46	66.51/0.34
	EEGNet	61.19/0.26	58.72/0.22	72.24/0.45	66.20/0.33
	FBCNet	60.86/0.25	58.36/0.22	71.18/0.44	65.53/0.31
	ATCNet	59.34/0.23	57.78/0.20	69.13/0.39	63.78/0.27
	SWT	66.13/0.33	63.58/0.28	74.58/0.49	69.79/0.40
	MCL-SWT	**66.48/0.33**	**64.52/0.30**	**75.62/0.51**	**73.27/0.47**
Max Acc/ Kappa	Shallow	67.28/0.34	65.36/0.32	75.18/0.51	71.69/0.45
	Deep	66.05/0.33	65.35/0.32	73.73/0.48	71.68/0.45
	EEGNet	65.51/0.32	64.51/0.30	73.31/0.47	71.16/0.44
	FBCNet	66.46/0.33	63.89/0.29	72.93/0.46	70.98/0.43
	ATCNet	64.11/0.29	62.18/0.28	70.84/0.43	68.22/0.36
	SWT	66.82/0.34	64.81/0.30	75.49/0.51	74.12/0.48
	MCL-SWT	**67.37/0.35**	**65.49/0.31**	**76.37/0.53**	**75.49/0.51**

Table 2. Classification performance (Accuracy and Kappa value) of subject-independent MI recognition on OpenBMI dataset. The best results are highlighted in bold.

		Shallow	Deep	EEGNet	FBCNet	ATCNet	MCL-SWT
Avg Acc/ Kappa	Fold1	81.61/0.63	79.56/0.59	80.17/0.60	79.00/0.58	80.89/0.62	82.56/0.65
	Fold2	77.67/0.55	75.89/0.52	75.94/0.52	71.56/0.43	73.72/0.47	80.01/0.60
	Fold3	78.33/0.57	80.44/0.61	82.00/0.64	82.39/0.65	83.83/0.68	76.38/0.53
	Fold4	81.00/0.62	81.22/0.62	82.06/0.64	79.11/0.58	80.06/0.60	85.51/0.71
	Fold5	74.39/0.49	72.28/0.45	76.89/0.54	71.67/0.43	72.94/0.46	75.48/0.51
	Fold6	80.33/0.61	82.61/0.65	82.75/0.66	76.89/0.54	81.78/0.64	80.46/0.61
	Fold7	70.33/0.41	69.22/0.38	70.44/0.41	67.83/0.36	72.28/0.45	74.05/0.48
	Avg	77.67/0.55	77.32/0.55	78.61/0.57	75.49/0.51	77.93/0.56	**79.21/0.58**

Table 3. The paired-sample one-sided Student's t-test of the performance on Table 1.

MCL-SWT vs.	Shallow	Deep	EEGNet	FBCNet	ATCNet
p-Values	0.0036	0.0020	0.0005	0.0007	0.0001

4.2. Sensitivity Analysis of Hyperparameters

The number of temporal multi-head self-attention blocks and the number of heads in each block are two hyperparameters of the SWT model. This subsection investigates the influence of these hyperparameters on the model's performance, with the experimental results summarized in Table 4. Increasing the number of heads or blocks results in a greater number of model parameters and attention scores, thereby increasing the model's complexity.

We can draw the following conclusions from Table 4: (a) the performance of the model deteriorates as the number of temporal multi-head self-attention blocks increases—this may be due to the limited number of training samples in MI EEG, which is insufficient to fully train a model with multiple self-attention blocks; (b) The number of self-attention heads has a minimal impact on the model's performance.

Table 4. The influence of the head number of self-attention and the number of multi-head self-attention block on the accuracy. The best results are highlighted in bold.

	4 Heads	8 Heads	10 Heads
1 block	**74.71/0.49**	**74.50/0.49**	**74.73/0.49**
2 block	74.30/0.48	74.20/0.48	74.45/0.49
3 block	73.16/0.46	73.67/0.47	73.26/0.46

We further investigate the impact of the hyperparameters w_o and w_m in the MCL. Figure 6 illustrates the model's performance with varying values of w_o and w_m on datasets IIa and IIb. The preprocessing step involved applying a uniform bandpass filter of "4–38 Hz". In Figure 6, W_m-2a represents the experimental results for different values of w_m on dataset IIa with w_o fixed, while W_o-2b presents the experimental results for varying w_o on dataset IIb with w_m fixed. The results suggest that values within the range $[0.1, 0.3]$ are effective for both datasets.

α and β are other two hyperparameters in the MCL training process. In the experiments, α and β were assigned different values, and their impact on the method's performance was evaluated. The specific experimental results are summarized in Table 5. Using a "4–38 Hz" bandpass filter for preprocessing, we examined the effect of various hyperparameter values on the method's performance by training the model on dataset IIa and testing it on new subjects from dataset IIb. The experimental results in Table 5 demonstrate the model's robustness to variations in hyperparameters.

Table 5. The performance evaluation on different hyperparameters in MCL. The best results are highlighted in bold.

hyperparameter values		$\beta = 1.2$	
	$\alpha = 0.2$	$\alpha = 0.4$	$\alpha = 0.8$
Average Accuracy	**72.54**	72.29	72.26
hyperparameter values		$\alpha = 0.2$	
	$\beta = 0.6$	$\beta = 1.2$	$\beta = 2.4$
Average Accuracy	71.95	**72.54**	72.02

Figure 5. Confusion matrices for SWT (**left**) and MCL-SWT (**right**) across nine subjects in dataset IIb.

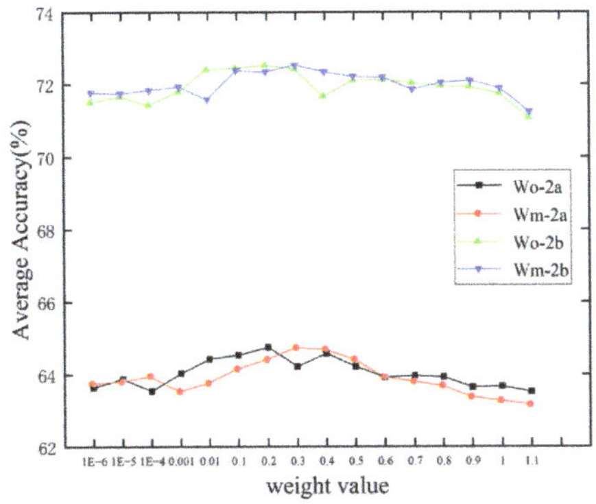

Figure 6. The influence of the hyperparameters w_o and w_m in the MCL.

4.3. Feature Visualization

To assess the impact of MCL on feature distribution, the features were visualized using t-SNE [40], with the results for subjects 4, 5, and 8 from dataset IIb shown in Figure 7. In these visualizations, the original EEG signals are represented as circles, while the mirror EEG signals are depicted as triangles. The left-hand MI is shown in red, and the right-hand MI is shown in blue.

From Figure 7, it is evident that MCL effectively separates features from different MI tasks and clusters features from the same MI task. Without MCL, the feature distributions of the left-hand and right-hand MI tasks exhibit significant overlap, making it challenging to distinguish between them. However, with MCL applied, the features from the same task (e.g., left-hand MI) are grouped more tightly, while features from opposite tasks are pushed further apart. This demonstrates the ability of MCL to enhance feature separability by leveraging mirror EEG data to align positive samples and enforce the distinction between negative sample pairs. By incorporating mirror EEG signals, MCL also enhances the model's ability to localize ERD/ERS phenomena accurately. This approach facilitates improved recognition of subject-independent MI EEG signals, confirming the effectiveness of MCL-SWT in addressing challenges associated with individual variability in EEG feature distributions.

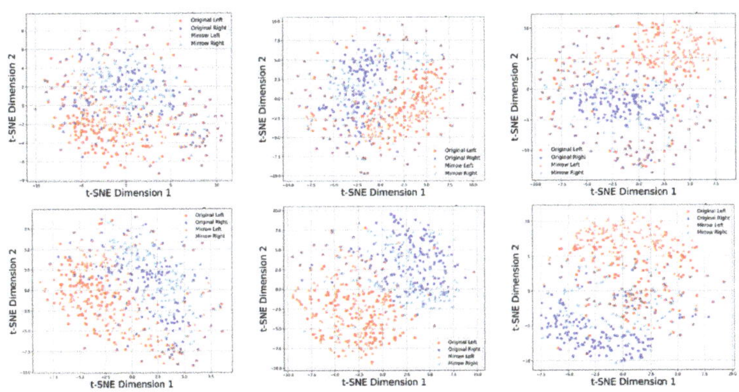

(a) **(b)** **(c)**

Figure 7. Feature distribution visualization of SWT without MCL (**top**) and with MCL (**bottom**) on datasets IIb. (**a**) Feature distribution of Subject 4. (**b**) Feature distribution of Subject 5. (**c**) Feature distribution of Subject 8.

4.4. Ablation Experiment on Mirror Contrastive Loss

Equation (12) consists of two components: the first term, $\mathcal{L}_o$, represents the contrastive loss within the original EEG signals, while the second term, $\mathcal{L}_m$, corresponds to the contrastive loss between the original EEG signals and their mirror counterparts. To evaluate the effectiveness of each component, ablation experiments were conducted on datasets IIa and IIb, with the results presented in Figure 8. For preprocessing, a "4–38 Hz" band-pass filter was applied. As shown in Figure 8, both $\mathcal{L}_o$ and $\mathcal{L}_m$ individually contribute to improving the average classification accuracy of the SWT model. Moreover, the combination of $\mathcal{L}_o$ and $\mathcal{L}_m$ further enhances the model's performance, demonstrating the complementary benefits of incorporating both components.

To assess the performance across different subjects, Tables 6–8 present the results of our proposed MCL method for each subject on datasets IIa, IIb and OpenBMI. Uniform preprocessing using a "4–38 Hz" bandpass filter on datasets IIa and IIb. Preprocessing method using the "8–30 Hz" bandpass filter on the OpenBMI dataset. Table 8 present the results of our proposed MCL method for each subject on OpenBMI datasets. A uniform preprocessing approach using a "8–30 Hz" bandpass filter was applied. The results indicate that the impact of each component varies across individual subjects. However, on average, both components contribute to performance improvement. Furthermore, the combination of $\mathcal{L}_o$ and $\mathcal{L}_m$ provides additional performance gains, consistent with the findings illustrated in Figure 8.

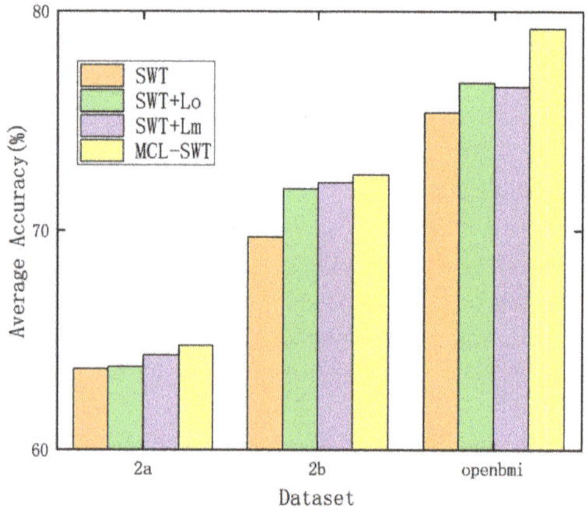

Figure 8. Ablation experiments of $\mathcal{L}_o$ and $\mathcal{L}_m$.

Table 6. Ablation experiments of $\mathcal{L}_o$ and $\mathcal{L}_m$ for each subject on dataset IIa. The best results are highlighted in bold. The check mark means the related loss is included.

SWT	$\mathcal{L}_o$	$\mathcal{L}_m$	sub1	sub2	sub3	sub4	sub5	sub6	sub7	sub8	sub9	Avg
✓			66.53	54.39	88.65	58.92	53.87	57.51	**62.35**	71.52	**59.39**	63.68
✓	✓		66.72	**58.73**	88.77	58.01	54.31	56.82	60.34	**74.03**	56.19	63.77
✓		✓	67.76	58.33	**89.97**	57.89	55.37	58.51	61.58	72.33	56.97	64.30
✓	✓	✓	**68.81**	57.26	89.83	**59.67**	**56.95**	**59.85**	59.18	73.18	57.94	**64.74**

Table 7. Ablation experiments of $\mathcal{L}_o$ and $\mathcal{L}_m$ for each subject on dataset IIb. The best results are highlighted in bold. The check mark means the related loss is included.

SWT	$\mathcal{L}_o$	$\mathcal{L}_m$	sub1	sub2	sub3	sub4	sub5	sub6	sub7	sub8	sub9	Avg
✓			65.83	53.12	54.75	80.23	73.74	67.60	73.30	83.50	75.32	69.71
✓	✓		**69.61**	54.15	55.50	81.63	78.71	68.70	**75.70**	86.23	76.85	71.90
✓		✓	68.36	53.97	56.76	**82.82**	**78.84**	**70.53**	73.76	85.54	**79.09**	72.19
✓	✓	✓	69.50	**55.69**	**57.13**	82.71	78.17	69.86	74.75	**86.93**	78.13	**72.54**

Table 8. Ablation experiments of $\mathcal{L}_o$ and $\mathcal{L}_m$ for each subject on OpenBMI dataset. The best results are highlighted in bold. The check mark means the related loss is included.

SWT	$\mathcal{L}_o$	$\mathcal{L}_m$	Fold1	Fold2	Fold3	Fold4	Fold5	Fold6	Fold7	Avg
✓			78.00	75.61	74.71	76.43	73.31	78.11	71.56	75.39
✓	✓		80.93	76.90	75.66	78.05	74.02	79.31	72.38	76.75
✓		✓	80.68	77.74	75.61	77.98	74.25	78.28	71.35	76.56
✓	✓	✓	**82.56**	**80.01**	**76.38**	**85.51**	**75.48**	**80.46**	**74.05**	**79.21**

4.5. Computing Complexity Analysis

The number of model parameters and inference time are key indicators of computational complexity and were evaluated in this section. Inference time was measured as the average duration of 1000 runs. All experiments were conducted on a server equipped with an Intel i7 10700K CPU and an NVIDIA GeForce RTX 3090 GPU. The results of the analysis are presented in Table 9, which shows that the MCL-SWT model achieves a good balance between parameter size and inference efficiency. While it has more parameters (155 M) than lightweight models like EEGNet and FBCNet (3 M), its inference time (8.36 ms) is significantly faster than FBCNet (37.64 ms) and ATCNet (15.37 ms). This highlights MCL-SWT's efficiency in maintaining high performance with low computational overhead, which shows the potential to be applied in online BCI system.

Table 9. Model complexity analysis.

	Shallow	Deep	EEGNet	FBCNet	ATCNet	MCL-SWT
Para num (M)	10	268	3	3	37	155
Infer time (ms)	0.56	1.42	2.48	37.64	15.37	8.36

5. Discussion

The utility of a brain-computer interface system is inextricably linked to the performance of its decoding module. For subject-independent EEG signal recognition, the variability of ERD/ERS spatial patterns across individuals poses a significant challenge to EEG signal decoding. We propose a mirror-contrastive learning-based sliding window transformer (MCL-SWT) to improve the spatial sensitivity of ERD/ERS by introducing mirror EEG signals and contrastive learning, and also to deal with inter-subject variability effectively. Additionally, we introduce the SWT mechanism to reduce the computational complexity compared to the global attention approach. Therefore, the proposed method improves subject-independent MI-based EEG signal recognition compared to existing models [16,28,38,39].

In the experiments, we can see that MCL-SWT achieves good results on the three datasets. The temporal SWT mechanism computes self-attention scores within a sliding window, effectively balancing improved model performance with acceptable computational complexity. We explored the impact of several key parameters on the model and showed that increasing the number of temporal multi-head self-attention blocks and the number of heads in each block leads to more model parameters and attention scores, increasing the complexity of the model and thus affecting classification performance. We also performed ablation experiments and feature visualizations where features from the same task (e.g., left MI) were more tightly grouped and features from the opposite task were pushed further apart with the application of MCL. The ability of MCL to enhance feature separability by using mirrored EEG data to align positive samples and force the distinction between negative sample pairs is demonstrated, enhancing the model's ability to accurately localise ERD/ERS phenomena. In the BCI field, where existing technologies are already quite mature, the room for enhancing the performance of high-precision algorithms and systems is inherently limited. Consequently, while the 2.82% and 2.17% accuracy improvements may appear modest numerically, from a practical application standpoint, these enhancements enable the system to significantly reduce misidentifications across extensive sample datasets.

However, although the SWT mechanism reduces computational complexity compared to global attention methods, its scalability to large-scale datasets or real-time applications requires further study. Future research could also focus on optimizing MCL-SWT for multi-class MI tasks or exploring its application in hybrid brain-computer interface (BCI) systems. Furthermore, most motor imagery EEG recognition methods are evaluated on EEG signals from healthy subjects, but motor imagery brain-computer interfaces are valuable in neurorehabilitation (e.g., post-stroke rehabilitation) [6,7]. Therefore, future research is planned on MI-EEG recognition algorithms for stroke patients.

Lastly, combining MCL with domain adaptation or transfer learning could further enhance its ability to generalize across subjects and even across datasets, paving the way for broader applicability in BCI applications.

6. Conclusions

In this paper, we propose a Mirror Contrastive Learning (MCL) method combined with a Sliding Window Transformer (SWT) model for subject-independent EEG signal recognition. By constructing mirror EEG signals, the model's sensitivity to the spatial location of ERD/ERS phenomena is enhanced. In addition, the temporal SWT mechanism computes self-attention scores within a sliding window, effectively balancing improved model performance with acceptable computational complexity. Detailed comparative experiments on different EEG datasets have been conducted with promising results. Furthermore, MCL, as a generalizable approach, has the potential to be integrated into various backbone networks for MI-EEG recognition.

Author Contributions: Methodology, Q.M. and J.L.; software, Q.M.; validation, Q.M. and Y.Z.; writ-ing—original draft preparation, Q.M.; writing—review and editing, H.Z. and J.L.; visualization, H.Z.; supervision, W.Y. and X.H.; funding acquisition, J.L. All authors have read and agreed to the published version of the manuscript.

Funding: This work is jointly supported by the Scientific Research Program Founded by Shaanxi Provincial Education Department of China under Grant 23JK0556 and the National Natural Science Foundation of China (Grant Nos. 61906152, 62376213 and U21A20524).

Institutional Review Board Statement: In this study, we used two publicly available dataset.

Informed Consent Statement: Not applicable.

Data Availability Statement: In this study, we utilized publicly available datasets. These datasets can be accessed at: 25 April 2025 https://www.bbci.de/competition/download/competition_iv/BCICIV_2a_gdf.zip, https://www.bbci.de/competition/download/competition_iv/BCICIV_2b_gdf.zip and https://gigadb.org/dataset/view/id/100542/File_page/5.

Conflicts of Interest: The authors declare no conflicts of interest.

References

1. Wolpaw, J.R. Brain-computer interfaces (BCIs) for communication and control. In Proceedings of the 9th International ACM SIGACCESS Conference on Computers and Accessibility, Tempe, AZ, USA, 15–17 October 2007; pp. 1–2.
2. Shanechi, M.M. Brain–Machine Interfaces from Motor to Mood. *Nat. Neurosci.* **2019**, *22*, 1554–1564. [CrossRef]
3. Benson, P.J. Decoding brain-computer interfaces. *Science* **2018**, *360*, 615–616.
4. Arpaia, P.; Esposito, A.; Natalizio, A.; Parvis, M. How to successfully classify EEG in motor imagery BCI: A metrological analysis of the state of the art. *J. Neural Eng.* **2022**, *19*, 031002. [CrossRef] [PubMed]
5. Burwell, S.; Sample, M.; Racine, E. Ethical aspects of brain computer interfaces: A scoping review. *BMC Med. Ethics* **2017**, *18*, 1–11. [CrossRef]
6. Altaheri, H.; Muhammad, G.; Alsulaiman, M.; Amin, S.U.; Altuwaijri, G.A.; Abdul, W.; Bencherif, M.A.; Faisal, M. Deep learning techniques for classification of electroencephalogram (EEG) motor imagery (MI) signals: A review. *Neural Comput. Appl.* **2023**, *35*, 14681–14722. [CrossRef]
7. Zhang, S.; Ang, K.K.; Zheng, D.; Hui, Q.; Chen, X.; Li, Y.; Tang, N.; Chew, E.; Lim, R.Y.; Guan, C. Learning eeg representations with weighted convolutional siamese network: A large multi-session post-stroke rehabilitation study. *IEEE Trans. Neural Syst. Rehabil. Eng.* **2022**, *30*, 2824–2833. [CrossRef]
8. Luo, J.; Wang, Y.; Xia, S.; Lu, N.; Ren, X.; Shi, Z.; Hei, X. A shallow mirror transformer for subject-independent motor imagery BCI. *Comput. Biol. Med.* **2023**, *164*, 107254. [CrossRef]
9. Zhu, X.; Meng, M.; Yan, Z.; Luo, Z. Motor Imagery EEG Classification Based on Multi-Domain Feature Rotation and Stacking Ensemble. *Brain Sci.* **2025**, *15*, 50 . [CrossRef]
10. Fraiwan, L.; Lweesy, K.; Khasawneh, N.; Wenz, H.; Dickhaus, H. Automated sleep stage identification system based on time–frequency analysis of a single EEG channel and random forest classifier. *Comput. Methods Prog. Biomed.* **2012**, *108*, 10–19. [CrossRef]
11. Bishop, C.M.; Nasrabadi, N.M. *Pattern Recognition and Machine Learning*; Springer: Berlin/Heidelberg, Germany, 2006; Volume 4.
12. Lemm, S.; Blankertz, B.; Curio, G.; Muller, K.R. Spatio-spectral filters for improving the classification of single trial EEG. *IEEE Trans. Biomed. Eng.* **2005**, *52*, 1541–1548. [CrossRef]
13. Ang, K.K.; Chin, Z.Y.; Zhang, H.; Guan, C. Filter bank common spatial pattern (FBCSP) in brain-computer interface. In Proceedings of the 2008 IEEE International Joint Conference on Neural Networks (IEEE World Congress on Computational Intelligence), Hong Kong, China, 1–8 June 2008; IEEE: Piscataway, NJ, USA, 2008; pp. 2390–2397.
14. Gu, H.; Chen, T.; Ma, X.; Zhang, M.; Sun, Y.; Zhao, J. CLTNet: A Hybrid Deep Learning Model for Motor Imagery Classification. *Brain Sci.* **2025**, *15*, 124. [CrossRef] [PubMed]
15. Bouchane, M.; Guo, W.; Yang, S. Hybrid CNN-GRU Models for Improved EEG Motor Imagery Classification. *Sensors* **2025**, *25*, 1399. [CrossRef] [PubMed]
16. Schirrmeister, R.T.; Springenberg, J.T.; Fiederer, L.D.J.; Glasstetter, M.; Eggensperger, K.; Tangermann, M.; Hutter, F.; Burgard, W.; Ball, T. Deep learning with convolutional neural networks for EEG decoding and visualization. *Hum. Brain Mapp.* **2017**, *38*, 5391–5420. [CrossRef] [PubMed]
17. Jiao, Y.; Zhang, Y.; Chen, X.; Yin, E.; Jin, J.; Wang, X.; Cichocki, A. Sparse group representation model for motor imagery EEG classification. *IEEE J. Biomed. Health Inform.* **2018**, *23*, 631–641. [CrossRef]
18. Autthasan, P.; Chaisaen, R.; Sudhawiyangkul, T.; Rangpong, P.; Kiatthaveephong, S.; Dilokthanakul, N.; Bhakdisongkhram, G.; Phan, H.; Guan, C.; Wilaiprasitporn, T. MIN2Net: End-to-end multi-task learning for subject-independent motor imagery EEG classification. *IEEE Trans. Biomed. Eng.* **2021**, *69*, 2105–2118. [CrossRef]
19. Pfurtscheller, G.; Da Silva, F.L. Event-related EEG/MEG synchronization and desynchronization: Basic principles. *Clin. Neurophysiol.* **1999**, *110*, 1842–1857. [CrossRef]
20. Makeig, S.; Bell, A.; Jung, T.P.; Sejnowski, T.J. Independent component analysis of electroencephalographic data. *Adv. Neural Inf. Process. Syst.* **1995**, *8*, 145–151.
21. Kim, H.; Yoshimura, N.; Koike, Y. Characteristics of kinematic parameters in decoding intended reaching movements using electroencephalography (EEG). *Front. Neurosci.* **2019**, *13*, 1148. [CrossRef]
22. Zhao, H.; Zheng, Q.; Ma, K.; Li, H.; Zheng, Y. Deep representation-based domain adaptation for nonstationary EEG classification. *IEEE Trans. Neural Netw. Learn. Syst.* **2020**, *32*, 535–545. [CrossRef]
23. Xu, L.; Xu, M.; Ke, Y.; An, X.; Liu, S.; Ming, D. Cross-dataset variability problem in EEG decoding with deep learning. *Front. Hum. Neurosci.* **2020**, *14*, 103. [CrossRef]
24. Rodrigues, P.L.C.; Jutten, C.; Congedo, M. Riemannian procrustes analysis: Transfer learning for brain–computer interfaces. *IEEE Trans. Biomed. Eng.* **2018**, *66*, 2390–2401. [CrossRef] [PubMed]

25. Zhang, K.; Robinson, N.; Lee, S.W.; Guan, C. Adaptive transfer learning for EEG motor imagery classification with deep convolutional neural network. *Neural Netw.* **2021**, *136*, 1–10. [CrossRef] [PubMed]
26. Jia, Z.; Lin, Y.; Wang, J.; Yang, K.; Liu, T.; Zhang, X. MMCNN: A Multi-branch Multi-scale Convolutional Neural Network for Motor Imagery Classification. In *Machine Learning and Knowledge Discovery in Databases* ; Hutter, F., Kersting, K., Lijffijt, J., Valera, I., Eds.; Springer: Cham, Switzerland, 2021; pp. 736–751.
27. Liu, Z.; Lin, Y.; Cao, Y.; Hu, H.; Wei, Y.; Zhang, Z.; Lin, S.; Guo, B. Swin transformer: Hierarchical vision transformer using shifted windows. In Proceedings of the IEEE/CVF International Conference on Computer Vision, Montreal, BC, Canada, 11–17 October 2021; pp. 10012–10022.
28. Altaheri, H.; Muhammad, G.; Alsulaiman, M. Physics-informed attention temporal convolutional network for EEG-based motor imagery classification. *IEEE Trans. Ind. Inform.* **2022**, *19*, 2249–2258. [CrossRef]
29. Luo, J.; Mao, Q.; Shi, W.; Shi, Z.; Wang, X.; Lu, X.; Hei, X. Mirror contrastive loss based sliding window transformer for subject-independent motor imagery based EEG signal recognition. In Proceedings of the Human Brain and Artificial Intelligence, 4th International Workshop, HBAI 2024, held in Conjunction with IJCAI 2024, Jeju Island, Republic of Korea, 3 August 2024.
30. Ioffe, S.; Szegedy, C. Batch normalization: Accelerating deep network training by reducing internal covariate shift. In Proceedings of the International Conference on Machine Learning, Lille, France, 7–9 July 2015; PMLR: Breckenridge, CO, USA, 2015; pp. 448–456.
31. Ba, J.L.; Kiros, J.R.; Hinton, G.E. Layer normalization. *arXiv* **2016**, arXiv:1607.06450.
32. Hendrycks, D.; Gimpel, K. Gaussian error linear units (gelus). *arXiv* **2016**, arXiv:1606.08415.
33. Luo, J.; Shi, W.; Lu, N.; Wang, J.; Chen, H.; Wang, Y.; Lu, X.; Wang, X.; Hei, X. Improving the performance of multisubject motor imagery-based BCIs using twin cascaded softmax CNNs. *J. Neural Eng.* **2021**, *18*, 036024. [CrossRef]
34. Hadsell, R.; Chopra, S.; LeCun, Y. Dimensionality reduction by learning an invariant mapping. In Proceedings of the 2006 IEEE Computer Society Conference on Computer Vision And Pattern Recognition (CVPR'06), New York, NY, USA, 17–22 June 2006; IEEE: Piscataway, NJ, USA, 2006; Volume 2, pp. 1735–1742.
35. Tangermann, M.; Müller, K.R.; Aertsen, A.; Birbaumer, N.; Braun, C.; Brunner, C.; Leeb, R.; Mehring, C.; Miller, K.J.; Müller-Putz, G.R.; et al. Review of the BCI competition IV. *Front. Neurosci.* **2012**, *6*, 55.
36. Lee, M.H.; Kwon, O.Y.; Kim, Y.J.; Kim, H.K.; Lee, Y.E.; Williamson, J.; Fazli, S.; Lee, S.W. EEG dataset and OpenBMI toolbox for three BCI paradigms: An investigation into BCI illiteracy. *GigaScience* **2019**, *8*, giz002. [CrossRef]
37. Kingma, D.P.; Ba, J. Adam: A method for stochastic optimization. *arXiv* **2014**, arXiv:1412.6980.
38. Lawhern, V.J.; Solon, A.J.; Waytowich, N.R.; Gordon, S.M.; Hung, C.P.; Lance, B.J. EEGNet: A compact convolutional neural network for EEG-based brain–computer interfaces. *J. Neural Eng.* **2018**, *15*, 056013. [CrossRef]
39. Mane, R.; Chew, E.; Chua, K.; Ang, K.K.; Robinson, N.; Vinod, A.P.; Lee, S.W.; Guan, C. FBCNet: A multi-view convolutional neural network for brain-computer interface. *arXiv* **2021**, arXiv:2104.01233.
40. Van der Maaten, L.; Hinton, G. Visualizing data using t-SNE. *J. Mach. Learn. Res.* **2008**, *9*, 2579–2605.

brain sciences

MDPI

Article

Effects of Different Individuals and Verbal Tones on Neural Networks in the Brain of Children with Cerebral Palsy

Ryosuke Yamauchi [1,2,*], Hiroki Ito [1], Ken Kitai [1], Kohei Okuyama [1], Osamu Katayama [1,3], Kiichiro Morita [4], Shin Murata [1] and Takayuki Kodama [1]

1 The Graduate School of Health Science, Kyoto Tachibana University, Kyoto 607-8175, Japan; h901524007@st.tachibana-u.ac.jp (H.I.); h901523004@st.tachibana-u.ac.jp (K.K.); h901523002@st.tachibana-u.ac.jp (K.O.); katayama.o@ncgg.go.jp (O.K.); murata-s@tachibana-u.ac.jp (S.M.); kodama-t@tachibana-u.ac.jp (T.K.)
2 Otemae Rehabilitation Center with Physical Disabilities, Osaka Red Cross Hospital, Osaka 543-8555, Japan
3 National Center for Geriatrics and Gerontology, Center for Gerontology and Social Science, Obu 474-8511, Japan
4 Cognitive and Molecular Research Institute of Brain Diseases, Kurume University, Fukuoka 830-0011, Japan; kiichiro@kurume-u.ac.jp
* Correspondence: h901522007@st.tachibana-u.ac.jp

Abstract: Background/Objectives: Motivation is a key factor for improving motor function and cognitive control in patients. Motivation for rehabilitation is influenced by the relationship between the therapist and patient, wherein appropriate voice encouragement is necessary to increase motivation. Therefore, we examined the differences between mothers and other individuals, such as physical therapists (PTs), in their verbal interactions with children with cerebral palsy who have poor communication abilities, as well as the neurological and physiological effects of variations in the tone of their speech. **Methods**: The three participants were children with cerebral palsy (Participant A: boy, 3 years; Participant B: girl, 7 years; Participant C: girl, 9 years). Participants' mothers and the assigned PTs were asked to speak under three conditions. During this, the brain activity of the participants was measured using a 19-channel electroencephalogram. The results were further analyzed using Independent Component Analysis frequency analysis with exact Low-Resolution Brain Electromagnetic Tomography, allowing for the identification and visualization of neural activity in three-dimensional brain functional networks. **Results**: The results of the ICA frequency analysis for each participant revealed distinct patterns of brain activity in response to verbal encouragement from the mother and PT, with differences observed across the theta, alpha, and beta frequency bands. **Conclusions**: Our study suggests that the children were attentive to their mothers' inquiries and focused on their internal experiences. Furthermore, it was indicated that when addressed by the PT, the participants found it easier to grasp the meanings and intentions of the words.

Keywords: motivation; rehabilitation; electroencephalogram; cerebral palsy; communication; emotion

1. Introduction

Motivation during the performance of physiotherapy tasks has been identified as a key factor for improving motor function and cognitive control in patients [1–3]. Motivation is a personal characteristic that facilitates the achievement and maintenance of a goal. It is clear that patients' motivation towards rehabilitation has a significant impact on the therapeutic effectiveness, especially in pediatric rehabilitation [4]. In child development, the active participation and continuous commitment of patients to rehabilitation are indispensable for promoting functional recovery and social adaptation. Additionally, it has been reported that motivation tends to decline over time, and factors such as fatigue, external environment, and reward misalignment are involved in this decrease in motivation [5,6]. Therefore, it is especially important to increase and maintain the motivation of children for pediatric rehabilitation by using preferred tasks, rewards [7], building trust relationships with others, and increasing self-esteem [8,9]. It has been reported that involving family members in the rehabilitation process maximizes the effectiveness of interventions, highlighting that the relationship between the child, family, and therapist is essential [10]. Moreover, the individual–environment interaction is imperative in child development, and the acquisition of motor skills enhances exploratory behaviors and influences perceptual, cognitive, and language development [11]. Therefore, children with cerebral palsy, who have severe motor and intellectual disability (CP-SMID), have impaired mobility owing to motor disabilities. This limits spontaneous exploration, which consequently induces severe intellectual developmental delays, decreased interest in objects and people, and decreased motivation to move and exercise. Therefore, we believe that it is essential to motivate children to move, exercise, and develop an interest in objects.

Motivation for rehabilitation is influenced by the relationship between the therapist and patient [8], wherein appropriate voice encouragement is necessary to increase motivation [12]. Changes in motivation are reported to be associated with changes in emotions. Therefore, it is essential to promote emotional changes in children to enhance their motivation. As children with CP-SMID have severe mental and physical disabilities that frequently include communication disorders, it is difficult to objectively assess how these patients perceive the voices of others, such as their parents and therapists, and ascertain the effect that they have on their motivation. Therefore, it is difficult to objectively evaluate and share children's emotional changes with others. Kangaroo care, wherein a contact stimulus is given to a neonate who has not yet acquired communication abilities, has been shown to elicit similar physiological responses toward both the mother and father, as measured by heart rate variability (HRV) and apnea/periodicity measures [13]. Additionally, in the field of pediatric rehabilitation, it is essential to collaborate not only with the child who is the patient but also with their family during interventions [14]. It has also been reported that involving the family in rehabilitation enhances the effectiveness of interventions [15]. A study of infants' sensitivity to language stimuli revealed that infants displayed more sensitivity in their native language rather than in a second/foreign language [16]. These finding has shown that even in infants who have not yet acquired language skills, differences in the individuals providing stimuli and the content of those stimuli can affect emotional and cognitive functions. While autonomic nervous responses, such as heart rate and other physiological measures, have been recognized, there is a lack of research quantifying neural activity and brain networks in children at later stages of communication development. Furthermore, little is known about how differences in vocal tones from various individuals influence these neural networks, enhance emotions, and consequently impact motivation. In light of these considerations, we believe that quantitatively assessing neural activity in the brain using an electroencephalogram (EEG), which has excellent temporal resolution, is essential for clarifying the mental elements necessary for forming a foundation for communication skills in rehabilitation.

We hypothesized that the elements of an emotion-based approach, such as verbal encouragement directed at children in the language development stage (i.e., the mental effects of language), may influence the effectiveness of pediatric rehabilitation. This study aimed to examine how differences in different vocal tones from mothers and therapists affect the neural networks of children with poor communication skills, specifically those with cerebral palsy, using an electroencephalogram (EEG) to clarify the necessary mental elements for forming a foundation for communication skills in rehabilitation.

2. Materials and Methods

The present study is designed as an observational study, utilizing a cross-sectional research approach.

2.1. Participants

The inclusion criteria for the subjects in this study were individuals diagnosed with CP-SMIDs who were under 12 years old and did not have accompanying hearing impairments, treated at the authors' affiliated hospital. The exclusion criteria included the following:

1. Severe body movements that could interfere with EEG recording;
2. Hearing impairment that could affect the perception of verbal stimuli;
3. Medical conditions contraindicating the use of EEG equipment.

These criteria were chosen to ensure the quality and reliability of the EEG data collected.

2.2. Ethical Considerations

The study was conducted in accordance with the Declaration of Helsinki and approved by Kyoto Tachibana University Research Ethics Committee, (approval number: 22–25; date of approval: 19 August 2022). Informed consent was obtained from all subjects involved in the study. Written informed consent to participate in this research was obtained from the parents/legal guardians of all participants, in accordance with institutional standards.

2.3. Stimuli

Speaking directed at the pediatric participants was provided by their mothers and the children's physical therapists (PTs). The content included the following:

- 20 sentences of speaking based on asking;
- 20 sentences based on informing;
- 40 sentences based on a mixed approach that combined asking and informing.

Although the content and character count were identical across conditions, the tone differed:

- Asking sentences were delivered in a friendly manner;
- Informing sentences were delivered in a non-friendly manner;
- The mixed approach incorporated tones appropriate for both asking and informing.

The sentences were developed in collaboration with speech therapists and child psychologists to ensure age-appropriateness and relevance to the rehabilitation context. Mothers reviewed the sentences before measurement to confirm that the content did not evoke undue joy or stress in the child and that there were no elements that could negatively stimulate the child's emotions. The verbal prompts used in this study were designed to exclude content that would elicit emotional changes, such as those that would either delight the children or insult them. Instead, the focus was on altering the tone and intonation of the questions and instructions. The dialogue examples for different conditions are shown in Figure 1, with both Japanese originals and English translations provided.

Line（asking）

It's (sunny or cloudy or rainy) today. 今日は（晴れ or 曇り or 雨）だね
Yesterday was (sunny or cloudy or rainy). 昨日は（晴れてたね or 曇ってたね or 雨が降ってたね）
It's (cold or hot) today. 今日は（寒い or 暑い）ね
Yesterday was (cold or hot) too, right? 昨日も（寒かった or 暑かった）よね
We came here today (by bike or on foot or by car).今日は（自転車 or 徒歩 or 車）で来たよね
Spring is just around the corner. もう春だね
Summer is coming soon. もうすぐ夏だね
It's going to get hotter soon. もうすぐもっと暑くなるね
It's almost time to head home, isn't it? もうすぐ帰る時間だね
You're not feeling unwell, right? 体調は悪くないよね
You're looking good today. 今日も元気だね
You seem the same as always. いつもと変わらないね
You slept well last night, didn't you? 昨日はよく寝れたね
You had lunch, right? ご飯食べたよね
You're full, aren't you?. お腹いっぱいだね
It was delicious. おいしかったよね
You're listening well, aren't you? よく聞いてるね
Just a little more, right? もう少しだね
You're doing well, aren't you? しっかりできてるね
You're going to do your best with rehab, right? リハビリ頑張ろうね

Line（informing）

Today is (sunny or cloudy or rainy). 今日は（晴れ or 曇り or 雨）です
Yesterday (it was sunny or it was cloudy or it rained). 昨日は（晴れでした or 曇でした or 雨が降
Today is (cold or hot). 今日は（寒い or 暑い）です
Yesterday was also (cold or hot). 昨日も（寒かった or 暑かった）です
We are here today (bike or walk or drive). 今日は（自転車 or 徒歩 or 車）で来ました
Spring is approaching soon. もう春です
Summer is approaching soon.もうすぐ夏です
It will be hotter soon. もうすぐもっと暑くなります
It is almost time to leave for home.もうすぐ帰る時間です
You are not feeling unwell, are you? 体調は悪くないです
You are looking well today. 今日も元気です
You do not seem different from usual. いつもと変わりません
You slept well last night. 昨日はよく寝ました
You have had lunch, haven't you? ご飯食べました
You are full. お腹いっぱいです
It was delicious. おいしかったです
You are listening well. よく聞いてます
It is just a little more. もう少しです
You are doing well. しっかりできてます
You are doing your best with rehab.リハビリ頑張ります

Line（asking and informing）

It's (sunny or cloudy or rainy) today. 今日は（晴れ or 曇り or 雨）だね
Yesterday was (sunny or cloudy or rainy). 昨日は（晴れてたね or 曇ってたね or 雨が降ってたね）
Today is (sunny or cloudy or rainy). 今日は（晴れ or 曇り or 雨）です
Yesterday (it was sunny or it was cloudy or it rained). 昨日は（晴れでした or 曇でした or 雨が降りました）
Today is (cold or hot). 今日は（寒い or 暑い）です
It's (cold or hot) today. 今日は（寒い or 暑い）ね
Yesterday was (cold or hot) too, right? 昨日も（寒かった or 暑かった）よね
Yesterday was also (cold or hot). 昨日も（寒かった or 暑かった）です
We came here today (by bike or on foot or by car).今日は（自転車 or 徒歩 or 車）で来たよね
We are here today (bike or walk or drive). 今日は（自転車 or 徒歩 or 車）で来ました
Spring is just around the corner. もう春だね
Summer is coming soon. もうすぐ夏だね
It's going to get hotter soon. もうすぐもっと暑くなるね
Spring is approaching soon. もう春です
Summer is approaching soon.もうすぐ夏です
It will be hotter soon. もうすぐもっと暑くなります
It's almost time to head home, isn't it? もうすぐ帰る時間だね
It is almost time to leave for home.もうすぐ帰る時間です
You're not feeling unwell, right? 体調は悪くないよね
You are not feeling unwell, are you? 体調は悪くないです
You are looking well today. 今日も元気です
You're looking good today. 今日も元気だね
You seem the same as always. いつもと変わらないね
You do not seem different from usual. いつもと変わりません
You slept well last night, didn't you? 昨日はよく寝れたね
You slept well last night. 昨日はよく寝ました
You had lunch, right? ご飯食べたよね
You're full, aren't you?. お腹いっぱいだね
It was delicious. おいしかったよね
You have had lunch, haven't you? ご飯食べました
You are full. お腹いっぱいです
It was delicious. おいしかったです
You are listening well. よく聞いてます
You're listening well, aren't you? よく聞いてるね
Just a little more, right? もう少しだね
It is just a little more. もう少しです
You are doing well. しっかりできてます
You are doing your best with rehab.リハビリ頑張ります
You're doing well, aren't you? しっかりできてるね
You're going to do your best with rehab, right? リハビリ頑張ろうね

Figure 1. Dialogue examples for different conditions [Asking condition, informing condition, and mixed condition (asking and informing)]. The verbal prompts were conducted in Japanese with English translations provided.

2.4. Procedure

The measurement protocol is shown in Figure 2. Each session proceeded as follows:

1. Control condition: 1 min EEG recording in silence, followed by 1 min recording with white noise;
2. 5 min break;
3. Mother's speaking conditions: asking, informing, and mixed, with 5 min breaks between each condition;
4. 10 min break;
5. Repeat of control condition;
6. PT's speaking conditions: asking, informing, and mixed, with 5 min breaks between each condition.

The measurements were conducted over two days, with the order of the mothers' and PTs' speaking reversed between the first and second measurements. The order of asking, informing, and mixed conditions was randomized across sessions to control for order effects. The measurement environment was a quiet, temperature-controlled room in the hospital, designed to minimize external distractions and ensure participant comfort.

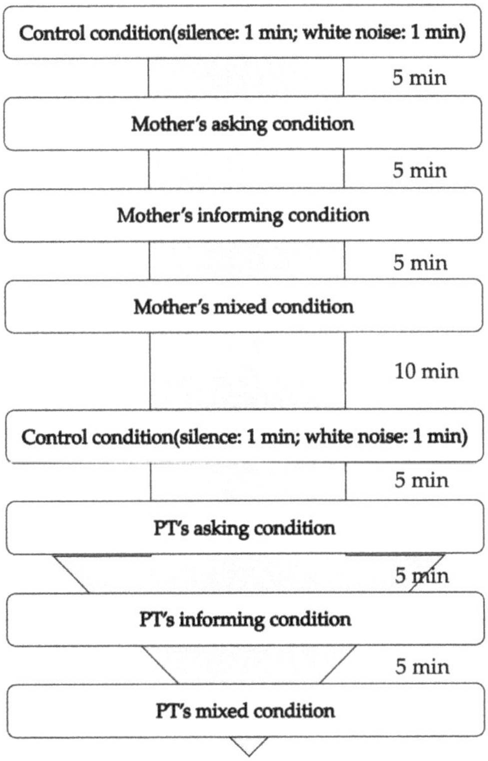

Figure 2. Measurement protocol. Twenty sentences were used for the asking condition, twenty sentences for the informing condition, and forty sentences for the mixed condition. Each sentence was read aloud at three-second intervals, with the asking condition and informing condition lasting approximately three minutes, and the mixed condition lasting about six minutes. Each condition was randomly administered in the first and second measurements.

2.5. EEG Recording and Analysis

Brain neural activity was recorded using the Polymatepro6100 bio-signal recording device (Miyuki Giken Co., Ltd., Tokyo, Japan), based on the international 10–20 system. Measurements were taken from 19 locations: Fp1, Fp2, F3, F4, F7, F8, Fz, C3, C4, Cz, P3, P4, Pz, O1, O2, T3, T4, T5, and T6. Electrode impedances were kept below 5 kΩ. The reference electrodes were placed on the earlobes. The frequency bands analyzed were theta (θ) (4–7 Hz), Alpha (α) (8–13 Hz), and Beta (β) (14–20 Hz). To accurately extract neural activity during verbal encouragement, the muscle activity of the speaker's suprahyoid muscles was measured as a trigger, allowing synchronization with the child's EEG measurements. The bandpass filter in the EEG measurement was 0.5–30 Hz, and the sampling rate was 1000 Hz. EEG data analysis was conducted using the Electro Magnetic Source Estimation program (EMSE) (Cortech Solutions, Inc., version 5.5.2). Data for each condition were extracted based on triggers, isolating the waveforms for 2 s after the end of each verbal encouragement stimulus. The data for each condition were then averaged to create the analysis dataset. Independent Component Analysis (ICA) was performed using Matrix Laboratory (MATLAB) (v.R2023b Update 6, MathWorks, 2023). ICA was chosen for its ability to separate EEG data into statistically independent components, allowing for effective artifact removal. Components identified as noise (e.g., eye blinks and muscle artifacts) were visually inspected and removed.

The ICA frequency analysis was conducted using Transposed fICA Networks within exact Low-Resolution Brain Electromagnetic Tomography (eLORETA, version 20160611). This method allows for the three-dimensional visualization of neural activity and the identification of brain functional networks. The fICA networks jointly represent space and frequency interactions, producing three sets of images (θ-wave, α-wave, and β-wave bands). This approach enables the identification of regions working together within each condition and captures coupling between frequencies.

2.6. Statistical Analysis

Due to the small sample size, formal statistical tests were not performed. Instead, descriptive analyses were conducted to explore patterns in the EEG data across conditions and participants. The activity values of the regions showing the maximum and minimum activity values were extracted from the activity values calculated by the eLORETA analysis for qualitative comparison.

3. Results

3.1. Participant Baseline Characteristics

The characteristics of the three participants are summarized in Table 1.

Table 1. Participant characteristics.

	Participant Characteristics		
	A	**B**	**C**
Age (years)	3	7	9
Gender (boy/girl)	boy	girl	girl
Disease	cerebral palsy (CP)		
Classified	severe psychosomatic disorder		
Verbal Communication	vocalizing but struggling to produce meaningful words		
Motor Level	difficulty turning over; unable to move voluntarily	difficulty turning over; unable to move voluntarily	able to crawl a few meters

Among the three participants, one was a boy (Participant A, age: 3 years) and two were girls (Participant B, age: 7 years; Participant C, age: 9 years). All participants had CP-SMID, classified as having severe psychosomatic disorders. Moreover, they all faced challenges in verbal communication, vocalizing but struggling to produce meaningful words. The motor level of Participant A was such that he had difficulty turning over and was unable to move voluntarily. The motor level of Participant B was similar to Participant A, as she had difficulty turning over and was unable to move voluntarily. However, the motor level of Participant C had developed to the point where she could crawl a few meters.

3.2. EEG Analysis

The results for the main brain activity areas in each frequency band for subjects A, B, and C under each condition are summarized in Table 2.

Table 2. Summary of the main brain activity areas in each frequency band under each condition.

Participant		Condition	Activity (Power Value)	θ-Wave	α-Wave	β-Wave
A	PT's	informing condition	amplification (max)	somatosensory association areas (1.64×10^0)	somatosensory association areas (1.54×10^0)	right superior temporal gyrus (1.27×10^0)
			reduction (minimum)	right inferior frontal gyrus (-1.75×10^0)	visual association area (-1.45×10^0)	
		asking condition	amplification (max)	frontal eye fields (1.68×10^0)	right visual association area (7.79×10^0)	frontal eye fields (1.02×10^0)
			reduction (minimum)	somatosensory association areas (1.67×10^0)	frontal poles (-8.78×10^0)	somatosensory association areas (-1.40×10^0)
		informing within mixed condition	amplification (max)	right frontal pole (8.63×10^{-1})	right visual association area (7.26×10^{-1})	somatosensory association areas (1.33×10^0)
			reduction (minimum)	somatosensory association areas (-8.54×10^{-1})	supplementary motor areas (-6.7×10^{-1})	frontal eye fields (-1.29×10^0)
		asking within mixed condition	amplification (max)	supplementary motor areas (1.32×10^0)	supplementary motor areas (1.06×10^0)	
			reduction (minimum)	orbitofrontal cortex (-1.39×10^0)	orbitofrontal cortex (-1.03×10^0)	somatosensory association areas (-1.54×10^0)
	mother's	informing condition	amplification (max)	frontal eye fields (7.76×10^{-1})	somatosensory association areas (7.78×10^{-1})	somatosensory association areas (-1.32×10^0)
			reduction (minimum)	visual association areas (-9.43×10^0)	supplementary motor areas (-7.93×10^0)	left frontal pole (-1.46×10^0)
		asking condition	amplification (max)	somatosensory association areas (1.08×10^0)	visual association areas (1.15×10^0)	somatosensory association areas (1.45×10^0)
			reduction (minimum)	frontal eye fields (-1.38×10^0)	frontal eye fields (-9.59×10^0)	Inferior temporal gyrus (-1.40×10^0)
		informing within mixed condition	amplification (max)	right somatosensory association area (1.33×10^0)	right visual association area (1.51×10^0)	supplementary motor areas (1.66×10^0)
			reduction (minimum)	right dorsolateral prefrontal cortex (-1.50×10^0)	left orbitofrontal cortex (-8.17×10^0)	right orbitofrontal cortex (-1.36×10^0)
		asking within mixed condition	amplification (max)	supplementary motor areas (2.10×10^0)	supplementary motor areas (1.40×10^0)	
			reduction (minimum)		frontal poles (-1.25×10^0)	orbitofrontal cortex (-1.98×10^0)

Table 2. *Cont.*

Participant		Condition	Activity (Power Value)	θ-Wave	α-Wave	β-Wave
B	PT's	informing condition	amplification (max)		dorsal posterior cingulate cortex (2.1×10^0)	
			reduction (minimum)	dorsolateral prefrontal cortex (-3.64×10^0)	supplementary motor areas (-2.46×10^0)	frontal poles (-3.13×10^0)
		asking condition	amplification (max)	orbitofrontal cortices (2.01×10^0)	somatosensory association areas (1.56×10^0)	frontal poles (1.65×10^0)
			reduction (minimum)	supplementary motor areas (-1.93×10^0)	supplementary motor areas (-1.69×10^0)	supplementary motor areas (-1.57×10^0)
		informing within mixed condition	amplification (max)	somatosensory association (2.18×10^0)	right frontal pole (1.36×10^0)	frontal poles (3.16×10^0)
			reduction (minimum)	orbitofrontal cortex (-2.67×10^0)	visual association areas (-2.45×10^0)	somatosensory association areas (-3.30×10^0)
		asking within mixed condition	amplification (max)	frontal pole (8.07×10^{-1})	right frontal pole (2.09×10^0)	somatosensory association areas (1.26×10^0)
			reduction (minimum)	dorsolateral prefrontal cortex (-8.12×10^{-1})		frontal pole (-2.39×10^0)
	mother's	informing condition	amplification (max)	left inferior frontal gyrus (1.88×10^0)	left visual association area (1.25×10^0)	right middle temporal gyrus (1.97×10^0)
			reduction (minimum)	somatosensory association areas (-2.73×10^0)	left orbitofrontal cortex (-1.42×10^0)	left somatosensory association area (-2.53×10^0)
		asking condition	amplification (max)	right middle temporal gyrus (2.29×10^0)	right dorsolateral prefrontal cortex (1.52×10^0)	left somatosensory association area (2.58×10^0)
			reduction (minimum)		right visual association areas (-1.32×10^0)	left frontal pole (-2.82×10^0)
		informing within mixed condition	amplification (max)			left frontal eye fields (2.16×10^0)
			reduction (minimum)	left frontal pole (-4.23×10^0) right fusiform gyrus (-3.97×10^0)	visual association areas (-2.22×10^0)	visual association areas (-3.04×10^0)
		asking within mixed condition	amplification (max)	left frontal pole (1.21×10^0)	somatosensory areas (2.98×10^0)	middle temporal gyri (1.66×10^0)
			reduction (minimum)		frontal poles (-2.50×10^0)	right frontal pole (-1.89×10^0)

Table 2. *Cont.*

Participant		Condition	Activity (Power Value)	θ-Wave	α-Wave	β-Wave
C	PT's	informing condition	amplification (max)	left frontal pole (2.13×10^0)	right dorsolateral prefrontal cortex (1.79×10^0)	right dorsolateral prefrontal cortex (1.28×10^0)
			reduction (minimum)	left visual association area (-2.11×10^0)	right visual association area (-1.42×10^0)	left somatosensory association area (-1.25×10^0)
		asking condition	amplification (max)	somatosensory association areas (1.37×10^0)		somatosensory association areas (1.33×10^0)
			reduction (minimum)	left orbitofrontal cortex(-1.41×10^0)	orbitofrontal cortex (-2.52×10^0)	middle temporal gyrus (-1.38×10^0)
		informing within mixed condition	amplification (max)	superior temporal gyrus (7.87×10^{-1})		right somatosensory association area (1.35×10^0)
			reduction (minimum)	somatosensory association areas (-7.29×10^{-1})	left frontal pole (-2.36×10^0)	orbitofrontal cortices (-2.01×10^0)
		asking within mixed condition	amplification (max)			ventral prefrontal cortex (1.84)
			reduction (minimum)	orbitofrontal cortices (-1.1×10^0)	right frontal pole (-9.80×10^{-1})	
	mother's	informing condition	amplification (max)		orbitofrontal cortices (1.91×10^0)	ventral prefrontal cortex (1.32×10^0)
			reduction (minimum)	somatosensory association area (-1.88×10^0)		
		asking condition	amplification (max)	right frontal pole (1.71×10^0)	left somatosensory association area ($8 68 \times 10^{-1}$)	right dorsolateral prefrontal cortex (1.29×10^0)
			reduction (minimum)		supplementary motor areas (-1.01×10^0)	
		informing within mixed condition	amplification (max)	right inferior temporal gyrus (5.82×10^{-1})	frontal poles (1.40×10^0)	right frontal pole (8.96×10^{-1})
			reduction (minimum)	left visual association area (-6.37×10^{-1})	right visual association area (-1.04×10^0)	right angular gyrus (-1.55×10^0)
		asking within mixed condition	amplification (max)	somatosensory association area (1.27×10^0)		frontal eye fields (1.44×10^0)
			reduction (minimum)	left frontal pole (-1.16×10^0)	frontal poles (-7.69×10^{-1})	left orbitofrontal cortex (-1.55×10^0)

3.2.1. Participant A

The results of the frequency analysis by ICA for Participant A are shown in Figure 3.

condition	informing	informing within mixed	asking	asking within mixed	informing	informing within mixed	asking	asking within mixed
	PT	PT	PT	PT	mother	mother	mother	mother
θ -wave	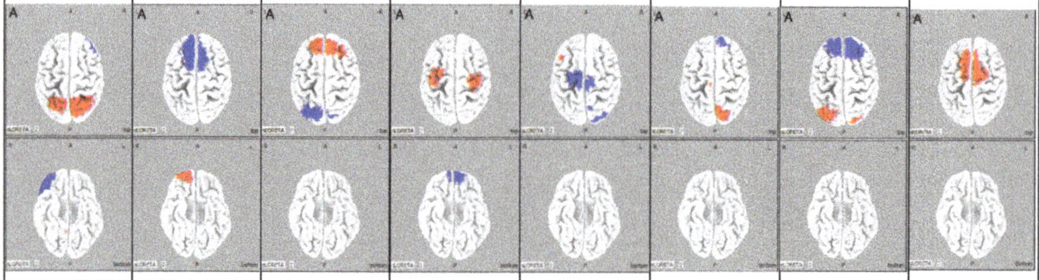							
α-wave	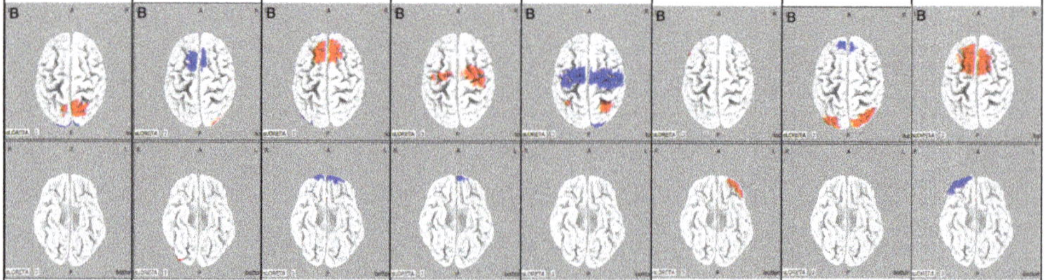							
β-wave	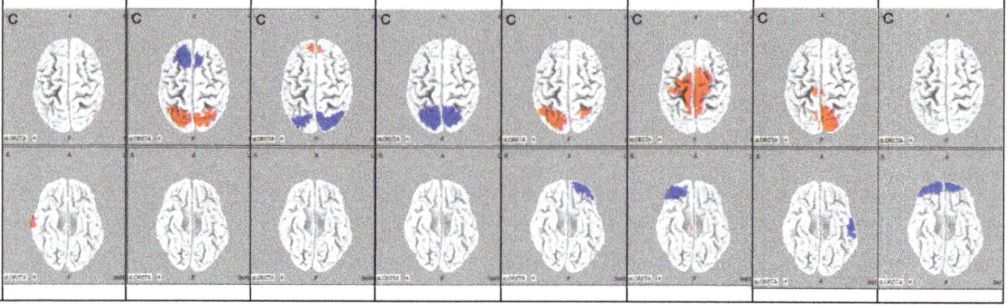							

Figure 3. Participant A: The θ wave band for each condition is designated as A, the α wave band as B, and the β wave band as C. The columns are classified by frequency, and the rows are classified by condition. Regions of attenuated activity are shown in blue, whereas regions of amplified activity are shown in red.

The results of the ICA frequency analysis for Participant A indicated the following findings. In the PT informing condition, in the θ wave band, there was an amplification in activity in the bilateral somatosensory association areas and a reduction in activity in the right inferior frontal gyrus; in the α wave band, there was an amplification in activity in the bilateral somatosensory association areas and a reduction in activity in the bilateral visual association area; in the β wave band, there was an amplification in activity in the right superior temporal gyrus and the right middle temporal gyrus. In the mother informing condition, in the θ wave band, there was an amplification in activity in the bilateral frontal eye fields, along with a reduction in activity in the bilateral supplementary motor areas and bilateral visual association areas; in the α wave band, there was an amplification in activity in the bilateral somatosensory association areas and a reduction in activity in the bilateral supplementary motor areas and bilateral visual association areas; in the β wave band, there was an amplification in activity in the bilateral somatosensory association areas and the left superior temporal gyrus, with a reduction in activity in the left frontal pole.

In the PT informing within mixed condition, in the θ wave band, there was an amplification in activity in the right frontal pole and a reduction in activity in the bilateral somatosensory association areas; in the α wave band, there was an amplification in activity in the right visual association area and a reduction in activity in the bilateral supplementary motor areas; and in the β wave band, there was an amplification in activity in the bilateral somatosensory association areas and a reduction in activity in the bilateral frontal eye fields. In the mother informing within mixed condition, in the θ wave band, there was a reduction in activity in the right somatosensory association area and a reduction in activity in the right dorsolateral prefrontal cortex; in the α wave band, there was an amplification in activity in the left orbitofrontal cortex; and in the β wave band, there was an amplification in activity in the bilateral supplementary motor areas and a reduction in activity in the right orbitofrontal cortex.

In the PT asking condition, in the θ wave band, there was an amplification in activity in the bilateral frontal eye fields and a reduction in activity in the bilateral somatosensory association areas; in the α wave band, there was an amplification in activity in the bilateral supplementary motor areas and a reduction in activity in the bilateral frontal poles and the left visual association area; and in the β wave band, there was an amplification in activity in the frontal eye fields and a reduction in activity in the somatosensory association areas. In the mother asking condition, in the θ wave band, there was an amplification in activity in the bilateral somatosensory association areas and a reduction in activity in the frontal eye fields; in the α wave band, there was an amplification in activity in the bilateral visual association areas and a reduction in activity in the frontal eye fields; and in the β wave band, there was an amplification in activity in the somatosensory association areas and a reduction in activity in the inferior temporal gyrus.

In the PT asking within mixed condition, in the θ wave band, there was an amplification in activity in the supplementary motor areas and a reduction in activity in the orbitofrontal cortex; in the α wave band, there was an amplification in activity in the supplementary motor areas and a reduction in activity in the orbitofrontal cortex; and in the β wave band, a reduction in activity was observed in the bilateral somatosensory association areas. In the mother asking within mixed condition, in the θ wave band, there was an amplification in activity in the supplementary motor areas; in the α wave band, there was an amplification in activity in the supplementary motor areas and a reduction in activity in the frontal poles; and in the β wave band, there was a reduction in activity in the orbitofrontal cortex.

Summarizing the results for Participant A, under the PT informing condition, somatosensory information processing increased, whereas visual information processing was suppressed. Conversely, under the mother informing condition, motor preparation was carried out quickly. Additionally, in the PT informing condition, frontal eye field activity increased, leading to controlled attention, whereas under the mother informing condition, somatosensory association area activity increased, directing attention toward bodily sensations. Under the PT informing within mixed condition, somatosensory information processing was suppressed, whereas decision-making increased. Similarly, under the mother informing within mixed condition, somatosensory information processing was suppressed, and motor preparation was carried out quickly. In the PT asking condition, somatosensory information processing was suppressed, whereas goal-directed behavior increased. Conversely, under the mother asking condition, somatosensory information processing increased. Under the PT asking within mixed condition, sensory information processing was suppressed, whereas concentration on attention tasks increased. However, in the mother asking within mixed condition, motor preparation was carried out quickly.

3.2.2. Participant B

The results of the frequency analysis by ICA for Participant B are shown in Figure 4.

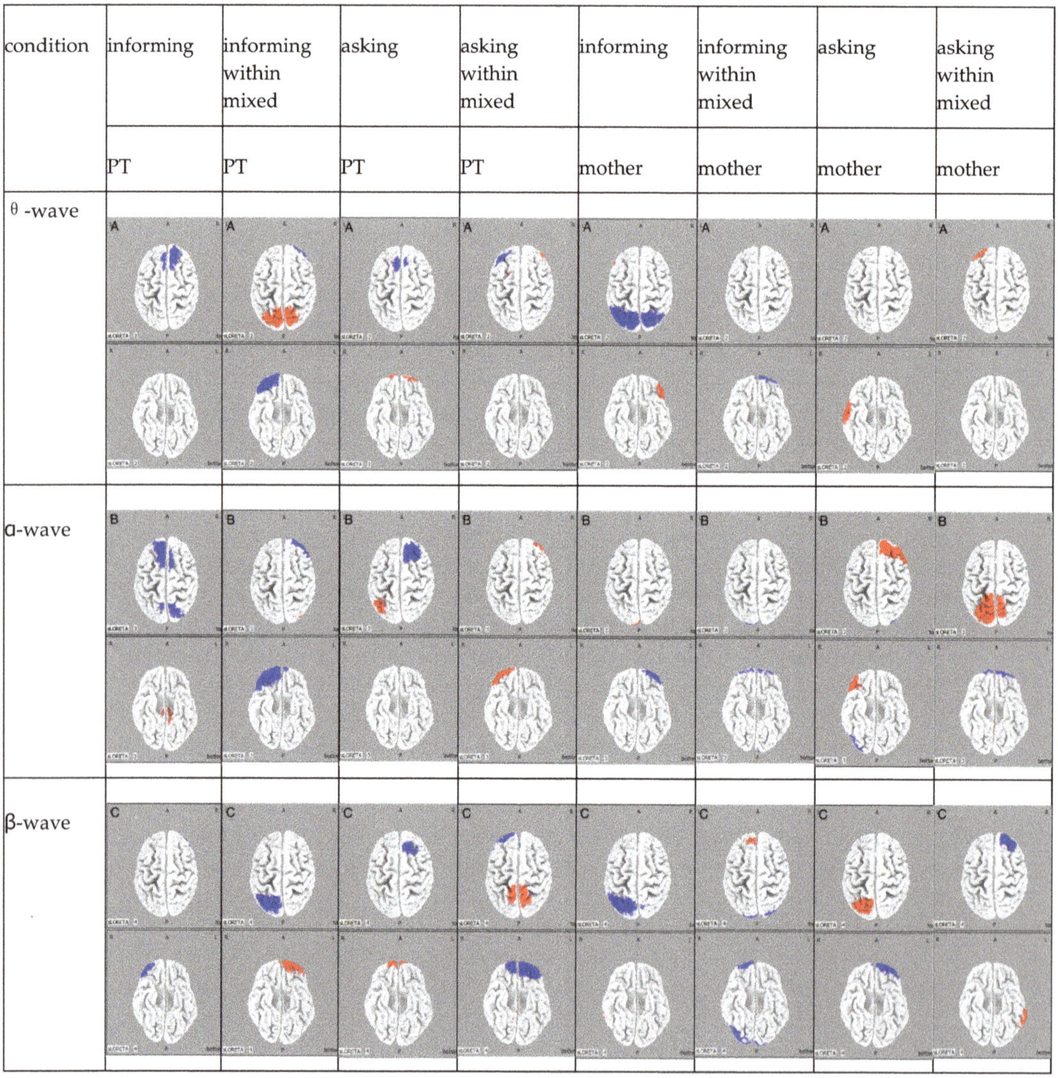

Figure 4. Participant B: The θ wave band for each condition is designated as A, the α wave band as B, and the β wave band as C. The columns are classified by frequency, and the rows are classified by condition. Regions of attenuated activity are shown in blue, whereas regions of amplified activity are shown in red.

The results of the ICA frequency analysis for Participant B indicated the following findings. In the PT informing condition, in the θ wave band, there was a reduction in activity in the dorsolateral prefrontal cortex; in the α wave band, there was an increase in activity in the bilateral dorsal posterior cingulate cortex, along with a reduction in activity in the bilateral somatosensory association areas and the bilateral supplementary motor areas; and in the β wave band, a reduction in activity was observed in the frontal poles. In the mother informing condition, in the θ wave band, there was an amplification in activity in the left inferior frontal gyrus, along with a reduction in activity in the bilateral somatosensory association areas; in the α wave band, there was an amplification in activity in the left visual association area and a reduction in activity in the left orbitofrontal cortex; and in the β wave band, there was an amplification in activity in the right middle temporal gyrus and a reduction in activity in the left somatosensory association area.

In the PT informing within mixed condition, in the θ wave band, there was an amplification in activity in the bilateral somatosensory association areas and a reduction in activity in the orbitofrontal cortex; in the α wave band, there was an amplification in activity in the right frontal pole and a reduction in activity in the visual association areas; and in the β wave band, there was an amplification in activity in the frontal poles and a reduction in activity in the somatosensory association areas. In the mother informing within mixed conditio, in the θ wave band, there was a reduction in activity in the left frontal pole and the right fusiform gyrus; in the α wave band, there was a reduction in activity in the bilateral orbitofrontal cortices and the bilateral visual association areas; and in the β wave band, there was an increase in activity in the left frontal eye fields and a reduction in activity in the right frontal pole and the bilateral visual association areas.

In the PT asking condition, in the θ wave band, there was an amplification in activity in the bilateral orbitofrontal cortices and a reduction in activity in the bilateral supplementary motor areas; in the α wave band, there was an amplification in activity in the somatosensory association areas and a reduction in the supplementary motor areas; and in the β wave band, there was an amplification in activity in the bilateral frontal poles and a reduction in the supplementary motor areas. In the mother asking condition, in the θ wave band, there was an amplification in activity in the right middle temporal gyrus; in the α wave band, there was an amplification in activity in the right dorsolateral prefrontal cortex, along with a reduction in activity in the right visual association areas; and in the β wave band, there was an amplification in activity in the left somatosensory association area and a reduction in activity in the left frontal pole.

In the PT asking within mixed condition, in the θ wave band, there was an amplification in activity in the frontal pole and a reduction in activity in the dorsolateral prefrontal cortex; in the α wave band, there was an amplification in activity in the right frontal pole; and in the β wave band, there was an amplification in activity in the somatosensory association areas and a reduction in activity in the frontal pole. In the mother asking within mixed condition, in the θ wave band, there was an amplification in activity in the left frontal pole; in the α wave band, there was an amplification in activity in the bilateral somatosensory areas and a reduction in activity in the bilateral frontal poles; and in the β wave band, there was an amplification in activity in the bilateral middle temporal gyri and a reduction in activity in the right frontal pole.

Summarizing the results for Participant B, under the PT informing condition, activity in the somatosensory association area was suppressed, whereas activity in the dorsal posterior cingulate cortex increased, indicating that sensory information processing was carried out selectively. Conversely, in the mother informing condition, activity increased in the visual association area and decreased in the orbitofrontal cortex, suggesting that visual information processing was prioritized. Additionally, in the PT informing condition, activity in the frontal pole increased, whereas activity in the supplementary motor area decreased, indicating that cognitive function was enhanced while motor preparation was suppressed. In the mother informing condition, activity increased in the middle temporal gyrus and the somatosensory association area, suggesting that attention was focused on language comprehension and bodily sensations. Under the PT informing within mixed condition, we observed a decrease in activity in the somatosensory association areas and an increase in activity in the frontal pole. This indicates suppression in attention to somatosensory information, whereas highlights an enhancement in decision-making. In the mother informing within mixed condition, simultaneous decrease in activity in the right frontal pole and an increase in activity in the left orbitofrontal cortex were observed. This indicates that higher cognitive processing was suppressed while decision-making was enhanced. Under the PT asked condition, activity in the frontal pole increased, whereas activity in the supplementary motor area decreased, indicating enhanced cognitive functions and reduced attention to movement. In the mother asking condition, activity increased in the somatosensory association area and decreased in the frontal pole, suggesting that sensory processing was enhanced while cognitive processing was diminished. Under the PT asking within mixed condition, activity increased in the somatosensory association area and decreased in the frontal pole, indicating that sensory processing was enhanced while cognitive processing was suppressed. However, in the mother asking within mixed condition, activity increased in the middle temporal gyri and decreased in the frontal pole, suggesting that language comprehension was enhanced while cognitive processing was suppressed.

3.2.3. Participant C

The results of the frequency analysis by ICA for pParticipant C are shown in Figure 5.

The results of the ICA frequency analysis for Participant C indicated the following findings. In the PT informing condition, in the θ wave band, there was an amplification in activity in the left frontal pole and a reduction in activity in the left visual association area; in the α wave band, there was an amplification in activity in the right dorsolateral prefrontal cortex and a reduction in activity in the right visual association area; and in the β wave band, there was an amplification in activity in the right dorsolateral prefrontal cortex and a reduction in activity in the left somatosensory association area. In the mother informing condition, in the θ wave band, there was a reduction in activity in the somatosensory association area; in the α wave band, there was an amplification in activity in the bilateral orbitofrontal cortices; and in the β wave band, there was an amplification in activity in the ventral prefrontal cortex.

In the PT informing within mixed condition, in the θ wave band, there was an amplification in activity in the superior temporal gyrus and a reduction in activity in the somatosensory association areas; in the α wave band, there was a reduction in activity in the left frontal pole; and in the β wave band, there was an amplification in activity in the right somatosensory association area and a reduction in activity in the bilateral orbitofrontal cortices. In the mother informing within mixed condition, in the θ wave band, there was an amplification in activity in the right inferior temporal gyrus and a reduction in activity in the left visual association area; in the α wave band, there was an amplification in activity in the bilateral frontal poles and a reduction in activity in the right visual association area; and in the β wave band, there was an amplification in activity in the right frontal pole and a reduction in activity in the right angular gyrus.

In the PT asking condition, in the θ wave band, there was an amplification in activity in the bilateral somatosensory association areas and a reduction in activity in the left orbitofrontal cortex; in the α wave band, there was a reduction in activity in the orbitofrontal cortex; and in the β wave band, there was an amplification in activity in the somatosensory association areas and a reduction in activity in the middle temporal gyrus. In the mother asking condition, in the θ wave band, there was an amplification in activity in the right frontal pole; in the α wave band, there was an amplification in activity in the left somatosensory association area and a reduction in activity in the bilateral supplementary motor areas; and in the β wave band, there was an amplification in activity in the right dorsolateral prefrontal cortex.

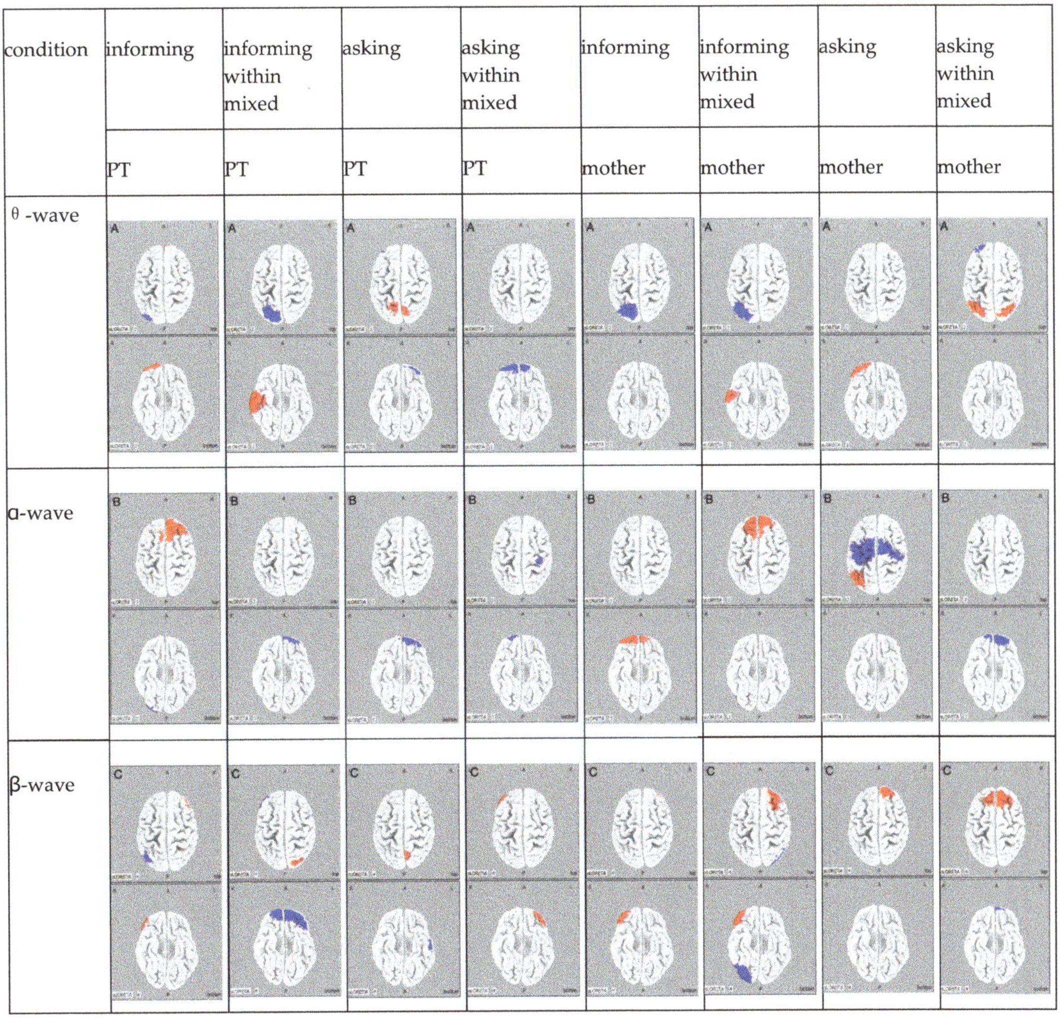

Figure 5. Participant C: The θ wave band for each condition is designated as A, the α wave band as B, and the β wave band as C. The columns are classified by frequency, and the rows are classified by condition. Regions of attenuated activity are shown in blue, whereas regions of amplified activity are shown in red.

In the PT asking within mixed condition, in the θ wave band, there was a reduction in activity in the bilateral orbitofrontal cortices; in the α wave band, there was a reduction in activity in the right frontal pole and the right somatosensory association area; and in the β wave band, there was an amplification in activity in the ventral prefrontal cortex. In the mother asking within mixed condition, in the θ wave band, there was an amplification in activity in the bilateral somatosensory association areas and a reduction in activity in the left frontal pole; in the α wave band, there was a reduction in activity in the bilateral frontal poles; and in the β wave band, there was an amplification in activity in the bilateral frontal eye fields and a reduction in activity in the left orbitofrontal cortex.

Summarizing the results for Participant C, under the PT informing condition, there was an increase in activity in the dorsolateral prefrontal cortex and a decrease in activity in the visual association cortex, suggesting that attention control was enhanced while visual information processing was suppressed. Under the mother informing condition, an increase in activity in the orbitofrontal cortex and ventromedial prefrontal cortex indicated that emotional processing was enhanced. Additionally, in the mother informing within mixed condition, an early increase in activity in the inferior temporal gyrus suggested that rapid language processing was taking place. In the PT informing condition, an increase in activity in the somatosensory association area and a decrease in activity in the orbitofrontal cortex indicated that attention was directed toward bodily sensations while emotional processing was suppressed. Under the mother informing condition, a decrease in activity in the supplementary motor area suggested that motor preparation was suppressed. In the PT informing within mixed condition, an increase in activity in the somatosensory association area and a decrease in activity in the orbitofrontal cortex indicated that sensory information processing was enhanced while behavioral inhibition was present. Under the mother informing within mixed condition, an increase in activity in the frontal pole and a decrease in activity in the angular gyrus suggested that decision-making was enhanced while responses to physical sensations were suppressed. Under the PT asking condition, an increase in activity in the somatosensory association areas and a decrease in activity in the middle temporal gyrus suggested that sensory information processing was enhanced while language processing was suppressed. Under the mother asking condition, a decrease in activity in the supplementary motor areas and an increase in activity in the dorsolateral prefrontal cortex indicated a reduction in the intention to move and an increase in emotional processing. In the PT asking within mixed condition, a decrease in activity in the somatosensory association areas and an increase in activity in the ventral prefrontal cortex indicated a reduction in sensory information processing and an increase in social behavior. In the mother asking within mixed condition, an amplification in activity in the frontal eye fields and a reduction in activity in the orbitofrontal cortex suggested an increase in emotional regulation and a decrease in social judgment.

4. Discussion

The ICA frequency analysis results for each participant revealed distinct brain activity patterns in response to verbal encouragement from the mother and PT, with variations observed across the θ (4–7 Hz), α (8–13 Hz), and β (14–30 Hz) frequency bands. These findings align with previous studies demonstrating the involvement of these frequency bands in cognitive processes such as attention, working memory, sensory processing, and motor control [17,18].

For Participant A, the PT informing condition showed an increase in β band activity in the somatosensory association cortex and a decrease in θ band activity in the inferior frontal gyrus, suggesting a focus on sensory information processing and a reduction in attention and working memory load. This aligns with prior research indicating the role of β oscillations in sensory processing and the inferior frontal gyrus in attention and working memory [19,20]. In the mother informing condition, α band activity increased in the frontal eye field, whereas β band activity decreased in the supplementary motor area and visual association area, indicating suppression of motor preparation and visual information processing in favour of auditory stimuli. These results are consistent with studies highlighting the role of alpha oscillations in auditory processing and the involvement of the frontal eye field, supplementary motor area, and visual association area in motor preparation and visual processing [21–23]. In the PT informing within mixed condition, activity increased in the right frontal pole within the θ wave band and decreased in the frontal eye fields within the β wave band, aligning with prior research on θ waves in decision-making and emotional processing and the prefrontal cortex in working memory [24–26]. Similarly, in the mother informing within mixed condition, α wave activity increased in the orbitofrontal cortex, whereas θ wave activity decreased in the somatosensory association areas, consistent with previous studies on α wave involvement in decision-making and emotional processing and sensory processing suppression in the somatosensory association areas [27,28]. For the PT asking condition, β wave activity increased in the frontal eye fields, whereas somatosensory association area activity decreased, supporting research on the orbitofrontal cortex's role in goal-directed behavior and the suppression of sensory information processing in the somatosensory association areas [18,29]. In the mother asking condition, θ wave activity decreased in the orbitofrontal cortex, whereas β wave activity increased, which is consistent with previous findings on the orbitofrontal cortex's role in attention tasks within the θ wave band and the somatosensory association areas' involvement in concentration and understanding within the β wave band [18,30]. For the PT asking within mixed condition, α wave activity decreased in the orbitofrontal cortex, alongside a reduction in β wave activity in the somatosensory association areas, in line with studies on the orbitofrontal cortex's role in internal cognitive processes and the β wave band's association with concentration and understanding [18,31]. In the mother asking within mixed condition, α wave activity increased in the supplementary motor area, whereas β wave activity decreased in the orbitofrontal cortex, mirroring research on the supplementary motor area's role in movement preparation and the orbitofrontal cortex's function in emotional suppression [26,32].

These findings suggest that during the informing condition, Participant A exhibited increased attention and engaged in external cognitive activities that support cognitive function. Conversely, during the asking condition, the child may have focused on intrinsic cognitive processes, prioritizing internal cognitive engagement over external prompts.

Participant B exhibited distinct brain activity patterns in response to verbal encouragement across conditions. In the PT informing condition, θ band activity decreased in the dorsolateral prefrontal cortex, α band activity increased in the dorsal posterior cingulate cortex, and β band activity decreased in the somatosensory association cortex and supplementary motor area. This pattern suggests suppressed working memory and attention to external stimuli, enhanced sensory information processing and attentional focus, and reduced interest in bodily sensory input and motor intent. These results align with previous research on the dorsolateral prefrontal cortex's role in working memory and attention [23], the dorsal posterior cingulate cortex's involvement in sensory processing and attention [33], and the somatosensory association cortex and supplementary motor area's role in sensory integration and motor planning [34]. In the mother informing condition, θ band activity increased in the left inferior frontal gyrus, β band activity decreased in the somatosensory association cortex, α band activity increased in the left visual association area, and θ band activity decreased in the left orbitofrontal cortex. This suggests enhanced working memory, reduced focus on bodily sensations, perceptual learning, and decreased attention to internal cognitive processes. These findings are supported by prior research on the inferior frontal gyrus in working memory [35], the somatosensory association cortex in body sensation processing [18], the visual association area in perceptual learning [36], and the orbitofrontal cortex's role in attention and internal cognition [31]. In the PT informing within mixed condition, α band activity increased in the frontal pole, whereas β band activity decreased in the somatosensory association areas, consistent with the right frontal pole's role in enhancing attention and the attenuation of somatosensory association areas in sensory information suppression [18,37]. In the mother informing within mixed condition, θ band activity decreased in the frontal pole, whereas β band activity increased in the frontal eye fields, aligning with studies linking θ wave reductions to decreased attention to cognitive tasks and the frontal eye fields to goal-directed behavior [29,38]. In the PT asking condition, α band activity increased in the somatosensory association areas, and β band activity increased in the frontal pole, aligning with findings on somatosensory association area activity in relaxation and the frontal pole's role in cognitive enhancement [39,40]. In the mother asking condition, α band activity increased in the dorsolateral prefrontal cortex, whereas β band activity decreased in the frontal pole, consistent with research on the dorsolateral prefrontal cortex's involvement in decision-making and the frontal pole's role in suppressing higher cognitive processing [41,42]. In the PT asking within mixed condition, θ band activity decreased in the dorsolateral prefrontal cortex, whereas α band activity increased in the right frontal pole, supporting research on the dorsolateral prefrontal cortex's role in suppressing working memory and external attention and the right frontal pole's role in decision-making [23,40]. In the mother asking within mixed condition, θ band activity increased in the frontal pole, whereas β band activity increased in the middle temporal gyri, aligning with studies on the frontal pole's role in decision-making and emotional processing [24,25] and the middle temporal gyri's function in language comprehension [43].

These findings suggest that during the informing condition, Participant B focused on sensory and cognitive processing, efficiently integrating information whereas suppressing external stimuli. In contrast, during the asking condition, emotional and cognitive processes appeared to be integrated, facilitating efficient and socially oriented information processing.

The results for Participant C showed that in the PT informing condition, there was an increase in β band activity in the left frontal pole and a decrease in α band activity in the left visual association area, an increase in θ band activity in the right dorsolateral prefrontal cortex and a decrease in α band activity in the right visual association area, and an increase in β band activity in the right dorsolateral prefrontal cortex and a decrease in θ band activity in the left somatosensory association area. This pattern suggests an enhancement of goal-directed behavior, decision-making, and emotional processing, an inhibition of visual information processing, increased concentration, planning, and sustained attention, and a suppression of responses to sensory information. These results are consistent with previous research indicating the involvement of the frontal pole in goal-directed behavior and decision-making [23,25], the visual association area in visual processing [44], the dorsolateral prefrontal cortex in decision-making, concentration, planning, and sustained attention [45], and the somatosensory association area in sensory processing and inhibition [18]. In the mother informing condition, a decrease in θ band activity in the somatosensory association area, an increase in α band activity in the orbitofrontal cortex, and an increase in β band activity in the inferior frontal gyrus were observed, suggesting a reduction in attention to bodily sensations, enhanced internal cognitive processes, and increased empathy and social behavior. These findings align with studies showing the role of the somatosensory association area in body sensation processing [28,46], the orbitofrontal cortex in internal cognitive processes [31,47], and the ventromedial prefrontal cortex in empathy and social behavior [48]. In the PT informing within mixed condition, there was an amplification in activity in the superior temporal gyrus in the θ wave band and a decrease in activity in the orbitofrontal cortex in the β wave band, which aligns with prior research regarding the role of the superior temporal gyrus in language processing and the role of the orbitofrontal cortex in emotional processing [26,48]. Similarly, in the mother informing within mixed condition, there was an amplification in activity in the inferior temporal gyrus in the θ wave band and an increase in activity in the frontal pole in the α and β wave bands, findings that are consistent with previous studies highlighting the role of the inferior temporal gyrus in language processing and the role of the frontal pole in decision-making and cognitive function enhancement [40,49]. In the PT asking condition, a decrease in activity was observed in the orbitofrontal cortex in the θ wave band, whereas activity in the somatosensory association areas increased in the β wave band, and activity in the middle temporal gyrus decreased. This supports prior research indicating the role of the orbitofrontal cortex in enhancing concentration on tasks, the somatosensory association areas in attention to bodily sensations, and the inhibition of language and auditory processing by the middle temporal gyrus [18,30]. In the mother asking condition, an amplification in activity was noted in the frontal pole in the θ wave band and an increase in activity in the dorsolateral prefrontal cortex in the β wave band, which aligns with previous studies highlighting the role of the frontal pole in decision-making and emotional processing and the dorsolateral prefrontal cortex in emotional processing [24–26]. In the PT asking within mixed condition, there was a decrease in activity in the somatosensory association areas in the α wave band, whereas activity in the ventromedial prefrontal cortex increased in the β wave band. This aligns with prior research indicating the role of the somatosensory association areas in the inhibition of sensory processing and the ventromedial prefrontal cortex in empathy and social behavior [23,50]. Additionally, in the mother asking within mixed condition, there was a decrease in activity in the frontal pole in the α wave band, while activity in the frontal eye fields increased in the β wave band, which aligns with previous studies highlighting the involvement of the frontal pole in higher-order brain functions and planning and the role of the frontal eye fields in emotional regulation [27,51].

These findings suggest that although Participant C's attention was directed toward auditory stimuli across all conditions, the child inhibited emotional evaluations and enhanced cognitive functions to engage in language processing during the PT asking and informing conditions, as well as the mother informing condition. In contrast, during the mother asking condition, the child may have engaged in emotional evaluations and focused attention on the social connection with the mother. Additionally, although attention was directed toward language processing during the PT's inquiries and instructions and the mother's instructions in the mixed condition, it is suggested that during the mother asking within mixed condition, the child focused on specific responses to the prompts.

The cognitive processes engaged by the participants differed not only depending on whether they were listening to instructions from the PT or the mother but also across different frequency bands. These frequency-specific effects align with previous research suggesting that different frequency bands are associated with distinct cognitive functions [52,53]. The theta band appears to be involved in attention, working memory, and internal cognitive processes [54,55], whereas the alpha band is associated with sensory processing, perceptual learning, and social behavior [56,57]. However, the beta band, seems to be involved in sensory processing, motor preparation, and goal-directed behavior [58,59].

The complex interplay between different frequency bands in mediating the cognitive processes engaged during verbal encouragement underscores the importance of considering these frequency-specific effects when designing interventions for children with CP-SMID. Previous studies have demonstrated that tailored interventions targeting specific frequency bands can enhance cognitive functions in various populations [60]. By selectively modulating the frequency bands associated with the desired cognitive processes, it may be possible to optimize the effectiveness of rehabilitation interventions for children with CP-SMID. However, it is important to acknowledge some limitations of the current study.

The small sample size restricts the generalizability of the findings, and potential confounding factors such as sleep conditions, diet, relationship with the child, and environmental variables at the time of measurement were not sufficiently controlled [61,62]. Additionally, the observed individual differences in the results of this study can be attributed to the lack of control for age and cognitive function level. There are no appropriate assessment scales to evaluate cognitive function levels in children with severe multiple disabilities due to cerebral palsy, making it difficult to conduct such evaluations. Future studies should aim to replicate these results in larger samples while accounting for these variables as rigorously as possible, including age as a controlling factor. Additionally, the absence of coherence and time-frequency analyses in the present study limits a more comprehensive understanding of the functional connectivity between brain regions and the temporal dynamics of brain activity [63,64]. Incorporating these advanced analytical techniques in future research could yield valuable insights into the neural mechanisms underlying the effects of verbal encouragement on cognitive processing in children with CP-SMID.

Despite these limitations, the present study provides novel evidence of the differential effects of verbal encouragement from mothers and PTs on cognitive processing in children with CP-SMID, as well as the frequency-specific nature of these effects. Emotions serve as the foundation of motivation; therefore, changes in emotional states resulting from verbal prompts may influence children's motivation. Additionally, the impact of these prompts on cognitive functions could lead to a greater understanding of tasks and related concepts among children. Thus, it can be concluded that these findings hold significant implications for optimizing rehabilitation interventions and fostering cognitive development in this population. By tailoring verbal encouragement to target specific frequency bands and cognitive processes, it may be possible to enhance the efficacy of these interventions and support cognitive development more effectively. Future research should build upon these findings by examining the effects of verbal encouragement in larger samples, across different developmental stages and contexts, and by employing more advanced analytical techniques to capture the intricate neural dynamics underlying these effects comprehensively.

5. Conclusions

This study provides key insights into the neural mechanisms underlying the effects of verbal encouragement on cognitive processing in children with CP-SMID. The findings indicate that cognitive engagement in these children varies based on the individual providing encouragement, their tone of voice, and the associated frequency bands of brain activity. Specifically, maternal encouragement appears to facilitate internal information processing and promote behavioral engagement, whereas therapist-led encouragement may play a role in emotion regulation. The findings indicate that verbal encouragement stimulates the emotions that are associated with changes in motivation and enhances cognitive activity. Thus, verbal encouragement plays a significant role in improving motivation, suggesting its potential to enhance the effectiveness of rehabilitation interventions. Furthermore, theta band activity is implicated in attention, working memory, and internal cognitive processes; alpha band activity is associated with sensory processing, perceptual learning, and social behavior; and beta band activity is linked to sensory processing, motor preparation, and goal-directed behavior.

These findings have significant implications for optimizing rehabilitation interventions and fostering cognitive development in children with CP-SMID. Tailoring verbal encouragement to target specific frequency bands and cognitive processes—while accounting for the individual providing encouragement and their tone of voice—could enhance the efficacy of these interventions. Furthermore, we are confident that this study can provide more effective interactions not only in rehabilitation settings but also in everyday life and therapeutic contexts to support children's cognitive and emotional development. However, limitations such as the small sample size and potential confounding variables underscore the need for further research. Additionally, this study focused on whether the tone of verbal prompts is teaching or asking, encompassing a variety of content ranging from motivational to everyday topics. Therefore, it is necessary to further examine the content of verbal prompts in the future and to advance research based on both the content and tone that are effective for motivation.

Future studies should aim to replicate these findings in larger samples while rigorously controlling for confounding factors. Moreover, future studies should aim to replicate these findings in larger samples with more homogeneous age groups or by controlling for age as a variable to better understand the influence of age on the observed neural activity patterns. Additionally, incorporating coherence and time-frequency analyses could offer deeper insights into the functional connectivity between brain regions and the temporal dynamics of neural responses to verbal encouragement. The application of advanced neuroimaging techniques, such as functional near-infrared spectroscopy (fNIRS) [65,66], could further elucidate the underlying neural mechanisms.

Beyond methodological improvements, future research should explore the effects of verbal encouragement across different developmental stages and contexts. Investigating these mechanisms in typically developing children and those with other neurodevelopmental disorders may help identify shared and distinct neural activity patterns, guiding the development of tailored interventions. Moreover, examining verbal encouragement in varied settings, such as classrooms or social interactions, could provide valuable insights into the generalizability of these findings. Addressing these research gaps will contribute to evidence-based interventions that support cognitive development and overall well-being in children with CP-SMID.

Author Contributions: Conceptualization, R.Y., H.I., K.K., K.O., O.K., K.M., S.M. and T.K.; methodology, R.Y., H.I., K.K., K.O., O.K., K.M., S.M. and T.K.; validation, R.Y.; formal analysis, R.Y.; investigation, R.Y.; resources, R.Y.; data curation, R.Y.; writing—original draft, R.Y.; writing—review and editing, R.Y., K.K., K.O., O.K., K.M., S.M. and T.K.; visualization, R.Y., H.I. and T.K.; supervision, R.Y. and T.K.; project administration, R.Y. All authors have read and agreed to the published version of the manuscript.

Funding: This research received no external funding.

Institutional Review Board Statement: This study was approved by the Research Ethics Committee of Kyoto Tachibana University (approval number: 22–25, 19 August 2022).

Informed Consent Statement: Informed consent was obtained from all participants or parents/guardians of participants involved in the study.

Data Availability Statement: We are unable to share these research data. We do not have consent from the research participants, and the data contain personal information that we cannot share without authorization. It is necessary for us to respect the privacy of the research participants, and sharing their personal information without permission would not be appropriate. We appreciate your understanding.

Conflicts of Interest: The authors declare no conflicts of interest.

References

1. Matteucci, G.; Guyoton, M.; Mayrhofer, J.M.; Auffret, M.; Foustoukos, G.; Petersen, C.C.H.; El-Boustani, S. Cortical sensory processing across motivational states during goal-directed behavior. *Neuron* **2022**, *110*, 4176–4193.e10. [CrossRef]
2. Kouneiher, F.; Charron, S.; Koechlin, E. Motivation and cognitive control in the human prefrontal cortex. *Nat. Neurosci.* **2009**, *12*, 939–945. [CrossRef] [PubMed]
3. McNaughton, H.; Fu, V. Intrinsic motivation. *Pract. Neurol.* **2023**, *23*, 489–492. [CrossRef]

4. Tatla, S.K.; Sauve, K.; Virji-Babul, N.; Holsti, L.; Butler, C.; Van Der Loos, H.F.M. Evidence for outcomes of motivational rehabilitation interventions for children and adolescents with cerebral palsy: An American academy for cerebral palsy and developmental medicine systematic review. *Dev. Med. Child Neurol.* **2013**, *55*, 593–601. [CrossRef] [PubMed]
5. LePine, J.A.; LePine, M.A.; Jackson, C.L. Challenge and Hindrance Stress: Relationships with Exhaustion, Motivation to Learn, and Learning Performance. *J. Appl. Psychol.* **2004**, *89*, 883–891. [CrossRef]
6. Ryan, R.M.; Mims, V.; Koestner, R. Relation of reward contingency and interpersonal context to intrinsic motivation: A review and test using cognitive evaluation theory. *J. Personal. Soc. Psychol.* **1983**, *45*, 736–750. [CrossRef]
7. Halamish, V.; Madmon, I.; Moed, A. Motivation to learn. *Exp. Psychol.* **2019**, *66*, 319–330. [CrossRef]
8. Núñez, L.; Midgley, N.; Capella, C.; Alamo, N.; Mortimer, R.; Krause, M. The therapeutic relationship in child psychotherapy: Integrating the perspectives of children, parents and therapists. *Psychother. Res.* **2021**, *31*, 988–1000. [CrossRef]
9. van Egmond, M.C.; Hanke, K.; Omarshah, T.T.; Navarrete Berges, A.; Zango, V.; Sieu, C. Self-esteem, motivation and school attendance among sub-Saharan African girls: A self-determination theory perspective. *Int. J. Psychol.* **2020**, *55*, 842–850. [CrossRef]
10. Bratton, S.C.; Ray, D.; Rhine, T.; Jones, L. The Efficacy of Play Therapy with Children: A Meta-Analytic Review of Treatment Outcomes. *Prof. Psychol. Res. Pract.* **2005**, *36*, 376–390. [CrossRef]
11. Franchak, J.M. The ecology of infants' perceptual-motor exploration. *Curr. Opin. Psychol.* **2020**, *32*, 110–114. [CrossRef] [PubMed]
12. Turner, J.C.; Meyer, D.K.; Schweinle, A. The Importance of emotion in theories of motivation: Empirical, methodological, and theoretical considerations from a goal theory perspective. *Int. J. Educ. Res.* **2003**, *39*, 375–393. [CrossRef]
13. Vogl, J.L.; Dunne, E.C.; Liu, C.; Bradley, A.; Rwei, A.; Lonergan, E.K.; Hopkins, B.S.; Kwak, S.S.; Simon, C.D.; Rand, C.M.; et al. Kangaroo Father Care: A pilot feasibility study of physiologic, biologic, and psychosocial measures to capture the effects of father-infant and mother-infant skin-to-skin contact in the Neonatal Intensive Care Unit. *Dev. Psychobiol.* **2021**, *63*, 1521–1533. [CrossRef]
14. King, G.; Williams, L.; Hahn Goldberg, S. Family-oriented services in pediatric rehabilitation: A scoping review and framework to promote parent and family wellness. *Child Care Health Dev.* **2017**, *43*, 334–347. [CrossRef]
15. Khetani, M.A.; Albrecht, E.C.; Jarvis, J.M.; Pogorzelski, D.; Cheng, E.; Choong, K. Determinants of change in home participation among critically illchildren. *Dev. Med. Child Neurol.* **2018**, *60*, 793–800. [CrossRef] [PubMed]
16. Sundara, M.; Zhou, Z.L.; Breiss, C.; Katsuda, H.; Steffman, J. Infants' developing sensitivity to native language phonotactics: A meta-analysis. *Cognition* **2022**, *221*, 104993. [CrossRef]
17. Başar, E.; Başar-Eroglu, C.; Karakaş, S.; Schürmann, M. Gamma, alpha, Delta, and theta oscillations govern cognitive processes. *Int. J. Psychophysiol.* **2001**, *39*, 241–248. [CrossRef] [PubMed]
18. Engel, A.K.; Fries, P. Beta-band oscillations—Signalling the status quo? *Curr. Opin. Neurobiol.* **2010**, *20*, 156–165. [CrossRef]
19. Pfurtscheller, G.; Woertz, M.; Müller, G.; Wriessnegger, S.; Pfurtscheller, K. Contrasting behavior of beta event-related synchronization and somatosensory evoked potential after median nerve stimulation during finger manipulation in man. *Neurosci. Lett.* **2002**, *323*, 113–116. [CrossRef]
20. Cho, S.S.; Pellecchia, G.; Ko, J.H.; Ray, N.; Obeso, I.; Houle, S.; Strafella, A.P. Effect of continuous theta burst stimulation of the right dorsolateral prefrontal cortex on cerebral blood flow changes during decision making. *Brain Stimul.* **2012**, *5*, 116–123. [CrossRef]
21. Strauß, A.; Wöstmann, M.; Obleser, J. Cortical alpha oscillations as a tool for auditory selective inhibition. *Front. Hum. Neurosci.* **2014**, *8*, 350. [CrossRef]
22. Babu Henry Samuel, I.; Wang, C.; Hu, Z.; Ding, M. The frequency of alpha oscillations: Task-dependent modulation and its functional significance. *Neuroimage* **2018**, *183*, 897–906. [CrossRef] [PubMed]
23. Clements, G.M.; Bowie, D.C.; Gyurkovics, M.; Low, K.A.; Fabiani, M.; Gratton, G. Spontaneous alpha and theta oscillations are related to complementary aspects of cognitive control in younger and older adults. *Front. Hum. Neurosci.* **2021**, *15*, 621620. [CrossRef] [PubMed]
24. Hsieh, L.-T.; Ranganath, C. Frontal midline theta oscillations during working memory maintenance and episodic encoding and retrieval. *Neuroimage* **2014**, *85*, 721–729. [CrossRef]
25. Romeo, Z.; Spironelli, C. Theta oscillations underlie the interplay between emotional processing and empathy. *Heliyon* **2024**, *10*, e34581. [CrossRef] [PubMed]
26. Schmidt, R.; Herrojo Ruiz, M.H.; Kilavik, B.E.; Lundqvist, M.; Starr, P.A.; Aron, A.R. Beta oscillations in working memory, executive control of movement and thought, and sensorimotor function. *J. Neurosci.* **2019**, *39*, 8231–8238. [CrossRef]
27. Klimesch, W. EEG alpha and theta oscillations reflect cognitive and memory performance: A review and analysis. *Brain Res. Brain Res. Rev.* **1999**, *29*, 169–195. [CrossRef]
28. Caplan, J.B.; Madsen, J.R.; Schulze-Bonhage, A.; Aschenbrenner-Scheibe, R.; Newman, E.L.; Kahana, M.J. Human theta oscillations related to sensorimotor integration and spatial learning. *J. Neurosci.* **2003**, *23*, 4726–4736. [CrossRef]
29. Chikazoe, J.; Konishi, S.; Asari, T.; Jimura, K.; Miyashita, Y. Activation of right inferior frontal gyrus during response inhibition across response modalities. *J. Cogn. Neurosci.* **2007**, *19*, 69–80. [CrossRef]

30. Knyazev, G.G. EEG correlates of self-referential processing. *Front. Hum. Neurosci.* **2013**, *7*, 264. [CrossRef]
31. Compton, R.J.; Arnstein, D.; Freedman, G.; Dainer-Best, J.D.; Liss, A. Cognitive control in the intertrial interval: Evidence from EEG alpha power. *Psychophysiology* **2011**, *48*, 583–590. [CrossRef] [PubMed]
32. Spooner, R.K.; Wilson, T.W. Cortical theta–gamma coupling governs the adaptive control of motor commands. *Brain Commun.* **2022**, *4*, fcac249. [CrossRef] [PubMed]
33. Vanneste, S.; Joos, K.; Ost, J.; De Ridder, D.D. Influencing connectivity and cross-frequency coupling by real-time source localized neurofeedback of the posterior cingulate cortex reduces tinnitus related distress. *Neurobiol. Stress.* **2018**, *8*, 211–224. [CrossRef]
34. Kandel, E.R.; Schwartz, J.H.; Jessell, T.M.; Siegelbaum, S.A.; Hudspeth, A.J. *Principles of Neural Science*; McGraw-Hill: New York, NY, USA, 2012; Volume 5.
35. Jensen, O.; Tesche, C.D. Frontal theta activity in humans increases with memory load in a working memory task. *Eur. J. Neurosci.* **2002**, *15*, 1395–1399. [CrossRef]
36. Sigala, R.; Haufe, S.; Roy, D.; Dinse, H.R.; Ritter, P. The role of alpha-rhythm states in perceptual learning: Insights from experiments and computational models. *Front. Comput. Neurosci.* **2014**, *8*, 36. [CrossRef]
37. Berger, A.M.; Davelaar, E.J. Frontal alpha oscillations and attentional control: A virtual reality neurofeedback study. *Neuroscience* **2018**, *378*, 189–197. [CrossRef]
38. Nigbur, R.; Ivanova, G.; Stürmer, B. Theta power as a marker for cognitive interference. *Clin. Neurophysiol.* **2011**, *122*, 2185–2194. [CrossRef]
39. Brickwedde, M.; Krüger, M.C.; Dinse, H.R. Somatosensory alpha oscillations gate perceptual learning efficiency. *Nat. Commun.* **2019**, *10*, 263. [CrossRef] [PubMed]
40. Doya, K. Reinforcement learning: Computational theory and biological mechanisms. *HFSP J.* **2007**, *1*, 30–40. [CrossRef]
41. Hwang, K.; Ghuman, A.S.; Manoach, D.S.; Jones, S.R.; Luna, B. Frontal preparatory neural oscillations associated with cognitive control: A developmental study comparing young adults and adolescents. *Neuroimage* **2016**, *136*, 139–148. [CrossRef]
42. Stoll, F.M.; Wilson, C.R.E.; Faraut, M.C.M.; Vezoli, J.; Knoblauch, K.; Procyk, E. The Effects of Cognitive Control and Time on Frontal Beta Oscillations. *Cereb. Cortex* **2016**, *26*, 1715–1732. [CrossRef] [PubMed]
43. Piai, V.; Roelofs, A.; Rommers, J.; Maris, E. Beta oscillations reflect memory and motor aspects of spoken word production. *Hum. Brain Mapp.* **2015**, *36*, 2767–2780. [CrossRef]
44. Kravitz, D.J.; Saleem, K.S.; Baker, C.I.; Mishkin, M. A new neural framework for visuospatial processing. *Nat. Rev. Neurosci.* **2011**, *12*, 217–230. [CrossRef] [PubMed]
45. Tanji, J.; Hoshi, E. Role of the lateral prefrontal cortex in executive behavioral control. *Physiol. Rev.* **2008**, *88*, 37–57. [CrossRef] [PubMed]
46. Chen, T.L.; Babiloni, C.; Ferretti, A.; Perrucci, M.G.; Romani, G.L.; Rossini, P.M.; Tartaro, A.; Del Gratta, C. Human secondary somatosensory cortex is involved in the processing of somatosensory rare stimuli: An fMRI study. *Neuroimage* **2008**, *40*, 1765–1771. [CrossRef]
47. Beer, J.S.; John, O.P.; Scabini, D.; Knight, R.T. Orbitofrontal cortex and social behavior: Integrating self-monitoring and emotion-cognition interactions. *J. Cogn. Neurosci.* **2006**, *18*, 871–879. [CrossRef]
48. Moore, R.C.; Dev, S.I.; Jeste, D.V.; Dziobek, I.; Eyler, L.T. Distinct neural correlates of emotional and cognitive empathy in older adults. *Psychiatry Res.* **2015**, *232*, 42–50. [CrossRef]
49. Pu, Y.; Cheyne, D.; Sun, Y.; Johnson, B.W. Theta oscillations support the interface between language and memory. *Neuroimage* **2020**, *215*, 116782. [CrossRef]
50. Nelson, B.D.; Sarapas, C.; Robison-Andrew, E.J.R.; Altman, S.E.; Campbell, M.L.; Shankman, S.A. Frontal brain asymmetry in depression with comorbid anxiety: A neuropsychological investigation. *J. Abnorm. Psychol.* **2012**, *121*, 579–591. [CrossRef]
51. Mesa-Gresa, P.; Gil-Gómez, J.-A.; Lozano-Quilis, J.A.; Schoeps, K.; Montoya-Castilla, I. Electrophysiological correlates of the emotional response on brain activity in adolescents. *Biomed. Signal Process. Control* **2024**, *89*, 105754. [CrossRef]
52. Başar, E.; Başar-Eroglu, C.; Karakaş, S.; Schürmann, M. Are cognitive processes manifested in event-related gamma, alpha, theta and delta oscillations in the EEG? *Neurosci. Lett.* **1999**, *259*, 165–168. [CrossRef]
53. Jensen, O.; Kaiser, J.; Lachaux, J.-P. Human gamma-frequency oscillations associated with attention and memory. *Trends Neurosci.* **2007**, *30*, 317–324. [CrossRef]
54. Aftanas, L.I.; Golocheikine, S.A. Human anterior and frontal midline theta and lower alpha reflect emotionally positive state and internalized attention: High-resolution EEG investigation of meditation. *Neurosci. Lett.* **2001**, *310*, 57–60. [CrossRef] [PubMed]
55. Sauseng, P.; Klimesch, W.; Freunberger, R.; Pecherstorfer, T.; Hanslmayr, S.; Doppelmayr, M. Relevance of EEG alpha and theta oscillations during task switching. *Exp. Brain Res.* **2006**, *170*, 295–301. [CrossRef] [PubMed]
56. Klimesch, W.; Doppelmayr, M.; Russegger, H.; Pachinger, T.; Schwaiger, J. Induced alpha band power changes in the human EEG and attention. *Neurosci. Lett.* **1998**, *244*, 73–76. [CrossRef]
57. Toscani, M.; Marzi, T.; Righi, S.; Viggiano, M.P.; Baldassi, S. Alpha waves: A neural signature of visual suppression. *Exp. Brain Res.* **2010**, *207*, 213–219. [CrossRef] [PubMed]

58. Pfurtscheller, G.; Stancák, A., Jr.; Neuper, C. Event-related synchronization (ERS) in the alpha band—An electrophysiological correlate of cortical idling: A review. *Int. J. Psychophysiol.* **1996**, *24*, 39–46. [CrossRef]
59. Zhang, Y.; Chen, Y.; Bressler, S.L.; Ding, M. Response preparation and inhibition: The role of the cortical sensorimotor beta rhythm. *Neuroscience* **2008**, *156*, 238–246. [CrossRef]
60. Gruzelier, J.H. EEG-neurofeedback for optimising performance. I: A review of cognitive and affective outcome in healthy participants. *Neurosci. Biobehav. Rev.* **2014**, *44*, 124–141. [CrossRef]
61. Benca, R.M.; Obermeyer, W.H.; Larson, C.L.; Yun, B.; Dolski, I.; Kleist, K.D.; Weber, S.M.; Davidson, R.J. EEG alpha power and alpha power asymmetry in sleep and wakefulness. *Psychophysiology* **1999**, *36*, 430–436. [CrossRef]
62. Smith, E.E.; Zambrano-Vazquez, L.; Allen, J.J.B. Patterns of alpha asymmetry in those with elevated worry, trait anxiety, and obsessive-compulsive symptoms: A test of the worry and avoidance models of alpha asymmetry. *Neuropsychologia* **2016**, *85*, 118–126. [CrossRef] [PubMed]
63. Astolfi, L.; Cincotti, F.; Mattia, D.; Marciani, M.G.; Baccalà, L.A.; de Vico Fallani, F.; Salinari, S.; Ursino, M.; Zavaglia, M.; Babiloni, F. Assessing cortical functional connectivity by partial directed coherence: Simulations and application to real data. *IEEE Trans. Biomed. Eng.* **2006**, *53*, 1802–1812. [CrossRef] [PubMed]
64. Lachaux, J.P.; Rodriguez, E.; Martinerie, J.; Varela, F.J. Measuring phase synchrony in brain signals. *Hum. Brain Mapp.* **1999**, *8*, 194–208. [CrossRef]
65. Mihara, M.; Miyai, I.; Hattori, N.; Hatakenaka, M.; Yagura, H.; Kawano, T.; Okibayashi, M.; Danjo, N.; Ishikawa, A.; Inoue, Y.; et al. Neurofeedback using real-time near-infrared spectroscopy enhances motor imagery related cortical activation. *PLoS ONE* **2012**, *7*, e32234. [CrossRef]
66. Strait, M.; Scheutz, M. What we can and cannot (yet) do with functional near infrared spectroscopy. *Front. Neurosci.* **2014**, *8*, 117. [CrossRef]

brain sciences

MDPI

Article

Neural Oscillatory Mechanisms Underlying Step Accuracy: Integrating Microstate Segmentation with eLORETA-Independent Component Analysis

Kohei Okuyama [1,2,*], Kota Maeda [2], Ryosuke Yamauchi [2], Daichi Harada [3] and Takayuki Kodama [2,3]

[1] Department of Physical Therapy, School of Health Sciences, Bukkyo University, Kyoto 604-8418, Japan
[2] Graduate School of Health Sciences, Kyoto Tachibana University, Kyoto 607-8175, Japan; onebase824@gmail.com (K.M.); h901522007@st.tachibana-u.ac.jp (R.Y.); kodama-t@tachibana-u.ac.jp (T.K.)
[3] Department of Physical Therapy, Faculty of Health Sciences, Kyoto Tachibana University, Kyoto 607-8175, Japan; a903021139@st.tachibana-u.ac.jp
* Correspondence: kohei.1127@gmail.com

Abstract: Background/Objectives: Precise stepping control is fundamental to human mobility, and impairments increase fall risk in older adults and individuals with neurological conditions. This study investigated the cortical networks underlying stepping accuracy using mobile brain/body imaging with electroencephalography (EEG)-based exact low-resolution electromagnetic tomography-independent component analysis (eLORETA-ICA) and microstate segmentation analysis (MSA). Methods: Sixteen healthy male participants performed a precision stepping task while wearing a mobile EEG system. Step performance was quantified using error distance, measuring deviation between target and heel contact points. Preprocessed EEG data were analyzed using eLORETA-ICA and MSA, with participants categorized into high- and low-performing groups. Results: Seven microstate clusters were identified, with the anterior cingulate cortex (ACC) showing the highest microstate probability (21.15%). The high-performing group exhibited amplified theta-band activity in the ACC, enhanced activity in the precuneus and postcentral gyrus, and suppressed mu- and beta-band activity in the paracentral lobules. Conclusions: Stepping accuracy relies on a distributed neural network, with the ACC playing a central role in performance monitoring. We propose an integrated framework comprising the following systems: error monitoring (ACC), sensorimotor integration (paracentral lobules), and visuospatial processing (precuneus and occipital regions). These findings highlight the importance of neural oscillatory mechanisms in precise motor control and offer insights for rehabilitation strategies and fall prevention programs.

Keywords: anterior cingulate cortex; mobile brain/body imaging; EEG; eLORETA-ICA; error monitoring; microstate segmentation analysis; motor control; neural oscillatory mechanisms; step accuracy

1. Introduction

Stepping accuracy is crucial for maintaining balance and mobility, particularly during gait initiation, as it determines how effectively an individual can transition from a stationary position to walking [1]. Deficits in this initial step may compromise stability, leading to an increased risk of falls, particularly among older adults [2,3] and individuals with neurological disorders such as Parkinson's disease [4,5]. The initial step movement plays a crucial role in maintaining balance and adapting to environmental changes. A better understanding of the cortical mechanisms underlying step accuracy could provide insights into fall prevention strategies and rehabilitation interventions for at-risk populations, including individuals with Parkinson's disease and older adults.

In this context, stepping performance refers to the ability to place the foot accurately in space and time during gait [6–8], requiring precise motor coordination and cognitive control [9]. This process engages multiple brain regions, including the supplementary motor area, primary motor cortex, prefrontal cortex, primary somatosensory cortex, posterior parietal cortex (PPC), and anterior cingulate cortex (ACC) [10]. However, the specific neural mechanisms underlying step accuracy remain unclear.

Traditional assessments of gait and stepping performance rely on biomechanical methods, including force plates [11,12], pressure-sensitive mats [13], motion capture systems [14], and inertial measurement units [15]. While these methods provide valuable kinematic and kinetic data, they offer limited insight into the neural mechanisms governing step accuracy. Functional neuroimaging techniques, such as functional magnetic resonance imaging (fMRI), have enhanced our understanding of motor-related brain function but are restricted to simulated movement paradigms in supine positions, limiting their ecological validity for studying real-world locomotion [10]. Functional near-infrared spectroscopy provides greater mobility but lacks the temporal resolution required to analyze rapid neural dynamics during step initiation [16].

To overcome these limitations, mobile brain/body imaging (MoBI) has emerged as a promising approach. MoBI integrates mobile electroencephalography (EEG) with motion capture and physiological monitoring, enabling real-time investigation of brain dynamics during natural movement [17,18]. Despite early challenges such as motion artifacts and non-neural signal contamination, advancements in signal processing have significantly improved artifact removal [19,20], enabling MoBI to be applied to dynamic tasks, including juggling [21] and skateboarding [22]. These improvements have expanded the application of MoBI to the study of step accuracy, allowing for the real-time analysis of neural activity at the precise moment of step execution. In contrast to traditional methods, MoBI provides both high temporal resolution and ecological validity, making it uniquely suited to studying the neural correlates of step accuracy in natural walking conditions.

Previous studies have identified neural signatures associated with gait control. Increased theta-band activity in the ACC has been linked to error monitoring and balance adjustments [23,24], while reductions in mu- and beta-band activity in the sensorimotor cortex and PPC reflect motor planning and execution [25,26]. Additionally, delta-band fluctuations in motor-related regions have been associated with gait control and feedback processing [27,28]. However, most of these studies have focused on steady-state walking rather than the initial step movement, which is critical for understanding gait initiation and step accuracy. While previous MoBI studies have investigated cortical activity during continuous walking, little is known about the precise neural mechanisms underlying the initiation of stepping, particularly in relation to step accuracy. In contrast to prior studies that primarily focused on gross locomotor patterns, our study examined the fine-grained neural dynamics of step execution.

To address this gap, this study employed exact low-resolution electromagnetic tomography–independent component analysis (eLORETA-ICA) to examine cortical activity associated with step accuracy. eLORETA-ICA combines source localization with independent component analysis, enabling the decomposition of EEG signals across multiple frequency bands with high spatial resolution [29–31]. Additionally, microstate segmentation analysis (MSA) enables tracking of temporal network dynamics at millisecond precision, offering a detailed view of the cognitive and motor processes underlying stepping accuracy [32,33].

This study investigated how different cortical regions contribute to step accuracy during gait initiation. By analyzing EEG data using eLORETA-ICA and MSA, this research aimed to clarify the neural mechanisms underlying step accuracy and provide new insights into motor planning, execution, and error monitoring processes. Understanding these neural dynamics could help inform the development of targeted interventions for populations at risk of impaired stepping performance, including older adults and individuals with neurological disorders.

2. Materials and Methods

2.1. Participants

This study included 16 healthy male paid volunteers (mean age: 23.19 ± 2.66 years; mean height: 169.86 ± 4.51 cm). Participant characteristics are presented in Table 1. The exclusion criteria were female sex, individuals with dementia or suspected dementia (Mini-Mental State Examination [MMSE] score < 24), and those with orthopedic issues that could affect walking. The MMSE was used to exclude participants with potential cognitive impairments that could confound the assessment of stepping performance, which requires cognitive resources such as attention and executive function. This exclusion criterion was implemented to ensure that the observed neural activity primarily reflects motor control mechanisms in the absence of significant cognitive decline. Female participants were excluded owing to the known susceptibility of EEG activity to hormonal fluctuations, particularly those associated with the menstrual cycle, which can affect EEG stability and signal patterns [34,35]. Given the exploratory nature of this study, male participants with higher EEG stability were selected to minimize the potential variability in EEG measurements. Furthermore, the participants were required to abstain from caffeine, alcohol, and tobacco use for 24 h before the experiment and confirm that they were not currently taking any medications. They also reported their sleep duration from the previous night, as sleep can affect EEG measurements [36]. The dominant foot was identified as the foot used to kick a ball [37].

Table 1. Participant characteristics (n = 16).

Characteristic	Mean ± SD or n (%)
Age (years)	23.2 ± 2.7
Height (cm)	169.9 ± 4.5
Weight (kg)	62.4 ± 6.9
Dominant foot (Right/Left)	13 (81.3)/3 (18.7)
Sleep duration, previous night (hours)	6.2 ± 1.4
MMSE (points)	30 (100)

MMSE: Mini-Mental State Examination; SD: standard deviation.
All participants scored 30 points on the MMSE, indicating no cognitive impairment.

2.2. Experimental Methodology

This study was conducted between December 2023 and March 2024 in the Exercise Physiology Function Assessment Laboratory at Kyoto Tachibana University, Kyoto, Japan.

2.2.1. Stepping Task Protocol

Participants performed a precision stepping task while their brain activity, eye movements, and muscle activity were recorded. Figure 1 shows the experimental setup with a participant wearing the EEG cap and standing on the pressure-sensitive mat system.

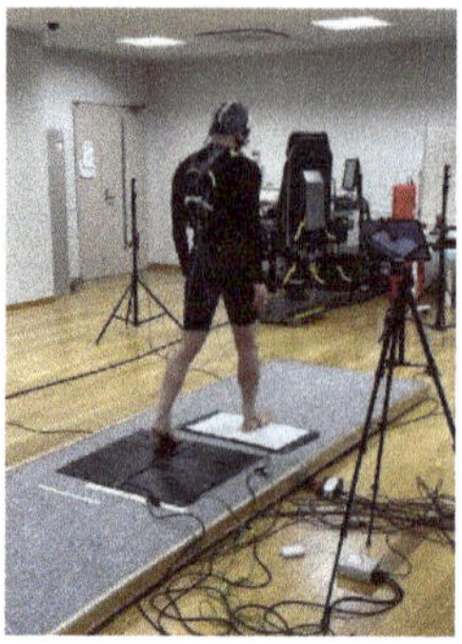

Figure 1. Experimental setup. The participant is wearing a 28-electrode EEG cap while standing on the pressure-sensitive mat system. The black mat in front of the participant is the initial position mat, and the white mat is the target position mat. Small markers were placed on the posterior and lateral sides of the calcaneus to precisely measure heel contact points. An iPad was used to record stepping movements. Participants performed 20 stepping trials, aiming to align their heels with the designated target point. EEG: electroencephalography.

For each trial, participants followed a standardized protocol as illustrated in Figure 2:

1. Initial Stance Phase (0–3 s): Participants stood in a stable, comfortable, and natural posture with their weight evenly distributed between both feet. An auditory cue signaled the beginning of the trial.
2. Preparation Phase (3–6 s): Participants visually identified the target point (marked with an 8 mm circular marker) and prepared their stepping. They were instructed to maintain their initial stance during this period.
3. Execution Phase (6–9 s): A second auditory cue prompted participants to take a single step forward with their dominant leg, aiming to align the calcaneal tuberosity with the designated target point. No specific instructions regarding step speed were given to ensure natural movement patterns. After heel contact, participants simultaneously performed a visual accuracy check by comparing their heel position to the target marker before returning to the starting position for the next trial.

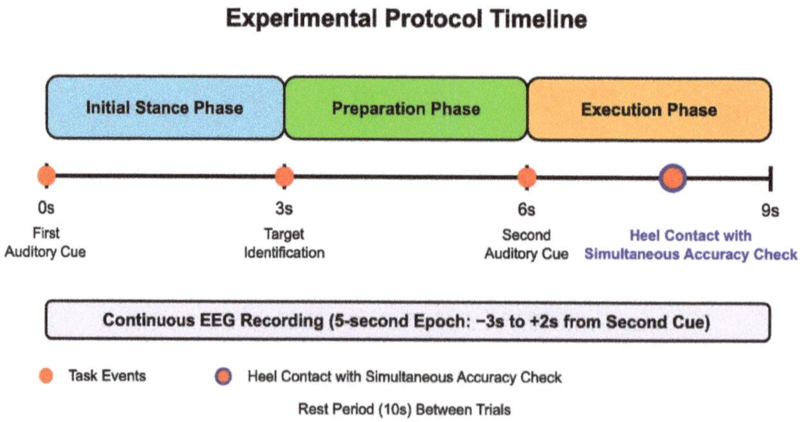

Figure 2. Experimental protocol timeline. The figure shows the temporal sequence of events during each stepping trial, including the initial stance, preparation, and execution phases. EEG recording epochs were extracted from 3 s before to 2 s after the second auditory cue.

A rest period of 10 s was provided between trials to prevent fatigue. Each participant completed 20 stepping trials with the same target location.

2.2.2. Data Acquisition Systems

EEG Recording

EEG data were recorded using active electrodes and the Polymate V AP5148 biological signal acquisition device (Miyuki Giken Co., Ltd., Tokyo, Japan). Measurements were performed at 28 electrode sites according to the international 10–10 system [38,39], including Fpz, Fz, Cz, Pz, Oz, FP1, FP2, F7, F8, F3, F4, C3, C4, C5, C6, P3, P4, P7, P8, O1, O2, T7, T8, CPz, CP1, CP2, PO3, and PO4. The ground electrode was placed on the spinous process of the seventh cervical vertebra, and the reference electrode was placed on the right earlobe. Recordings were sampled at 1 kHz, and electrode impedance was maintained below 10 kΩ. To minimize artifacts, participants were instructed to limit head movements and eye blinks during the stepping task.

Supplementary Physiological Recordings

In addition to EEG recordings, the following physiological signals were recorded simultaneously:

1. Electrooculography (EOG): Vertical EOG was recorded by placing electrodes above and below the dominant eye to detect eye movement and blink artifacts.
2. Electromyography (EMG): Surface EMG was recorded bilaterally from the tibialis anterior muscles. Electrode placement followed SENIAM's recommendations [40]. A bipolar derivation was used with an inter-electrode distance of 2 cm. The EMG signals were sampled at 1 kHz.

Pressure-Sensitive Mat System

For accurate detection of stepping timing, a pressure-sensitive mat system developed by Caretech Inc. (Japan) was employed. This system featured the following characteristics:

1. Sensor Configuration: Three pressure-sensitive mats were positioned as follows:
   - Two mats at the initial standing position (one for the right foot and one for the left foot);
   - One mat at the target position (around the heel contact area).
2. Data Acquisition and Synchronization: Signals from each mat sensor were directly synchronized with the EEG recording system. Trigger signals were collected at the same sampling rate as the EEG (1 kHz), with a temporal precision of ±5 ms.
3. Trigger Functions: The mat sensors generated trigger signals to accurately detect the following two key events:
   - Moment of unloading of the dominant foot (decrease in pressure on the initial position mat at step initiation);
   - Moment of heel contact (increase in pressure on the target position mat).

Video Recording System

Stepping movements were recorded using an iPad Air (Apple Inc., Cupertino, CA, USA; 11-inch, 4th generation; resolution: 1920 × 1080 pixels; frame rate: 30 frames per second). The device was positioned at a fixed height of 1.5 m and oriented as close to perpendicular to the plane of motion as possible to minimize perspective distortion.

2.2.3. Step Performance Assessment

Step performance was quantified using the error distance (ED), defined as the spatial deviation between the target point and the actual heel contact point. To ensure accurate measurement, the following procedure was employed:

1. Video Processing: Videos of the stepping trials were processed using the Fiji software (Fiji Is Just ImageJ, Version 1.54f) [38] to extract individual frames corresponding to heel contact moments.
2. Calibration: A 10 mm reference marker was placed in each video frame for accurate spatial calibration and distance measurement.
3. Reference Points: The heel contact position was defined using the following anatomical landmarks:
    - X-axis reference point: Dorsal (posterior) side of the calcaneal tuberosity;
    - Y-axis reference point: Lateral side of the calcaneal tuberosity.
4. Measurement: These reference points were marked with small (8 mm diameter) high-contrast circular markers to enhance visibility. ED was calculated as the Euclidean distance between the target point (xt, yt) and the heel contact point (xh, yh) using the following formula.

$$d = \sqrt{\left(X_{target} - X_{heel}\right)^2 + \left(Y_{target} - Y_{heel}\right)^2}$$

All measurements were recorded in millimeters. For each participant, the mean ED across all 20 trials was calculated and used for subsequent analysis.

2.3. Data Analysis

2.3.1. EEG Preprocessing

EEG data preprocessing was performed using the multimodal EEG analysis program EMSE Suite (CORTECH SOLUTIONS, Inc., Wilmington, NC, USA; Miyuki Giken Co., Ltd., Japan). The preprocessing pipeline comprised the following steps:

1. Filtering: Initially, continuous EEG data were filtered using the following parameters:
    - High-pass filter: 1.5 Hz (to remove low-frequency drift);
    - Low-pass filter: 30 Hz (to remove high-frequency noise);
    - Zero-phase digital filtering was implemented using the filter pipeline in EMSE to minimize phase distortion.
2. Power Line Noise Removal: A band-stop filter (notch filter) was applied to selectively remove power line noise:
    - Since the experiment was conducted in western Japan (Kyoto), where the commercial power frequency is 60 Hz, a band-stop filter centered at 60 Hz with a bandwidth of 2 Hz was applied.
3. EOG Artifact Removal: Eye movement and blink artifacts were removed using the template-based EOG artifact removal function implemented in EMSE. Using the simultaneously recorded EOG signal as a reference, typical eye movement and blink patterns were defined as templates, and these patterns were detected and removed from each EEG channel.
4. Segmentation: The preprocessed continuous data were segmented according to each stepping trial as follows:
    - From 3 s before to 2 s after the presentation of the second auditory cue (stepping initiation instruction);

- Each segment covered a 5 s time window, encompassing the preparation phase through post-movement completion.

5. Quality Control: The segmented data underwent quality verification through the following procedures:
   - Automatic rejection of epochs containing values exceeding ±100 μV;
   - Visual inspection by experienced researchers (K.O. and T.K.) to identify any remaining artifacts;
   - Re-referencing to average reference to minimize the influence of the reference electrode location.

2.3.2. eLORETA-ICA Analysis

Source analysis was performed on the preprocessed EEG data using the eLORETA software (version 20140711, http://www.uzh.ch/keyinst/loreta.htm, accessed on 1 August 2023) [41]. The following analytical procedures were employed:

1. Source Estimation: Preprocessed scalp EEG data were transformed into current density distributions in the brain using the eLORETA algorithm. eLORETA features minimal localization bias for source estimation at 6239 cortical gray matter voxels with 5 mm spatial resolution within a realistic head model [42].
2. Frequency-Domain ICA Application: The frequency-domain ICA (transposed fICA networks) module integrated into the eLORETA was directly applied to the preprocessed EEG data. This approach decomposes cortical electrical activity across multiple frequency bands into independent components (ICs), enabling simultaneous analysis of multiple frequency bands that would be difficult with conventional structural source analysis [29,31]. This module was configured to analyze the following four frequency bands:
   - Delta band (1.5–3.5 Hz);
   - Theta band (4–7.5 Hz);
   - Alpha/mu band (8–13 Hz, with the mu rhythm specifically referring to activity recorded over the sensorimotor areas) [43];
   - Beta band (13.5–30 Hz).

   This approach enables the extraction of functionally independent brain network components by applying ICA in the source space after decomposing signals into frequency bands.
3. Component Evaluation: Three experienced researchers (K.O., K.M., and T.K.) evaluated the resulting components and collaboratively classified them into neural activity and artifact components based on their spatial and spectral characteristics.

2.3.3. Microstate Segmentation Analysis

The identified ICs from eLORETA-ICA were further analyzed using the eLORETA clusters/microstates module integrated within the eLORETA software to assess patterns of neural activity during the stepping task. This analysis identifies quasi-stable patterns of brain activity (microstates) that represent distinct functional states. The procedure comprised the following steps:

1. Clustering Algorithm: The k-means clustering algorithm was applied to the source-space data to identify prototypical microstate maps.
2. Number of Clusters: Based on previous literature on gait-related brain activity [10], the number of clusters was set to seven, corresponding to brain regions previously identified as associated with gait and stepping motions. The optimal number of

clusters was verified using the cross-validation criterion, which confirmed that seven clusters provided the best balance between model complexity and explanatory power.

3. Microstate Parameter: For each identified microstate, its probability (the likelihood of each microstate occurring during the task) was calculated and used for subsequent analysis.

2.3.4. Statistical Analysis

After completing the EEG analysis using eLORETA-ICA, the participants were categorized into the following two groups based on their stepping performance:

1. High-Performing (HiP) Group: Participants with smaller ED values (below the median value), indicating more accurate stepping (n = 8).
2. Low-Performing (LoP) Group: Participants with larger ED values (above the median value), indicating less accurate stepping (n = 8).

The median split method was chosen based on its established use to robustly dichotomize continuous variables for group comparisons [44]. To ensure the robustness of our findings, we verified that the two groups significantly differed in their stepping performance using an independent samples *t*-test.

Statistical comparisons between the HiP and LoP groups were conducted using the statistics eLORETA SnPM 26 function (multiple paired *t*-tests with nonparametric randomization) included in the analysis program. This approach uses permutation tests to identify statistically significant differences between groups while controlling for multiple comparisons. The significance level was set at $p < 0.05$.

2.4. Ethical Considerations

Before the initiation of this study, all participants were provided with both written and verbal explanations of the purpose and content of the study through a research information sheet. All participants provided informed consent to participate in this study via signatures on a consent form. This study was conducted in accordance with the Declaration of Helsinki and was approved by the Research Ethics Committee of Kyoto Tachibana University (approval number: 23–44).

3. Results

3.1. Quality of EEG Data and Preprocessing Efficacy

Figure 3 presents representative EEG, EOG, and EMG recordings from a participant during the following three conditions: resting state (A); stepping task before preprocessing (B); and stepping task after preprocessing (C). The comparison between the raw and preprocessed stepping data demonstrates the efficacy of our artifact removal methodology. Movement-related artifacts visible in panel B were effectively eliminated in panel C while preserving the neurophysiological signals of interest. These preprocessed data were subsequently used for eLORETA-ICA analysis to identify neural components associated with stepping performance.

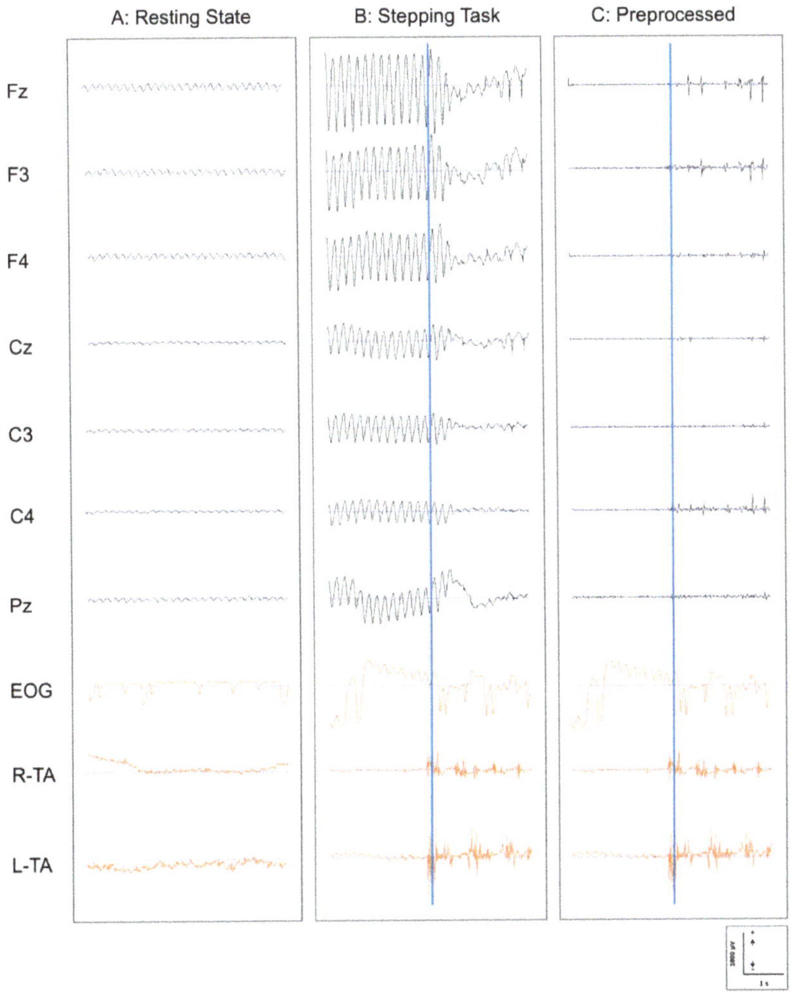

Figure 3. EEG, EOG, and EMG recordings during resting state and stepping task: before and after preprocessing. (**A**) Resting state: raw data during the resting condition; (**B**) stepping task: raw data during the stepping task showing significant movement artifacts; and (**C**) preprocessed: the same stepping task data after application of the artifact removal pipeline. Displayed channels include frontal (Fz, F3, and F4), central (Cz, C3, and C4), and parietal (Pz) EEG recordings (black); vertical EOG; and bilateral tibialis anterior EMG (R-TA and L-TA) (orange). Blue vertical lines in panels B and C indicate stepping trigger moments, marking the initiation of dominant foot movement. EEG signals are presented with conventional polarity (positive deflections upward). The preprocessing pipeline effectively removed movement artifacts while preserving relevant neural signals. Scale bars: 1000 μV (vertical) and 1 s (horizontal). EEG: electroencephalography; EMG: electromyography; EOG: electrooculography.

3.2. IC and Microstate Analysis

Using eLORETA-ICA, 13 ICs were identified from EEG recordings obtained during the stepping task. Of these, 10 ICs (IC-1, IC-2, IC-3, IC-5, IC-7, IC-9, IC-10, IC-11, IC-12, and IC-13) were associated with neural network activity, while the remaining ICs (IC-4, IC-6, and IC-8) were classified as artifacts (see Appendix A for visualization).

The activity was analyzed across four frequency bands (theta, alpha, mu, and beta) for each of the 10 ICs, and eLORETA microstate analysis grouped them into seven clusters (Clusters 1–7). These seven microstate clusters are presented in Figure 4, with each cluster representing a distinct functional network. The detailed anatomical locations of these clusters are presented in Table 2.

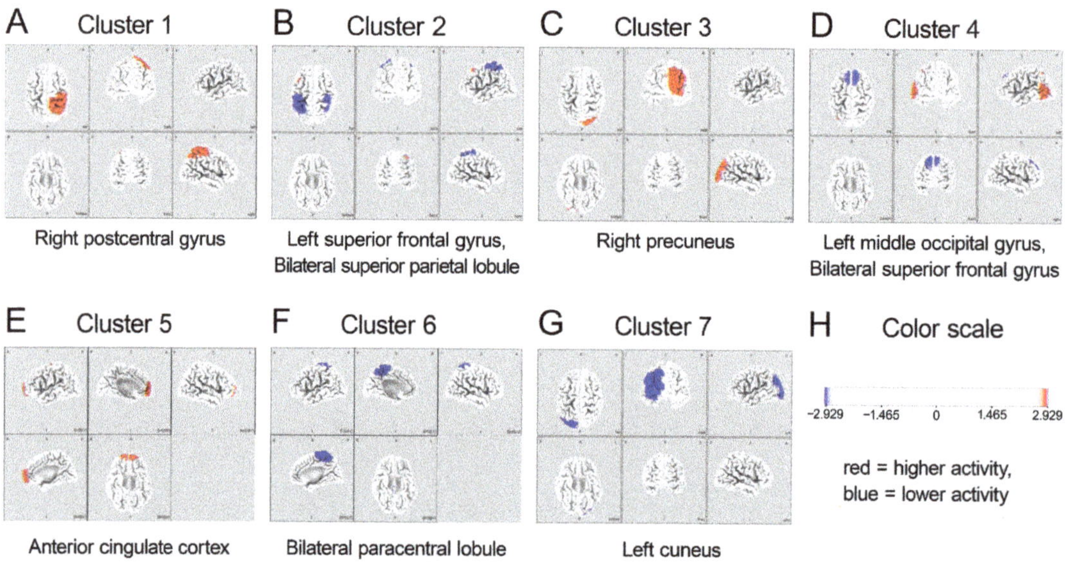

Figure 4. Seven clusters calculated using eLORETA microstates. Each panel (**A**–**G**) represents an independent microstate cluster: (**A**) Cluster 1: right postcentral gyrus; (**B**) Cluster 2: left superior frontal gyrus/bilateral superior parietal lobule; (**C**) Cluster 3: right precuneus; (**D**) Cluster 4: left middle occipital gyrus/bilateral superior frontal gyrus; (**E**) Cluster 5: anterior cingulate cortex; (**F**) Cluster 6: bilateral paracentral lobule; (**G**) Cluster 7: left cuneus. The color scale (**H**) represents t-values (t = 2.929), with red and blue indicating higher and lower activities, respectively. Brain images are displayed from various perspectives, with medial views chosen when either maximum or minimum activity is located in medial brain regions to optimally visualize both the activation and deactivation patterns of each cluster. eLORETA: exact low-resolution electromagnetic tomography.

Table 2. Areas of neural activity in the seven clusters.

	HiP, *n* = 8			**LoP, *n* = 8**		
Cluster	**Brain Lobe**	**Region**	**BA**	**Brain Lobe**	**Region**	**BA**
1	Right parietal lobe	Postcentral gyrus	3	No EEG activity		
2	Left frontal lobe	Superior frontal gyrus	8	Bilateral parietal lobes	Superior parietal lobule	7
3	Right parietal lobe	Precuneus	7	No EEG activity		
4	Left occipital lobe	Middle occipital gyrus	19	Bilateral frontal lobes	Superior frontal gyrus	8
5	Limbic lobe	Anterior cingulate cortex	32	No EEG activity		
6	No EEG activity			Bilateral parietal lobes	Paracentral lobule	4
7	No EEG activity			Left occipital lobe	Cuneus	19

HiP: high-performing group characterized by a small error distance between the target point and the heel contact point; LoP: low-performing group characterized by a large error distance between the target point and the heel contact point; BA: Brodmann area; EEG: electroencephalography.

The detailed frequency-band characteristics of each cluster are listed in Table 3. The microstate probability analysis revealed the relative contribution of each frequency band within the seven clusters. Notably, Cluster 5 (corresponding to the ACC) exhibited the highest overall microstate probability of 21.15%, with a particularly strong representation in the theta band (38.46%).

Table 3. The microstate probabilities for each frequency band in the seven clusters.

Cluster	Total	δ	θ	α (μ)	β
1	7.69	7.69	7.69	7.69	7.69
2	15.39	15.39	15.39	15.39	15.39
3	17.31	23.08	15.39	15.39	15.39
4	5.77	0.00	7.69	7.69	7.69
5	21.15	23.08	38.46	15.39	7.69
6	17.31	15.39	7.69	23.08	23.08
7	15.39	15.39	7.69	15.39	23.08

δ: delta; θ: theta; α: alpha; β: beta. Each value represents a percentage.

3.3. Comparison Between HiP and LoP Groups

Based on stepping performance quantified by ED, participants were categorized into HiP ($n = 8$; mean ED = 12.8 ± 2.6 mm) and LoP ($n = 8$; mean ED = 23.1 ± 4.1 mm) groups using a median-split method. The between-group difference in ED was statistically significant ($t(14) = 5.02$, $p < 0.001$), confirming the validity of this grouping approach.

When comparing neural activity patterns between the HiP and LoP groups, significant differences emerged across several brain regions. The HiP group demonstrated amplified theta-band activity in the ACC (Cluster 5, $p < 0.05$; Cohen's $d = 1.57$), enhanced activity in the precuneus (Cluster 3) and right postcentral gyrus (Cluster 1), and significant attenuation of mu- and beta-band activity in the bilateral paracentral lobules (Cluster 6, $p < 0.05$).

4. Discussion

In this study, we analyzed brain activity during initial step movement using eLORETA-ICA and MSA, comparing HiP and LoP stepping groups. Our analysis identified seven microstate clusters, with Cluster 5 (ACC) exhibiting the highest probability of theta band activity. The HiP group demonstrated amplified ACC theta-band activity, increased activity in the precuneus and right parietal lobe (postcentral gyrus), and the suppression of mu- and beta-band activity in the bilateral paracentral lobules.

4.1. Amplified ACC Theta-Band Activity: Role in Performance Monitoring

The ACC is widely recognized for its involvement in error detection and conflict monitoring, playing a crucial role in adjusting motor control systems [45,46]. Particularly noteworthy is the association between theta band activity and cognitive control functions [47]. Our finding of enhanced theta-band activity in the ACC in the HiP group extends these observations to the domain of stepping accuracy, providing novel insights into the neural mechanisms of precise locomotion.

Accurate stepping demands the continuous integration of visual and somatosensory information while rapidly detecting and correcting errors. The amplified ACC theta activity in the HiP group indicates more efficient performance monitoring and top-down control, a critical neural signature differentiating high performers from their counterparts. This observation directly addresses our primary research question regarding the neural mechanisms underlying stepping accuracy, establishing the ACC as a central hub for error monitoring during precision stepping tasks.

These results align with multiple lines of evidence from locomotor research. Studies on brain–computer interface (BCI)-controlled treadmill walking have reported increased low-frequency activity in the ACC associated with error monitoring and motor learning networks [27]. Similarly, investigations using split-belt treadmill adaptation paradigms have demonstrated increased theta power in the ACC during early adaptation phases [23], providing converging evidence for the role of ACC theta oscillations in locomotor adaptation. During perturbations to standing balance, the ACC functions as a hub for theta-band coupling, suggesting its broader roles in various aspects of balance control [24].

Our findings reveal that individuals with superior stepping accuracy engage in enhanced ACC-mediated monitoring processes, which likely facilitate the detection of minor deviations between intended and actual foot placement. This real-time error detection system appears crucial for the precise adjustments necessary to achieve accurate stepping. The theta oscillations may serve as a communication mechanism between monitoring systems and motor execution areas, enabling rapid corrections to ensure accurate foot placement.

While traditional neuroimaging studies using fMRI have emphasized the role of the supplementary motor area and primary motor cortex in gait control [48], our electrophysiological approach reveals the dynamic temporal characteristics of ACC involvement. These findings suggest that precise stepping relies not merely on motor execution capabilities but critically depends on cognitive monitoring processes mediated by the ACC.

4.2. Suppression of Mu- and Beta-Band Activity in Sensorimotor Regions: Enhanced Motor Preparation and Control

The precuneus plays a crucial role in visuospatial processing and motor imagery [6–8]. The increased precuneus activity we observed in the HiP group specifically links this region to stepping accuracy. Accurate stepping requires integrating visual information with body-centered coordinates to form a spatial map, suggesting the precuneus significantly contributes to this process.

Increased activity in the postcentral gyrus suggests that sensory inputs, particularly plantar sensations, are processed with higher precision in individuals with superior stepping accuracy. This finding aligns with those of studies by Koenraadt et al. [49] and Haefeli et al. [50], who reported increased cortical activity during precision stepping and obstacle avoidance; however, our study specifically identifies the relevant frequency bands and anatomical regions involved.

The HiP group showed suppression in mu- and beta-band activity in the bilateral paracentral lobules, consistent with findings by Wagner et al. [25,51,52] and Seeber et al. [53,54]. This reduction in mu- and beta-band activity represents a well-established neurophysiological marker of motor planning and execution [55]. The event-related desynchronization (ERD) observed in the mu and beta bands from motor preparation through execution reflects fundamental mechanisms of motor control [25,26]. Research on voluntary gait control using BCI has reported sustained alpha/mu-band ERD associated with intentional control [56], supporting our findings in precision stepping.

Beta-band activity plays a critical role in motor and cognitive control [25,57], suggesting that the suppression pattern observed in the HiP group reflects smoother lower-limb motor control. A recent study by Nordin et al. [55] demonstrated that at faster gait speeds, the sensorimotor cortex shows reduced alpha and beta EEG spectral power, aligning with our observations that reduced mu- and beta-band activity is associated with skilled motor performance, suggesting a common neural mechanism for both stepping accuracy and gait speed modulation.

4.3. Visual Processing in the Occipital Region: Enhanced Visuospatial Integration

We observed changes in frequency band activity in the occipital lobes (middle occipital gyrus and cuneus). These regions are crucial for both initial visual processing and the integration of higher-order visuospatial information [58,59], essential for accurate foot placement.

The stepping task required participants to visually identify a target point and precisely align their heel with it, demanding efficient visual processing to create an accurate spatial map of the target location relative to the body. The distinct patterns of occipital activity in the HiP group suggest more effective visual information processing and integration with motor planning areas.

Bradford et al. [60] demonstrated that electrocortical activity can distinguish between different walking conditions. Our findings advance this understanding by showing that occipital activity patterns differentiate between high and low stepping accuracy, suggesting that precise stepping relies on optimized visual processing to inform motor planning.

Wagner et al. [51] showed that walking down a virtual alley activates premotor and parietal areas, emphasizing visuospatial processing in locomotion. Our results demonstrate that occipital processing efficiency directly relates to stepping performance, highlighting the importance of visual processing in precise foot placement. This visual–motor connection is further supported by studies showing electrocortical correlates of gait adaptation to visual kinematic perturbations [56], suggesting a critical visual–motor integration pathway for precision locomotion.

4.4. A Proposed Neural Network Framework for Stepping Accuracy

Integrating our findings, we propose a comprehensive neural network framework for stepping accuracy that builds upon current models of motor control. This proposed framework comprises the following three interconnected neural systems working in concert:

1. Error Monitoring System: Centered on the ACC, this system generates theta oscillations that monitor performance in real-time, detecting discrepancies between intended and actual foot placement. The enhanced theta activity in the HiP group suggests more efficient error monitoring processes facilitating precise stepping, aligning with findings from Luu et al. [27] and Jacobsen and Ferris [23] on ACC theta activity in error monitoring and locomotor adaptation.
2. Sensorimotor Integration System: Encompassing the paracentral lobules and postcentral gyrus, this system features task-specific mu and beta desynchronization patterns supporting flexible motor execution. The greater suppression of these rhythms in the HiP group may enable more precise motor adjustments, corresponding with findings from Wagner et al. [25] and Nordin et al. [55] on active locomotion and adaptation.
3. Visuospatial Processing System: Including the precuneus and occipital regions, this system processes visual information and integrates it with body-centered coordinates to create a spatial map for guiding foot placement. Enhanced activity in these regions among high performers suggests more effective visuospatial processing, consistent with studies by Wagner et al. [51] on parietal regions in locomotion.

We hypothesize that these systems interact dynamically throughout stepping execution. The error monitoring system likely modulates the sensorimotor integration system through top-down control, while the visuospatial processing system provides essential spatial information guiding motor planning and execution. Enhanced coordination between these systems appears associated with superior stepping accuracy; however, further research is needed to validate this proposed framework and examine the causal relationships between these systems.

This integrated framework extends traditional motor control models by emphasizing the role of cognitive monitoring processes in precise motor execution. Rather than viewing motor control as primarily bottom-up, our findings suggest the importance of top-down cognitive control mechanisms in fine-tuning motor output. This perspective is supported by electrocortical evidence from BCI-controlled walking studies [27] and split-belt adaptation paradigms [23].

4.5. Study Limitations

Some limitations should be considered when interpreting our results. Our study included only young male participants, limiting generalizability to females, older adults, and individuals with neurological disorders. Sex differences in neural control mechanisms have been documented [22,23], underscoring the importance of examining whether the neural correlates of stepping accuracy are consistent across sexes. While we focused on the initial step movement to isolate neural mechanisms of step accuracy, future studies should examine how these mechanisms function during more complex walking tasks involving obstacle avoidance or dual-task conditions. The laboratory environment may not fully capture the neural dynamics of stepping in real-world environments with varying surfaces and distractions. Our sample size, while sufficient to detect significant differences between groups, could be expanded in future studies to enhance statistical power.

Future research should investigate these neural mechanisms in clinical populations with gait impairments. The frequency-specific oscillatory patterns identified could serve as potential biomarkers for fall risk assessment and as targets for neuromodulation or neurofeedback interventions aimed at enhancing stepping accuracy in vulnerable populations.

5. Conclusions

Our study provides novel insights into the neural oscillatory mechanisms underlying stepping accuracy. We demonstrated that accurate step execution relies on a distributed yet specialized neural network comprising (1) enhanced performance monitoring via ACC theta-band activity; (2) effective sensorimotor integration through mu- and beta-band suppression in motor areas; and (3) efficient visuospatial processing in parietal and occipital regions.

The frequency-specific oscillatory patterns identified have important clinical implications as potential biomarkers for assessing fall risk and as targets for interventions to improve stepping accuracy. Our methodological approach using eLORETA-ICA effectively addressed common challenges in mobile EEG research by separating neural signals from movement-related artifacts. By elucidating these neural mechanisms, our findings provide a foundation for developing more effective rehabilitation strategies and fall prevention programs targeting the specific neural systems involved in precise foot placement.

Author Contributions: Conceptualization, K.O. and T.K.; Methodology, K.O. and T.K.; Software, T.K.; Validation, K.O., K.M., R.Y. and T.K.; Formal Analysis, K.O.; Investigation, K.O. and K.M.; Resources, T.K.; Data Curation, K.O. and D.H.; Writing—Original Draft Preparation, K.O.; Writing—Review and Editing, T.K., K.M. and R.Y.; Visualization, K.O.; Supervision, T.K.; Project Administration, K.O. All authors have read and agreed to the published version of the manuscript.

Funding: This research received no external funding.

Institutional Review Board Statement: This study was conducted in accordance with the Declaration of Helsinki and was approved by the Research Ethics Committee of Kyoto Tachibana University (approval number: 23–44; approval date: 12 September 2023).

Informed Consent Statement: Informed consent for participation was obtained from all participants involved in this study.

Data Availability Statement: Some of the participant data presented in this study are available upon request from the corresponding author. However, due to privacy concerns, certain participants' data may not be offered.

Acknowledgments: We would like to express our sincere gratitude to all participants who generously contributed their time and valuable insights to this study. Their cooperation was instrumental in bringing this research to fruition. We also extend our appreciation to CARETECH for their expert collaboration in developing the tool central to this work.

Conflicts of Interest: The authors declare no conflicts of interest.

Abbreviations

The following abbreviations are used in this manuscript:

ACC	Anterior cingulate cortex
BCI	Brain–computer interface
ED	Error distance
EEG	Electroencephalography
eLORETA	Exact low-resolution electromagnetic tomography
EMG	Electromyography
EOG	Electrooculography
MMSE	Mini-mental state examination
ERD	Event-related desynchronization
fMRI	Functional magnetic resonance imaging
HiP group	High-performing group
ICA	Independent component analysis
ICs	Independent components
LoP group	Low-performing group
MoBI	Mobile brain/body imaging
MSA	Microstate segmentation analysis
PPC	Posterior parietal cortex

Appendix A. Images of Independent Components from eLORETA-ICA Analysis

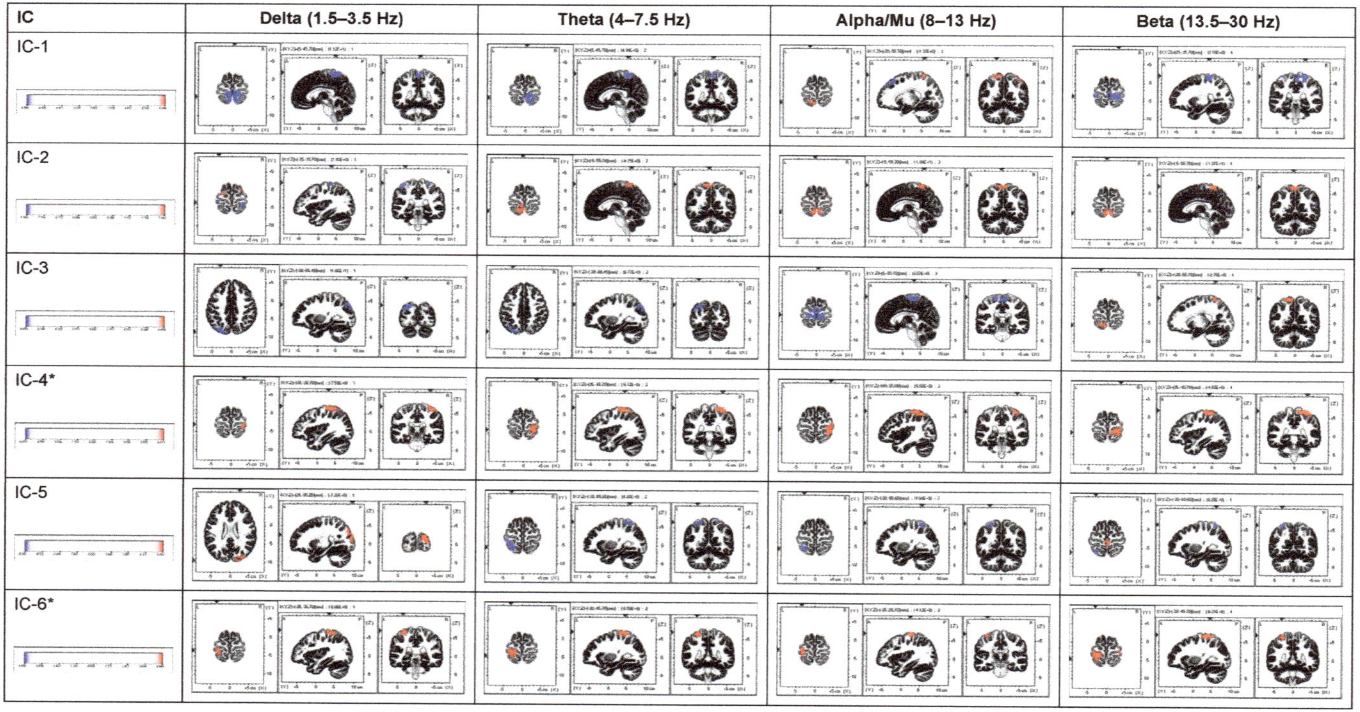

Figure A1. *Cont.*

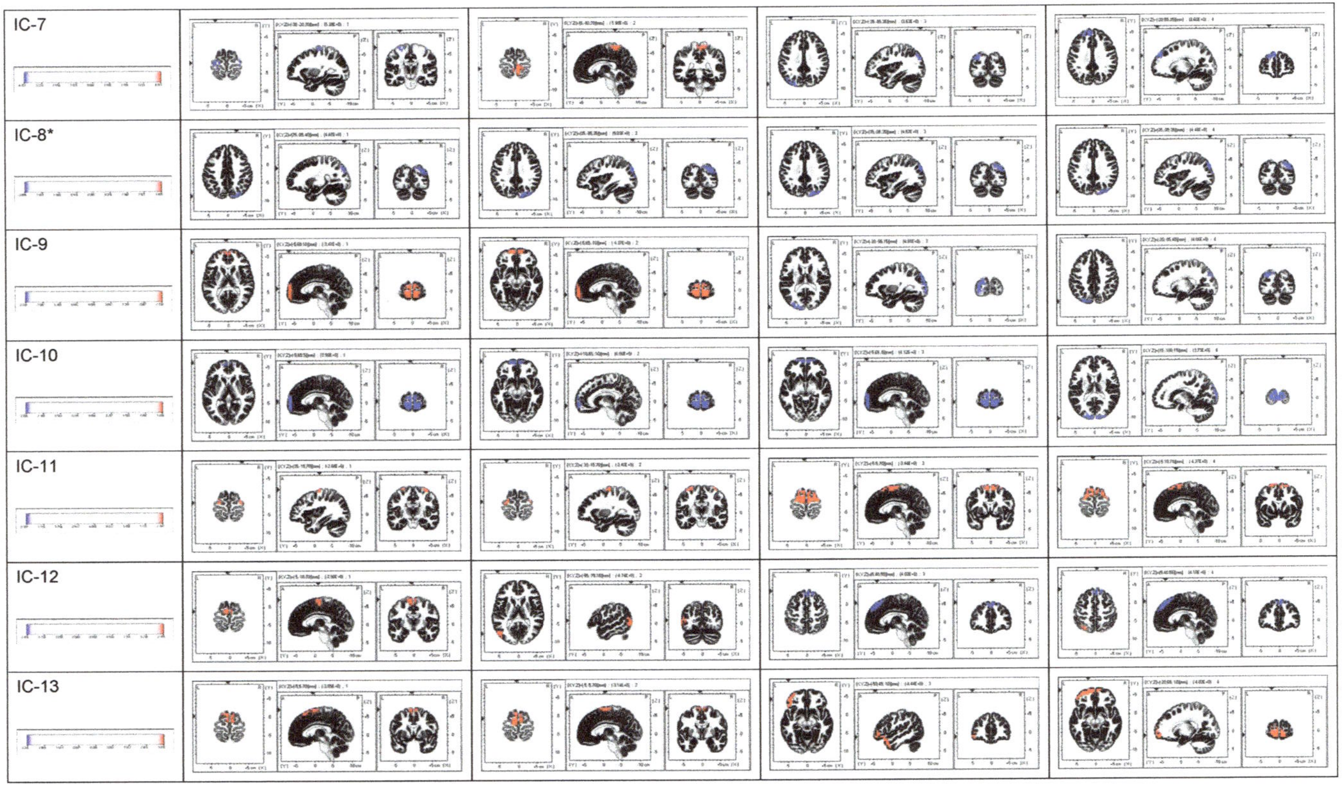

Figure A1. Images of 13 independent components (ICs) within their respective frequency bands obtained using eLORETA-ICA on EEG data during the stepping task. Each row represents a different IC, and columns represent the four frequency bands analyzed: Delta (1.5–3.5 Hz), Theta (4–7.5 Hz), Alpha/Mu (8–13 Hz), and Beta (13.5–30 Hz). Ten ICs (IC-1, IC-2, IC-3, IC-5, IC-7, IC-9, IC-10, IC-11, IC-12, and IC-13) were identified as physiological neural network activity, while three ICs (IC-4, IC-6, and IC-8) were classified as artifacts based on their spatial and spectral characteristics. In the color-coded maps, red represents increased power, whereas blue indicates decreased power as IC activity increases. * Note: IC-4, IC-6, and IC-8 were identified as artifacts and excluded from subsequent analyses.

References

1. Bruijn, S.M.; van Dieën, J.H. Control of human gait stability through foot placement. *J. R. Soc. Interface* **2018**, *15*, 20170816. [CrossRef]
2. Robinovitch, S.N.; Feldman, F.; Yang, Y.; Schonnop, R.; Leung, P.M.; Sarraf, T.; Sims-Gould, J.; Loughin, M. Video capture of the circumstances of falls in elderly people residing in long-term care: An observational study. *Lancet* **2013**, *381*, 47–54. [CrossRef]
3. Tisserand, R.; Robert, T.; Chabaud, P.; Bonnefoy, M.; Chèze, L. Elderly fallers enhance dynamic stability through anticipatory postural adjustments during a choice stepping reaction time. *Front. Hum. Neurosci.* **2016**, *10*, 613. [CrossRef]
4. Bloem, B.R.; Hausdorff, J.M.; Visser, J.E.; Giladi, N. Falls and freezing of gait in Parkinson's disease: A review of two interconnected, episodic phenomena. *Mov. Disord.* **2004**, *19*, 871–884. [CrossRef] [PubMed]
5. Nutt, J.G.; Bloem, B.R.; Giladi, N.; Hallett, M.; Horak, F.B.; Nieuwboer, A. Freezing of gait: Moving forward on a mysterious clinical phenomenon. *Lancet Neurol.* **2011**, *10*, 734–744. [CrossRef]
6. Chapman, G.J.; Hollands, M.A. Evidence that older adult fallers prioritise the planning of future stepping actions over the accurate execution of ongoing steps during complex locomotor tasks. *Gait Posture* **2007**, *26*, 59–67. [CrossRef]
7. Young, W.R.; Wing, A.M.; Hollands, M.A. Influences of state anxiety on gaze behavior and stepping accuracy in older adults during adaptive locomotion. *J. Gerontol. B Psychol. Sci. Soc. Sci.* **2012**, *67*, 43–51. [CrossRef]
8. Caetano, M.J.D.; Lord, S.R.; Brodie, M.A.; Schoene, D.; Pelicioni, P.H.S.; Sturnieks, D.L.; Menant, J.C. Executive functioning, concern about falling and quadriceps strength mediate the relationship between impaired gait adaptability and fall risk in older people. *Gait Posture* **2018**, *59*, 188–192. [CrossRef]
9. Takakusaki, K. Functional neuroanatomy for posture and gait control. *J. Mov. Disord.* **2017**, *10*, 1–17. [CrossRef]
10. Hamacher, D.; Herold, F.; Wiegel, P.; Hamacher, D.; Schega, L. Brain activity during walking: A systematic review. *Neurosci. Biobehav. Rev.* **2015**, *57*, 310–327. [CrossRef]
11. Yiou, E.; Teyssèdre, C.; Artico, R.; Fourcade, P. Comparison of base of support size during gait initiation using force-plate and motion-capture system: A bland and Altman analysis. *J. Biomech.* **2016**, *49*, 4168–4172. [CrossRef] [PubMed]
12. Khanmohammadi, R.; Talebian, S.; Hadian, M.R.; Olyaei, G.; Bagheri, H. Time and frequency domain analysis of gait initiation in younger and older adults. *J. Aging Phys. Act.* **2017**, *25*, 212–217. [CrossRef] [PubMed]
13. Jeon, E.-T.; Cho, H.-Y. A novel method for gait analysis on center of pressure excursion based on a pressure-sensitive mat. *Int. J. Environ. Res. Public Health* **2020**, *17*, 7845. [CrossRef]
14. Lee, M.; Kim, J.; Son, J.; Kim, Y. Kinematic and kinetic analysis during forward and backward walking. *Gait Posture* **2013**, *38*, 674–678. [CrossRef]
15. Ogata, T.; Hashiguchi, H.; Hori, K.; Hirobe, Y.; Ono, Y.; Sawada, H.; Inaba, A.; Orimo, S.; Miyake, Y. Foot trajectory features in gait of Parkinson's disease patients. *Front. Physiol.* **2022**, *13*, 726677. [CrossRef]
16. Bishnoi, A.; Holtzer, R.; Hernandez, M.E. Brain activation changes while walking in adults with and without neurological disease: Systematic review and meta-analysis of functional near-infrared spectroscopy studies. *Brain Sci.* **2021**, *11*, 291. [CrossRef]
17. Makeig, S.; Gramann, K.; Jung, T.-P.; Sejnowski, T.J.; Poizner, H. Linking brain, mind and behavior. *Int. J. Psychophysiol.* **2009**, *73*, 95–100. [CrossRef]
18. Song, S.; Nordin, A.D. Mobile electroencephalography for studying neural control of human locomotion. *Front. Hum. Neurosci.* **2021**, *15*, 749017. [CrossRef]
19. Gramann, K.; Ferris, D.P.; Gwin, J.; Makeig, S. Imaging natural cognition in action. *Int. J. Psychophysiol.* **2014**, *91*, 22–29. [CrossRef]
20. Ladouce, S.; Donaldson, D.I.; Dudchenko, P.A.; Ietswaart, M. Mobile EEG identifies the re-allocation of attention during real-world activity. *Sci. Rep.* **2019**, *9*, 15851. [CrossRef]
21. Kim, H.; Miyakoshi, M.; Iversen, J.R. Approaches for hybrid coregistration of marker-based and markerless coordinates describing complex body/object interactions. *Sensors* **2023**, *23*, 6542. [CrossRef] [PubMed]
22. Callan, D.E.; Torre-Tresols, J.J.; Laguerta, J.; Ishii, S. Shredding artifacts: Extracting brain activity in EEG from extreme artifacts during skateboarding using ASR and ICA. *Front. Neuroergonomics* **2024**, *5*, 1358660. [CrossRef]
23. Jacobsen, N.A.; Ferris, D.P. Electrocortical activity correlated with locomotor adaptation during split-belt treadmill walking. *J. Physiol.* **2023**, *601*, 3921–3944. [CrossRef]
24. Peterson, S.M.; Ferris, D.P. Group-level cortical and muscular connectivity during perturbations to walking and standing balance. *Neuroimage* **2019**, *198*, 93–103. [CrossRef]
25. Wagner, J.; Makeig, S.; Gola, M.; Neuper, C.; Müller-Putz, G. Distinct β band oscillatory networks subserving motor and cognitive control during gait adaptation. *J. Neurosci.* **2016**, *36*, 2212–2226. [CrossRef]
26. Luu, T.P.; Brantley, J.A.; Nakagome, S.; Zhu, F.; Contreras-Vidal, J.L. Electrocortical correlates of human level-ground, slope, and stair walking. *PLoS ONE* **2017**, *12*, e0188500. [CrossRef]
27. Luu, T.P.; Nakagome, S.; He, Y.; Contreras-Vidal, J.L. Real-time EEG-based brain-computer interface to a virtual avatar enhances cortical involvement in human treadmill walking. *Sci. Rep.* **2017**, *7*, 8895. [CrossRef]

28. Li, J.; Dimitrakopoulos, G.N.; Thangavel, P.; Chen, G.; Sun, Y.; Guo, Z.; Yu, H.; Thakor, N.; Bezerianos, A. What are spectral and spatial distributions of EEG-EMG correlations in overground walking? An exploratory study. *IEEE Access* **2019**, *7*, 143935–143946. [CrossRef]
29. Aoki, Y.; Ishii, R.; Pascual-Marqui, R.D.; Canuet, L.; Ikeda, S.; Hata, M.; Imajo, K.; Matsuzaki, H.; Musha, T.; Asada, T.; et al. Detection of EEG-resting state independent networks by ELORETA-ICA method. *Front. Hum. Neurosci.* **2015**, *9*, 31. [CrossRef]
30. Aoki, Y.; Takahashi, R.; Suzuki, Y.; Pascual-Marqui, R.D.; Kito, Y.; Hikida, S.; Maruyama, K.; Hata, M.; Ishii, R.; Iwase, M.; et al. EEG resting-state networks in Alzheimer's disease associated with clinical symptoms. *Sci. Rep.* **2023**, *13*, 3964. [CrossRef]
31. Katayama, O.; Stern, Y.; Habeck, C.; Coors, A.; Lee, S.; Harada, K.; Makino, K.; Tomida, K.; Morikawa, M.; Yamaguchi, R.; et al. Detection of neurophysiological markers of cognitive reserve: An EEG study. *Front. Aging Neurosci.* **2024**, *16*, 1401818. [CrossRef] [PubMed]
32. Férat, V.; Seeber, M.; Michel, C.M.; Ros, T. Beyond broadband: Towards a spectral decomposition of EEG microstates. *Hum. Brain Mapp.* **2022**, *43*, 3047–3061. [CrossRef] [PubMed]
33. Zhang, K.; Shi, W.; Wang, C.; Li, Y.; Liu, Z.; Liu, T.; Li, J.; Yan, X.; Wang, Q.; Cao, Z.; et al. Reliability of EEG microstate analysis at different electrode densities during propofol-induced transitions of brain states. *Neuroimage* **2021**, *231*, 117861. [CrossRef] [PubMed]
34. Becker, D.; Creutzfeldt, O.D.; Schwibbe, M.; Wuttke, W. Changes in physiological, EEG and psychological parameters in women during the spontaneous menstrual cycle and following oral contraceptives. *Psychoneuroendocrinology* **1982**, *7*, 75–90. [CrossRef]
35. de Souza, R.F.L.; Mendes, T.M.A.S.; de Araujo Lima, L.A.B.; Brandão, D.S.; Laplagne, D.A.; de Sousa, M.B.C. Effect of the menstrual cycle on electroencephalogram alpha and beta bands during motor imagery and action observation. *Front. Hum. Neurosci.* **2022**, *16*, 878887. [CrossRef]
36. Gibbings, A.; Ray, L.B.; Berberian, N.; Nguyen, T.; Shahidi Zandi, A.; Owen, A.M.; Comeau, F.J.E.; Fogel, S.M. EEG and behavioural correlates of mild sleep deprivation and vigilance. *Clin. Neurophysiol.* **2021**, *132*, 45–55. [CrossRef]
37. Goda, A.; Izumi, M.; Kido, H.; Yamashita, H.; Murata, S. Differences in foot morphology and center of gravity sway between dominant and nondominant legs in young healthy adults. *Jpn. J. Health Promot. Phys. Ther.* **2024**, *11*, 79–83. [CrossRef]
38. Chatrian, G.E.; Lettich, E.; Nelson, P.L. Ten percent electrode system for topographic studies of spontaneous and evoked EEG activities. *Am. J. EEG Technol.* **1985**, *25*, 83–92. [CrossRef]
39. Oostenveld, R.; Praamstra, P. The five percent electrode system for high-resolution EEG and ERP measurements. *Clin. Neurophysiol.* **2001**, *112*, 713–719. [CrossRef]
40. Hermens, H.J.; Freriks, B.; Disselhorst-Klug, C.; Rau, G. Development of recommendations for SEMG sensors and sensor placement procedures. *J. Electromyogr. Kinesiol.* **2000**, *10*, 361–374. [CrossRef]
41. Pascual-Marqui, R.D.; Lehmann, D.; Koukkou, M.; Kochi, K.; Anderer, P.; Saletu, B.; Tanaka, H.; Hirata, K.; John, E.R.; Prichep, L.; et al. Assessing interactions in the brain with exact low-resolution electromagnetic tomography. *Philos. Trans. A Math. Phys. Eng. Sci.* **2011**, *369*, 3768–3784. [CrossRef] [PubMed]
42. Fuchs, M.; Kastner, J.; Wagner, M.; Hawes, S.; Ebersole, J.S. A standardized boundary element method volume conductor model. *Clin. Neurophysiol.* **2002**, *113*, 702–712. [CrossRef] [PubMed]
43. Pfurtscheller, G.; Neuper, C. Motor imagery activates primary sensorimotor area in humans. *Neurosci. Lett.* **1997**, *239*, 65–68. [CrossRef]
44. Iacobucci, D.; Posavac, S.S.; Kardes, F.R.; Schneider, M.J.; Popovich, D.L. The median split: Robust, refined, and revived. *J. Consum. Psychol.* **2015**, *25*, 690–704. [CrossRef]
45. Botvinick, M.M.; Cohen, J.D.; Carter, C.S. Conflict monitoring and anterior cingulate cortex: An update. *Trends Cogn. Sci.* **2004**, *8*, 539–546. [CrossRef]
46. Rolls, E.T. The cingulate cortex and limbic systems for emotion, action, and memory. *Brain Struct. Funct.* **2019**, *224*, 3001–3018. [CrossRef]
47. Cavanagh, J.F.; Frank, M.J. Frontal theta as a mechanism for cognitive control. *Trends Cogn. Sci.* **2014**, *18*, 414–421. [CrossRef]
48. Suzuki, M.; Miyai, I.; Ono, T.; Oda, I.; Konishi, I.; Kochiyama, T.; Kubota, K. Prefrontal and premotor cortices are involved in adapting walking and running speed on the treadmill: An optical imaging study. *Neuroimage* **2004**, *23*, 1020–1026. [CrossRef]
49. Koenraadt, K.L.M.; Roelofsen, E.G.J.; Duysens, J.; Keijsers, N.L.W. Cortical control of normal gait and precision stepping: An FNIRS Study. *Neuroimage* **2014**, *85 Pt 1*, 415–422. [CrossRef]
50. Haefeli, J.; Vögeli, S.; Michel, J.; Dietz, V. Preparation and Performance of Obstacle Steps: Interaction between Brain and Spinal neuronal activity. *Eur. J. Neurosci.* **2011**, *33*, 338–348. [CrossRef]
51. Wagner, J.; Solis-Escalante, T.; Scherer, R.; Neuper, C.; Müller-Putz, G. It's how you get there: Walking down a virtual alley activates premotor and parietal areas. *Front. Hum. Neurosci.* **2014**, *8*, 93. [CrossRef]
52. Wagner, J.; Martinez-Cancino, R.; Delorme, A.; Makeig, S.; Solis-Escalante, T.; Neuper, C.; Mueller-Putz, G. High-density EEG mobile brain/body imaging data recorded during a challenging auditory gait pacing task. *Sci. Data* **2019**, *6*, 211. [CrossRef] [PubMed]

53. Seeber, M.; Scherer, R.; Wagner, J.; Solis-Escalante, T.; Müller-Putz, G.R. EEG beta suppression and low gamma modulation are different elements of human upright walking. *Front. Hum. Neurosci.* **2014**, *8*, 485. [CrossRef]
54. Seeber, M.; Scherer, R.; Wagner, J.; Solis-Escalante, T.; Müller-Putz, G.R. High and low gamma EEG oscillations in central sensorimotor areas are conversely modulated during the human gait cycle. *Neuroimage* **2015**, *112*, 318–326. [CrossRef]
55. Nordin, A.D.; Hairston, W.D.; Ferris, D.P. Faster gait speeds reduce alpha and beta EEG spectral power from human sensorimotor cortex. *IEEE Trans. Biomed. Eng.* **2020**, *67*, 842–853. [CrossRef]
56. Luu, T.P.; He, Y.; Brown, S.; Nakagame, S.; Contreras-Vidal, J.L. Gait adaptation to visual kinematic perturbations using a real-time closed-loop brain-computer interface to a virtual reality avatar. *J. Neural Eng.* **2016**, *13*, 036006. [CrossRef]
57. Yuan, H.; Liu, T.; Szarkowski, R.; Rios, C.; Ashe, J.; He, B. Negative covariation between task-related responses in alpha/beta-band activity and BOLD in human sensorimotor cortex: An EEG and FMRI study of motor imagery and movements. *Neuroimage* **2010**, *49*, 2596–2606. [CrossRef]
58. Vanni, S.; Tanskanen, T.; Seppä, M.; Uutela, K.; Hari, R. Coinciding early activation of the human primary visual cortex and anteromedial cuneus. *Proc. Natl. Acad. Sci. USA* **2001**, *98*, 2776–2780. [CrossRef]
59. Neuner, I.; Arrubla, J.; Werner, C.J.; Hitz, K.; Boers, F.; Kawohl, W.; Shah, N.J. The default mode network and EEG regional spectral power: A simultaneous FMRI-EEG study. *PLoS ONE* **2014**, *9*, e88214. [CrossRef]
60. Bradford, J.C.; Lukos, J.R.; Ferris, D.P. Electrocortical activity distinguishes between uphill and level walking in humans. *J. Neurophysiol.* **2016**, *115*, 958–966. [CrossRef]

brain sciences

MDPI

Article

Exploring Machine Learning Classification of Movement Phases in Hemiparetic Stroke Patients: A Controlled EEG-tDCS Study

Rishishankar E. Suresh [1,2], M S Zobaer [2,3,*], Matthew J. Triano [1,3], Brian F. Saway [1,3], Parneet Grewal [2,4] and Nathan C. Rowland [1,2,3,4]

1 College of Medicine, Medical University of South Carolina, Charleston, SC 29425, USA; sureshr@musc.edu (R.E.S.); triano@musc.edu (M.J.T.); saway@musc.edu (B.F.S.); rowlandn@musc.edu (N.C.R.)
2 MUSC Institute for Neuroscience Discovery (MIND), Medical University of South Carolina, Charleston, SC 29425, USA; grewalp@musc.edu
3 Department of Neurosurgery, Medical University of South Carolina, Charleston, SC 29425, USA
4 Department of Neurology, Medical University of South Carolina, Charleston, SC 29425, USA
* Correspondence: zobaer@musc.edu; Tel.: +1-843-792-7700

Abstract: Background/Objectives: Noninvasive brain stimulation (NIBS) can boost motor recovery after a stroke. Certain movement phases are more responsive to NIBS, so a system that auto-detects these phases would optimize stimulation timing. This study assessed the effectiveness of various machine learning models in identifying movement phases in hemiparetic individuals undergoing simultaneous NIBS and EEG recordings. We hypothesized that transcranial direct current stimulation (tDCS), a form of NIBS, would enhance EEG signals related to movement phases and improve classification accuracy compared to sham stimulation. Methods: EEG data from 10 chronic stroke patients and 11 healthy controls were recorded before, during, and after tDCS. Eight machine learning algorithms and five ensemble methods were used to classify two movement phases (hold posture and reaching) during each of these periods. Data preprocessing included z-score normalization and frequency band power binning. Results: In chronic stroke participants who received active tDCS, the classification accuracy for hold vs. reach phases increased from pre-stimulation to the late intra-stimulation period (72.2% to 75.2%, $p < 0.0001$). Late active tDCS surpassed late sham tDCS classification (75.2% vs. 71.5%, $p < 0.0001$). Linear discriminant analysis was the most accurate (74.6%) algorithm with the shortest training time (0.9 s). Among ensemble methods, low gamma frequency (30–50 Hz) achieved the highest accuracy (74.5%), although this result did not achieve statistical significance for actively stimulated chronic stroke participants. Conclusions: Machine learning algorithms showed enhanced movement phase classification during active tDCS in chronic stroke participants. These results suggest their feasibility for real-time movement detection in neurorehabilitation, including brain–computer interfaces for stroke recovery.

Keywords: chronic stroke; machine learning; electroencephalogram; noninvasive brain stimulation; transcranial direct current stimulation

1. Introduction

Chronic stroke affects over 10 million people in the US and remains a major source of disability worldwide [1,2]. Brain–computer interfaces (BCIs) are a suggested method to improve the quality of life for affected individuals given their potential to detect stroke severity, sense ongoing motor behaviors, and assist with longitudinal recovery [3–8]. To advance BCIs for chronic stroke towards clinical practice, many groups are interested in creating a simple BCI embedded with machine learning (ML) that can be deployed at the population scale [9]. Yet it is currently unknown what neural input features and ML approaches are optimized for this task.

Electroencephalography (EEG) recordings are a noninvasive and an accessible method of sampling brain activity for neural computations in BCI use. Moreover, EEG can be paired with noninvasive brain stimulation (NIBS) to enhance certain features of control signal classification [10]. Transcranial direct current stimulation (tDCS), a form of NIBS, has been used in stroke rehabilitation [11] and has the specific advantage of continuous and simultaneous stimulation and EEG recording [10,12]. Anodal tDCS applied to the motor cortex has also been shown to affect BCI performance [12]. While mental rehearsal has evolved into the standard paradigm for BCI studies [13–17], particularly in those with tetraplegia, the vast majority of participants with stroke retain some level of motor capability. Accordingly, many have questioned whether ML-assisted EEG classification in stroke participants performing real movements requires the modeling of a substantively different parameter space [18–20]. As an example, Mebarkia et al. [21] found that layering three support vector machines (SVMs) (i.e., multi-voting) was necessary to exceed 90% accuracy in classifying left versus right hand movements in 3D space; however, this BCI architecture has only been tested in healthy participants.

The parameter space for EEG BCIs designed for stroke rehabilitation is extensive. First, the spectral content of EEG recordings in healthy controls is significantly different from that of individuals post-stroke [22–24]. For instance, mu (8–12 Hz) and beta (13–30 Hz) power are attenuated following a stroke, yet event-related desynchronization (ERD) and synchronization (ERS) of mu and beta power are frequently used to provide the feedback signal for decoding movement intention [25–27]. For those with severe motor deficits, ERD/ERS can be effectively absent, and thus, alternate control signals must be identified, though it is not clear what approach should be used to select surrogate markers. Second, vascular compromise gives rise to hemispheric asymmetry (i.e., ipsi- vs. contralesional), and this asymmetry is reflected as imbalanced oscillatory patterns following stroke [28–30]. Whether these factors preclude ML classification of brain states during real movement is a critical question to address prior to the development of a clinically accepted BCI for post-stroke hemiparesis.

In this study, we aimed to test the overall hypothesis that EEG data recorded during different movement phases in individuals with chronic stroke can be accurately classified using ML. We further explored whether transcranial direct current stimulation (tDCS), which can be activated continuously during EEG recording, boosts the ML classification accuracy for certain movement phases compared to sham stimulation. As a control, we employed an identical ML pipeline in healthy participants.

2. Materials and Methods

2.1. Sample Size Calculation

In a previous study comparing Parkinson's (n = 10) and essential tremor (n = 8) patients, we observed an absolute difference in cortical beta event-related desynchronization (ERD) during movement preparation of 0.24 with a standard deviation of 0.19 (PD -0.31 ± 0.06, ET -0.071 ± 0.05, $\log10(\mu V)^2/Hz$, mean $\pm$ SE, $p = 0.0061$) [31]. Based on this experience, with an observed absolute difference in cortical beta ERD of 0.24 $\log10(\mu V)^2/Hz$ and an assumed standard deviation to be as observed in our preliminary data for the PD group, we estimated that a sample size of 20 subjects for both the HC and CS groups would be sufficient to yield a power of 91% assuming a moderate intra-class correlation of 0.7. The sample size was divided in half for this pilot study, which was focused on the feasibility of the methods, with plans to carry out a fully powered study as an immediate follow-up to this manuscript.

2.2. Participants

Ten chronic stroke (CS) participants with upper-extremity hemiparesis and eleven healthy controls (HCs) were included in this study, which was approved by the Medical University of South Carolina Institutional Review Board (Pro#00087153; date: 16 July 2022) (Figure 1A). Hemiparetic patients were chosen as the vast majority of patients with chronic stroke retain some degree of motor function. Chronic stroke was defined as greater than six months from a sentinel infarct as determined by a fellowship-trained stroke neurologist (PG). Only participants who could raise and reach with the affected upper extremity were included. Participants with a history of seizure, intracranial mass, infection or cerebrovascular malformation, cranial defect, radiotherapy to the brain, scalp injury or disease, pregnancy, or the inability to consent to the study were excluded. The mean time after stroke for CS participants was 98.8 ± 36.7 months. The CS and HC groups did not differ with regard to sex ($\chi^2_{[1]} < 0$; $p > 0.05$). The mean age was higher in CS (63.3 ± 10.2 years; range = 30 to 78 years) than HC participants (46.3 ± 11.3 years; range = 24 to 73 years, $t_{[18]} = 2.31$, $p = 0.0319$). Six CS participants had left-sided infarcts, while four had right-sided infarcts.

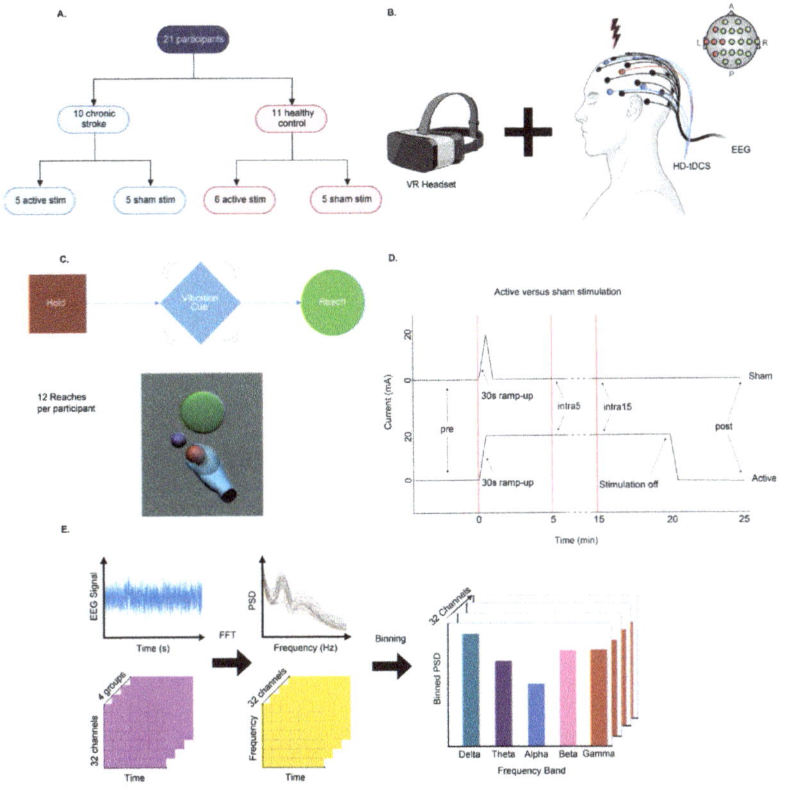

Figure 1. *Cont.*

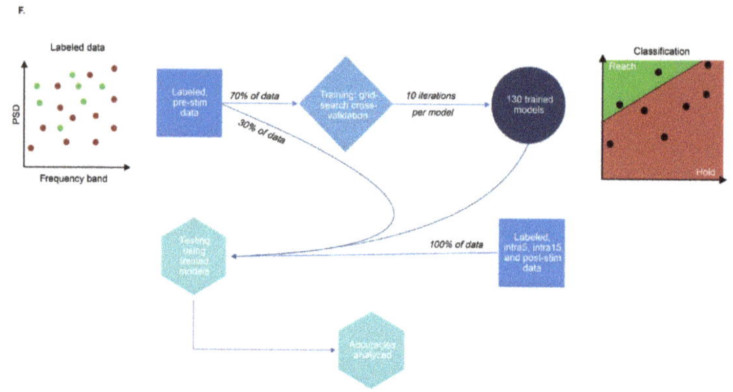

Figure 1. (**A**) A total of 10 participants with chronic stroke and 11 healthy controls were randomized into active and sham stimulation groups. (**B**) The participants were fitted with EEG electrodes and HD-tDCS delivering anodal stimulation to the ipsilesional side (contralateral to the motor deficit); for healthy participants, laterality was randomized. (**C**) During the stimulation phase, the participants performed a VR motor task involving reaches toward a virtual blue sphere target (3 cm radius) placed 0.3–0.5 m away. The task had three steps: hold, prep (Cue), and move. An example VR scene is shown. (**D**) Sham participants received 30 s of ramp-up current, then no stimulation, while active stimulation included 30 s ramp-up followed by 20 min of stimulation. Twelve reaches were performed at each time period. EEG was recorded at pre, 5 min (intra5), 15 min (intra15), and post-stimulation (post). (**E**) The raw EEG signal was recorded from all channels. After normalization, the power spectral density was calculated and binned across frequency bands to complete feature extraction. (**F**) Thirteen ML models were trained using 70% of the pre-stim data and then tested on 30% of the pre-stim data and 100% of the data from other time periods. This was performed to simulate ex vivo training of an onboard BCI.

2.3. Randomization and Single Blinding

Study participants in each CS or HC cohort were assigned to either a stimulation or sham group (Figure 1A). Each participant was assigned to the opposite group (i.e., stimulation or sham) as the previously recruited participant to maintain balance. Participants were blinded to their treatment assignment. For the CS cohort, 5 participants were assigned to active stim, and 5 were assigned to sham; for the HC cohort, 6 were assigned to active stim, and 5 were assigned to sham (Figure 1A). The sham group underwent a placebo procedure, in which current was applied in increasing steps for 30 s to induce a tingling sensation in the scalp. This mimicked the procedure in the stimulation group that involved an identical 30 s of current ramp before applying 20 min of DC current. Research staff operating the tDCS device were aware of the group assignment. Randomization and blinding were applied to both CS and HC cohorts in an identical fashion.

2.4. Task and Dataset

During a single recording session, the participants were fitted with a total of 20 Ag-AgCl EEG electrodes using the 10–20 International EEG system (DC115, Rhythmlink International LLC, Columbia, SC, USA) (Figure 1B). Next, a 5-electrode, center-surround, anodal high-density (HD) transcranial direct current stimulation (tDCS) montage was arranged on the scalp contralateral to the affected arm (model 2001tE and 4×1-C3ASKU 4×1-C3ASoterix Medical; 4×1 HD-tDCS/HD-tES adaptor, Soterix Medical, Inc, Woodbridge, NJ, USA). The central anodal tDCS electrode was positioned near the C3 or C4 EEG electrode, depending on the laterality of the upper extremity performing the task. The C3 and C4 electrodes represent the site of the primary motor cortex (M1) ipsilateral to the stroke lesion for participants with stroke. The surrounding cathodal tDCS electrodes were arranged in a square configuration in relation to the central anode. All tDCS electrodes were placed adjacent to, but out of direct contact with, nearby EEG electrodes (Figure 1B). Next, an Oculus Rift (Model C4-A, Menlo Park, CA, USA) virtual reality (VR) headset was carefully fitted over the EEG-tDCS arrangement, and a wireless controller containing an accelerometer was placed in the affected hand. For healthy participants, the hand used for the task was randomized. EEG signals were sampled at 1024 Hz (Natus® Neuroworks®, Pleasanton, CA, USA). An arm reaching task designed in the Unity® (version 2022.2, Unity Technologies, San Francisco, CA, USA) programming environment began with a holding position requiring a slightly outstretched upper limb to remain in place for several seconds prior to the appearance of a colored sphere. Once the sphere appeared, a reaching movement was made by the participant to virtually "touch" the sphere and return to the holding position (the total hand distance traveled ranged from 15 to 50 cm). The duration of the holding position was randomly varied between 2 and 5 s. Additionally, the location of the sphere was randomly varied between different locations within the virtual environment to reduce learning of the task over time. In between the hold and reach periods, a 0.5 s preparatory (or "prep") cue was delivered in the form of a vibratory pulse of the controller. The participants were not explicitly instructed to attend to this cue. The trial ended after 12 consecutive hold, prep, and reach cycles were completed (the average trial duration for all 12 reaches was approximately 3 min) (Figure 1C). The number of trials was selected empirically to protect against fatigue for mild-to-moderate hemiparetic individuals. For analysis purposes, each movement cycle was divided into the following states, or "epochs": the hold epoch, occurring prior to the prep cue, and the reach epoch occurring from initiation of movement until returning to the holding position. After an initial trial during the pre-stimulation period, the tDCS system was activated and within 30 s reached a maximum current delivery of 2.0 mA. Two additional trials of 12 reach cycles were performed at 5 and 15 min each after tDCS activation. Next, at 20 min after tDCS activation, the current was switched off. Five minutes after the deactivation of tDCS, a fourth and final trial was performed (Figure 1D). Thus, all participants performed a total of four trials of 12 reaches each. As per the published tDCS literature [10,32,33], the participants in the sham control group had tDCS activated for 30 s before ramping down to zero over an additional 30 s. This short period of stimulation mimics the 30 s of scalp tingling induced by active tDCS [33]. The HC group underwent the same randomization and procedures. The entire experiment for each participant lasted approximately one hour. The participants were given the option to take breaks to prevent fatigue. VR and EEG signals were labeled using synchronized TTL pulses across all data streams.

2.5. Preprocessing and Feature Extraction

Depending on the configuration, BCIs for stroke may have low computational power, which limits access to other critical signals such as electrooculography (EOG) and electromyography (EMG). An additional constraint observed by Winkler et al. and McDermott et al. [34,35] is that artifact rejection can attenuate BCI performance due to signal loss. Thus, we omitted resampling and limited signal preprocessing to bandpass filtering (a 1 Hz high-pass filter and 50 Hz low-pass filter were employed to remove environmental noise [36] and a portion of EOG and EMG artifacts, respectively [37]). All data were z-score transformed over the entire EEG signal for each channel per participant. Normalization was performed via the method shown in Equation (1), where x, $\overline{x}$, and σ represent the raw, mean, and standard deviation of the input values [38].

$$x_{norm} = \frac{x - \overline{x}}{\sigma} \quad (1)$$

Recordings were divided into 1 s epochs to balance point-to-point variations in the signal while maintaining adequate temporal resolution. Power spectral densities (PSDs) were calculated for each epoch using Welch's method, which applies the discrete Fourier transform (DFT) to several contiguous windowed subsets of the original signal [39]. Hann windowing was used to generate windowed segments with 50% overlap, and the number of Fast Fourier Transform (FFT) segments was set to the sample rate to maintain a spectral resolution of 1 Hz, as described above (Figure 1E). PSD values were subsequently resolved into the following frequency bands: delta (1–4 Hz), theta (4–8 Hz), alpha (8–12 Hz), beta (12–30 Hz), and gamma (30–50 Hz). As the bands did not exceed 50 Hz, a 60 Hz notch filter was not applied.

Extracted features from all participants in each of the four experimental groups (i.e., CS active stimulation, CS sham stimulation, HC active stimulation, and HC sham stimulation) were compiled into one dataset per group for model training. The labeled dataset in each experimental group was split 70:30 for model training and testing with grid search cross-validation used for hyperparameter tuning. The model training dataset consisted of 5100 samples with feature dimensions of (6, 5100), including 5 frequency band features and 1 electrode feature; the model testing dataset contained 2300 samples with dimensions of (6, 2300). Model training consisted of the classification of extracted features into 2 pre-labeled classes: the "hold" phase, during which participants were stationary, and the "reach" phase, during which participants extended their affected or dominant hand towards a VR target. Model training was carried out 10 times for each model per trial, and the average training times, with standard deviations, were recorded. Thus, models were trained for each experimental group rather than individual participants. From this dataset, a total of 4030 models were trained with a total computing time of 103 h. Training was performed on a 6th Gen Intel Xeon (R) Gold 6226R processor at 2.90 GHz with 64 cores and 187.5 GB RAM in serial processing alongside two NVIDIA RTX A5000 GPUs. Trained models were then tested on the remaining 30% of pre-stimulation data to obtain accuracies for the pre-stimulation time period. For subsequent time periods, 100% of the data were classified using these pre-trained models without further training, and these are the accuracies reported in the text (Figure 1F). All analyses were conducted in Python 3.9 using the NumPy [40], SciPy [40], and Scikit-Learn libraries [41].

2.6. Machine Learning Implementation

The classification accuracy of movement phases was tested among 13 different machine learning algorithms on the same dataset. These were selected to explore the effect of the disease state, stimulation state, time period, frequency band, and EEG electrode location. The time periods examined were pre-stimulation, intra-stimulation at 5 and 15 min after stimulation began, and post-stimulation, hereafter referred to as "Pre", "Intra5", "Intra15", and "Post". Features were tested in limited combinations to (1) reduce the exponential increase in model permutations and (2) to narrow the parameter space with a focus on clinical interpretability. A set of 8 primary algorithms was selected to identify an optimal model with respect to accuracy, computational demand, and training time. Additionally, we created 5 ensemble models using these 8 primary algorithms with varied weights to assess their combined accuracy. The following algorithms were chosen for this study: logistic regression (LR), linear discriminant analysis (LDA), decision trees (DT), Naïve Bayes (NB), K-nearest neighbors (KNN), random forest (RF), AdaBoost, and XGBoost.

LR was chosen due to its prior use in motor imagery classification derived from EEG signals [42]. LDA classifiers model the distribution of each class and were included due to their ability to perform dimensionality reduction and minimize the training time [43]. DT and RF were chosen to detect complex patterns in binarized data that may lead to a higher classification accuracy. The Naïve Bayes classifier applies Bayes' theorem to calculate the probability of an observation belonging to a given class based on the assumption that the data are distributed in a Gaussian manner; NB was included due to its minimal training time and mathematical simplicity [43]. KNN was included to determine if the movement phases exhibited clustering behavior in the associated feature space, as this would reveal insights beyond improved classification accuracy. The boosting algorithms XGboost and AdaBoost were included to detect the importance of incorrectly classified data points.

The hyperparameter search process for each classifier was defined in the following way: For LR, a univariate grid search was performed on the parameter C for values 2x for $15 \leq x < 35$. For LDA, comparisons were conducted for accuracies achieved by singular value decomposition (SVD), least squares (LSQR), and eigen solvers. Although all models demonstrated similar accuracies, SVD was chosen since it does not compute the covariance matrix and, therefore, has a shorter run time. For NB classification, a Gaussian implementation was used due to the Gaussian nature of epoched EEG data. A parameter representing the variance was computed using 10x for $-15 \leq x < 0$. For KNN, the number of neighbors considered was varied for $3 \leq x < 10$. For DT, the minimum weight fraction of each leaf node was empirically determined to be 0. A grid search was performed to optimize the minimum number of samples required to split individual nodes (varied for $2 \leq x < 11$) and the maximum depth allowed for trees (varied for $2 \leq x < 30$). For RF, the minimum number of samples per leaf was optimally determined to be 1, the method of determining the maximum samples to split a node was set to the square root of the total number of samples, and the maximum depth of each tree was set to 30. A grid search was performed to optimize the minimum number of samples required to split individual nodes (varied from $2 \leq x < 5$); the number of individual trees was varied over the set 5n for $5 \leq n < 21$. Feature importance was calculated by taking each feature's average depth of use and weighing the average from one relative to the other features' depths. The earlier a feature was used in a tree, the more important it was considered. For AdaBoost, NB and DT were contrasted as base estimators, and the number of individual estimators was varied by 25x for $8 \leq x \leq 16$. For XGBoost, the number of individual estimators was varied by 25x for $50 \leq x \leq 400$, and the max tree depth varied by 5x for $5 \leq x \leq 30$. The "Hist" method, as implemented by XGBoost 2.0.3 [44], was chosen for the Tree Method hyperparameter to reduce the training time. For voting classifiers, an ensemble of pre-trained models for LR, LDA, DT, RF, NB, and KNN was first created. One hard-voting classifier and four soft-voting classifiers among these were then utilized. Soft-voting classifiers used the following weight methods: uniform weights (termed "uni"), weights determined by the individual models' training set accuracy (termed "train"), uniform weights determined by the highest model accuracy (termed "hard"), and weights predetermined based upon the empirical global accuracy of all models within the ensemble (termed "global"). The ensemble labeled "me" was weighted based on the mean of several selected base estimators that appeared to perform better during the initial classification tests used to classify hold vs. reach. Effect weights, hyperparameters, and computation times were saved for each training. Prediction results were stored as text files labeled by the electrode and feature. Note that all accuracies depicted are the classification accuracies of the validation (i.e., testing) set; no samples in the validation set were used to train any of the models.

2.7. Statistical Analysis

Statistical analysis was conducted using the rstatix package for R (version 4.3.1) [45]. Combining data using two to three features for each model (e.g., grouping all electrodes and frequencies) resulted in groups sufficiently large to satisfy the central limit theorem, which were, therefore, treated as parametric data. For comparisons between multiple groups, student's t-test and repeated measures ANOVA (one and two way) were used. Sphericity was tested using Mauchly's test and, when violated, was corrected using the Greenhouse–Geisser correction. Bonferroni adjustment was applied for multiple comparisons. Specific comparisons are highlighted in the text with the appropriate test statistics; full results are given Supplementary Table S6. Data visualizations were generated using the ggpubr package for R [46]. The default threshold for significance was set at $p < 0.05$ for all tests. All error bars represent the standard error of the mean. Effect sizes are reported in Supplementary Table S11.

3. Results

3.1. Baseline Classification Accuracies

To investigate the baseline classification characteristics in our cohort, we compared the accuracy of each algorithm in distinguishing healthy from chronic stroke participants. This baseline analysis was carried out using all electrodes, frequency bands, and movement phases grouped together. This is also an important criterion our models need to meet, since BCIs, like all neuromodulatory devices, are designed to be disease specific. During the pre-stimulation time period, we observed that the mean accuracy of classifying HC and CS participants was 83.4% for the active group and 71.1% for the sham group ($t_{[22.1]} = -3.60$, $p = 0.0016$, Figure 2A), suggesting that some baseline variation persisted between the active and sham groups after randomization. This difference was notable in 12 out of 13 algorithms, the exception being linear discriminant analysis (LDA). To account for this variation, we normalized the accuracies for each group using their respective pre-stimulation accuracy (Figure 2C), and the resulting curves showed wide deviations between the groups in the classification accuracy over time. As expected, a two-way repeated measures ANOVA revealed a statistically significant effect of the stimulation group as a function of accuracy ($F_{[1,24]} = 4.502$, $p = 0.044$, see Supplementary Table S6 for full statistical results, n.b., red italicized text). Notably, normalized classification accuracies between the groups were most different at the post-stimulation state (active: 98.6%, sham: 82.5%, $t_{[23.8]} = -3.25$, $p = 0.0013$). These results confirm that differences in the classification of disease states between active and sham tDCS are statistically significant, consistent with our overall hypothesis.

3.2. Movement Phase

To investigate the accuracy of each algorithm in discriminating hold vs. reach movement phases, we trained ML models using all electrodes and frequency bands separated by disease and stimulation state over time (Figure 3). We observed that the mean accuracy of classifying hold vs. reach was 72.2% for the CS active group, 68.6% for the CS sham group, 71.6% for the HC active group, and 79.6% for the HC sham group at the pre-stimulation time period. For the CS active group, the mean classification accuracy increased during the early and late stimulation periods, peaking at intra15 (pre-stim: 72.2% vs. intra15: 75.3%, $t_{[17.1]} = 9.20$, $p < 0.0001$, Figure 3A,B, see Supplementary Table S6). Additionally, the mean classification accuracy between hold vs. reach was higher at intra15 for the CS active versus the CS sham cohort (active: 75.3%, sham: 71.5%, $t_{[23.7]} = -9.73$, $p < 0.0001$, Figure 3C). A two-way repeated measures ANOVA confirmed a significant effect on the accuracy as a function of the stimulation group ($F_{[1,24]} = 20.174$, $p = 0.00015$), as well as a significant effect of time ($F_{[3,72]} = 234.59$, $p < 0.0001$) and group–time interaction ($F_{[3,72]} = 111.599$, $p < 0.0001$). These results suggest divergence in the classification accuracy as a function of active tDCS while controlling for the disease state. Interestingly, the mean classification accuracy between hold vs. reach was significantly different at all time periods except intra15 for the CS active versus the HC active cohort (CS active: 75.3%, HC active:

74.5%, $t_{[24]} = 2.27$, $p = 0.1$, see Supplementary Table S6). Thus, late active tDCS exacerbates movement phase classification differences with sham tDCS (i.e., stimulation state) and attenuates classification differences between healthy and stroke participants (i.e., disease state).

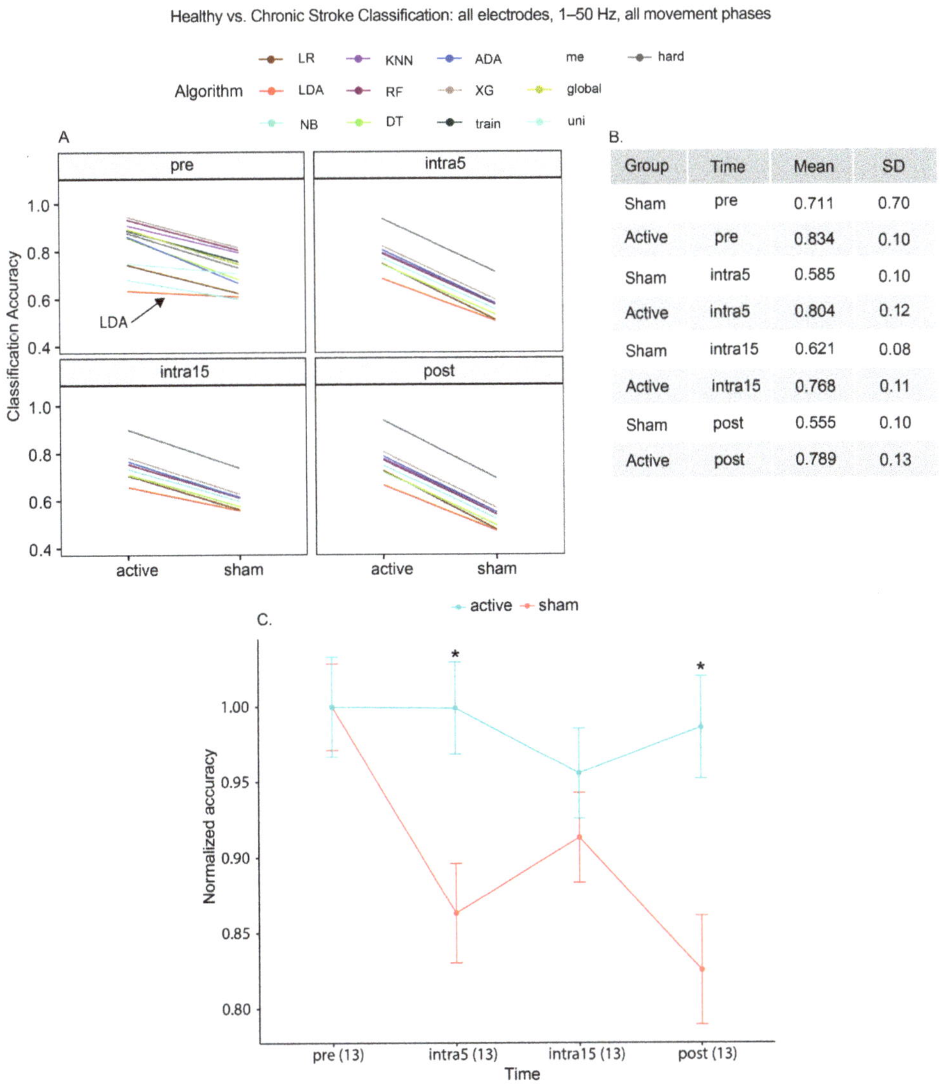

Group	Time	Mean	SD
Sham	pre	0.711	0.70
Active	pre	0.834	0.10
Sham	intra5	0.585	0.10
Active	intra5	0.804	0.12
Sham	intra15	0.621	0.08
Active	intra15	0.768	0.11
Sham	post	0.555	0.10
Active	post	0.789	0.13

Figure 2. (**A**) Classification of disease state (healthy versus chronic stroke) using all frequencies from 1 to 50 Hz and all electrodes. Although most algorithms detected differences prior to stimulation, LDA was not affected. (**B**) Mean accuracies. (**C**) To account for baseline differences at the pre-stimulation time periods, we normalized the accuracies to the pre-stimulation accuracy for each group. Asterisks (*) indicate a significant difference between active and sham groups. We observed a significantly increased classification accuracy for the active stim group at the intra5 and post-stimulation time periods, with accuracies converging at intra15. Values in parentheses represent the number of algorithms per group at each time period.

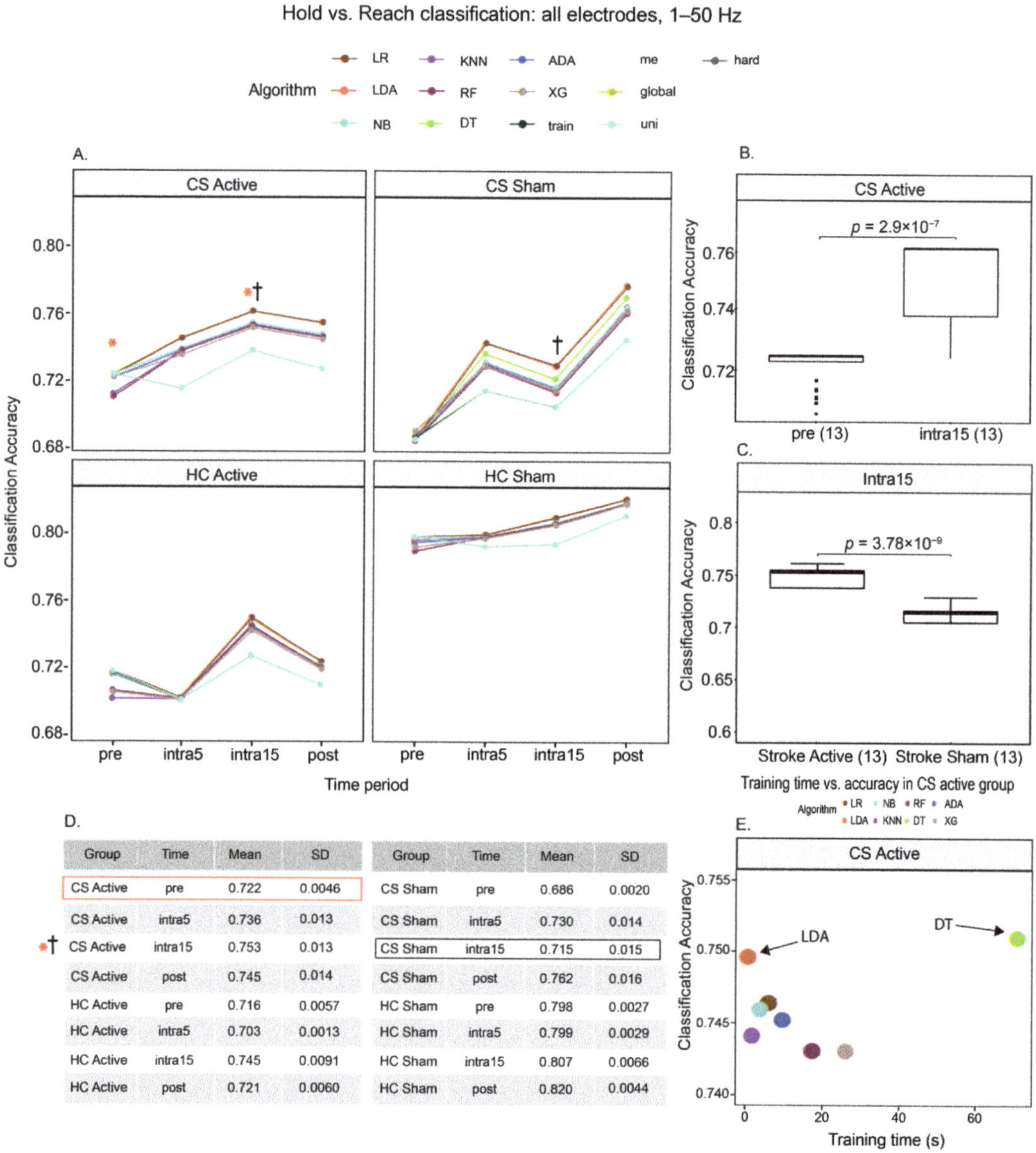

Group	Time	Mean	SD	Group	Time	Mean	SD
CS Active	pre	0.722	0.0046	CS Sham	pre	0.686	0.0020
CS Active	intra5	0.736	0.013	CS Sham	intra5	0.730	0.014
*† CS Active	intra15	0.753	0.013	CS Sham	intra15	0.715	0.015
CS Active	post	0.745	0.014	CS Sham	post	0.762	0.016
HC Active	pre	0.716	0.0057	HC Sham	pre	0.798	0.0027
HC Active	intra5	0.703	0.0013	HC Sham	intra5	0.799	0.0029
HC Active	intra15	0.745	0.0091	HC Sham	intra15	0.807	0.0066
HC Active	post	0.721	0.0060	HC Sham	post	0.820	0.0044

Figure 3. (**A**) Hold versus reach movement classification using all frequencies of 1–50 Hz and all electrodes. In the CS active group, the classification accuracy increased after stimulation and peaked at intra15 (*, asterisk). Furthermore, the accuracy at this time period was significantly higher than in the CS sham group at the same time (†, cross). (**B**) Movement classification was higher at intra15 than pre in the CS active group. (**C**) Movement classification was higher for intra15 CS active than intra15 CS sham. Values in parentheses represent the number of algorithms per group at each time period. (**D**) Mean accuracies. Values in parentheses represent the number of algorithms per group at each time period. (**E**) We observed a tradeoff between the training time and accuracy, as LDA produced the highest accuracy with a short training time compared to the other models with the exception of DT, which obtained a slightly higher mean accuracy but required the longest training time.

In the CS active group, LDA performed hold vs. reach classification with a mean accuracy of 74.6% at the shortest mean training time of 0.9 s per iteration. By comparison, DT performed hold vs. reach classification with a 75.1% accuracy at the longest average training time of 1 min 11.1 s per model.

3.3. Frequency Band

To investigate the accuracy of each algorithm in discriminating hold vs. reach phases by frequency band, we created ML models using all electrodes recorded in the CS active group only (Figure 4). We observed that the mean accuracy of classifying hold vs. reach was consistently higher in the stimulation and post-stimulation time periods for all frequency bands (Figure 4A,C; note that for the pre-stimulation period, the RF classification accuracy was below 65% for all frequency bands and is not shown in Figure 4A,C). When examining individual frequencies, only a low gamma classification accuracy significantly increased immediately after stimulation (gamma: pre = 71.7%, intra5 = 73.5%, $t_{[15.9]} = -2.88$, $p = 0.03$, Figure 4C; outliers omitted for clarity), although there is a broadband accuracy increase at intra15 observed for all bands when compared to pre (Figure 4C, with multiple comparison results in Supplementary Table S6). This increased accuracy persisted into the post-stimulation period (Figure 4C, with multiple comparison results in Supplementary Table S6). Interestingly, the largest differences in the classification accuracies by frequency band appeared between ML approaches rather than individual algorithms (Figure 4D, $F_{[6,253]} = 6.304$, $p < 0.0001$, Supplementary Table S6). While a two-way ANOVA between the ML method and frequency band across all time periods confirmed a significant effect on the accuracy as a function of the ML method ($F_{[6,225]} = 5.564$, $p < 0.0001$), it uncovered no significant effect of the frequency band ($F_{[4,225]} = 0.203$, $p = 0.936$) or method–frequency band interaction ($F_{[24,225]} = 0.046$, $p = 1.00$). Notwithstanding, we observed a sub-significant increase in the low gamma frequency band resolution, being most prominent in the ensemble methods ($F_{[4,95]} = 1.353$, $p = 0.256$, Figure 4D), whereas other methods appeared relatively indifferent to the frequency band choice.

3.4. Electrode Laterality

To investigate the accuracy of each algorithm in discriminating hold vs. reach phases by electrode laterality, we created ML models using all frequency bands grouped together. We labeled the electrode overlying the motor cortex (C3 or C4) contralateral to the hand used to perform the task as the ipsi-stimulated electrode. That is, if the right hand performed the task, then the C3 electrode was labeled as the ipsi-stimulated electrode while the C4 electrode was labeled as the contra-stimulated electrode and vice versa. We observed that the mean accuracy of classifying hold vs. reach was consistently higher in the contra-stimulated electrode for the CS sham and HC sham groups), with the exception of the pre-stimulation period in the HC sham cohort. In contrast, in the CS active group, the classification accuracy was highest in the ipsi-stimulated electrode during active stimulation periods only (intra5: contra = 59.3%, ipsi = 68.1%, $t_{[12]} = -7.01$, $p = 5.64 \times 10^{-5}$; intra15: contra = 60.6%, ipsi = 69.7%, $t_{[12]} = -4.71$, $p = 0.00204$, Figure 5A, see Supplementary Table S6). As expected, a two-way repeated measures ANOVA identified a significant effect of time ($F_{[3,72]} = 4.206$, $p = 0.032$) but not the electrode group ($F_{[1,24]} = 1.887$, $p = 0.182$), although a significant interaction between the electrode and time existed ($F_{[3,72]} = 7.196$, $p = 0.005$).

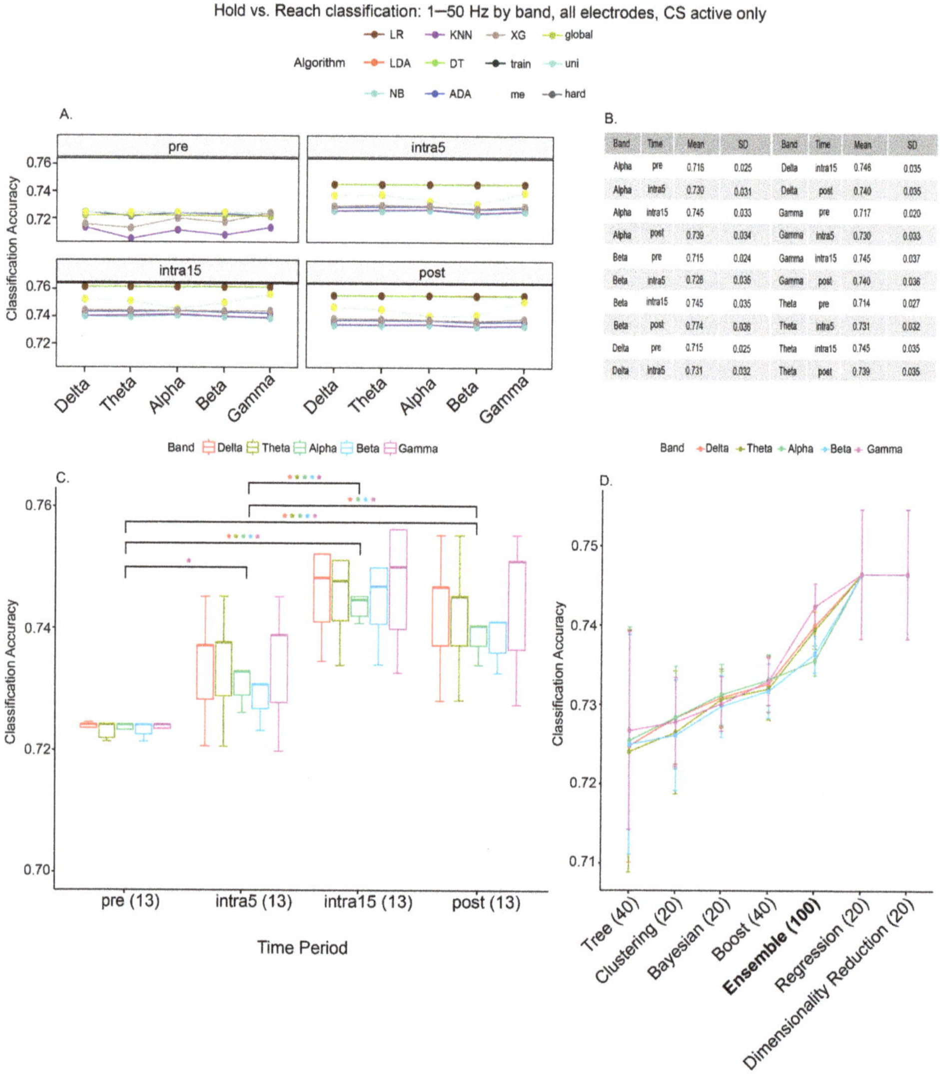

Band	Time	Mean	SD	Band	Time	Mean	SD
Alpha	pre	0.716	0.025	Delta	intra15	0.746	0.035
Alpha	intra5	0.730	0.031	Delta	post	0.740	0.035
Alpha	intra15	0.745	0.033	Gamma	pre	0.717	0.020
Alpha	post	0.739	0.034	Gamma	intra5	0.730	0.033
Beta	pre	0.715	0.024	Gamma	intra15	0.745	0.037
Beta	intra5	0.728	0.035	Gamma	post	0.740	0.036
Beta	intra15	0.745	0.035	Theta	pre	0.714	0.027
Beta	post	0.774	0.036	Theta	intra5	0.731	0.032
Delta	pre	0.715	0.025	Theta	intra15	0.745	0.035
Delta	intra5	0.731	0.032	Theta	post	0.739	0.035

Figure 4. (**A**) Hold versus reach movement classification using all frequencies of 1–50 Hz and all electrodes by the frequency band used (delta through gamma) in CS active participants only. Gamma PSD alone produced the highest classification accuracy for most models, although this was not statistically significant. (**B**) Mean accuracies. (**C**) When comparing the classification accuracy over time by band, we observe a broadband increase in the accuracy, seen in all bands at intra15. We also observe some differences in the band response, highlighted here using colored asterisks corresponding to each band. Values in parentheses represent the number of algorithms per individual frequency band at each time period. (**D**) When aggregating algorithms by method, we observed that ensemble methods (such as global voting, or hard voting) resolved frequency bands more so than other models, although this was not significant. Dimensionality reduction (LDA) and regression (LR) methods outperformed the others. Values in parentheses represent the number of algorithms per method at each time period.

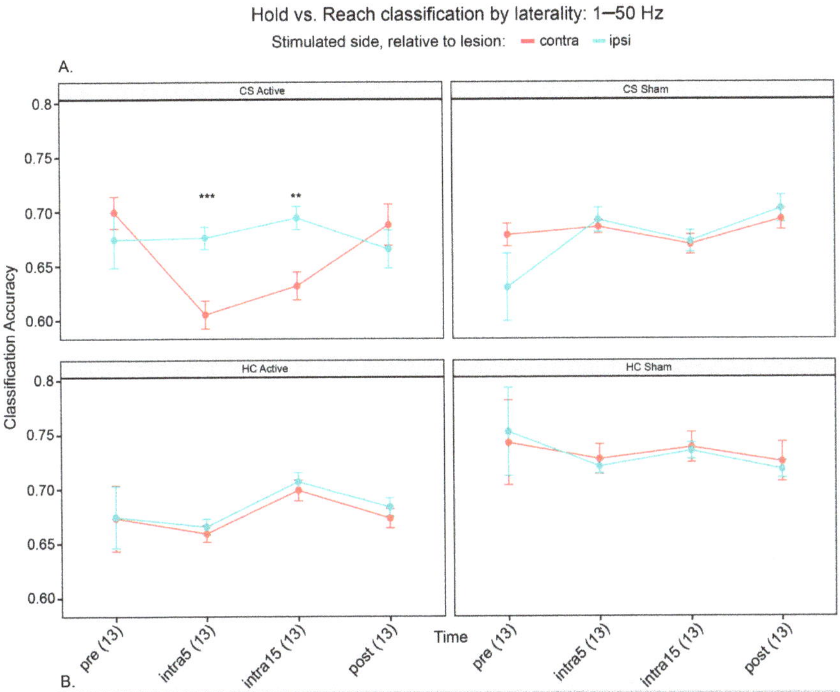

B.

Group	Time	Electrode	Mean	SD	Group	Time	Electrode	Mean	SD
CS Sham	pre	contra	0.678	0.037	CS Sham	intra15	contra	0.676	0.052
CS Sham	pre	ipsi	0.630	0.11	CS Sham	intra15	ipsi	0.657	0.11
CS Active	pre	contra	0.699	0.051	CS Active	intra15	contra	0.606	0.090
CS Active	pre	ipsi	0.674	0.089	CS Active	intra15	ipsi	0.697	0.080
HC Sham	pre	contra	0.742	0.14	HC Sham	intra15	contra	0.725	0.18
HC Sham	pre	ipsi	0.753	0.14	HC Sham	intra15	ipsi	0.719	0.071
HC Active	pre	contra	0.673	0.11	HC Active	intra15	contra	0.695	0.11
HC Active	pre	ipsi	0.674	0.099	HC Active	intra15	ipsi	0.692	0.12
CS Sham	intra5	contra	0.686	0.037	CS Sham	post	contra	0.695	0.062
CS Sham	intra5	ipsi	0.667	0.12	CS Sham	post	ipsi	0.679	0.14
CS Active	intra5	contra	0.593	0.079	CS Active	post	contra	0.715	0.059
CS Active	intra5	ipsi	0.681	0.075	CS Active	post	ipsi	0.682	0.088
HC Sham	intra5	contra	0.718	0.16	HC Sham	post	contra	0.722	0.18
HC Sham	intra5	ipsi	0.709	0.057	HC Sham	post	ipsi	0.703	0.082
HC Active	intra5	contra	0.652	0.094	HC Active	post	contra	0.664	0.12
HC Active	intra5	ipsi	0.658	0.097	HC Active	post	ipsi	0.669	0.11

Figure 5. (**A**) Hold versus reach movement classification using C3 or C4 electrodes and all frequencies of 1–50 Hz. Number of asterisks (*) represents significance level between contra- and ipsi-lesional accuracy. Here, we investigated the effect of recording electrode lesion laterality. We observed that contralesional classification accuracy significantly decreases during stimulation compared to ipsilesional accuracy in CS active participants; this is not seen in any of the other groups. Values in parentheses represent the number of algorithms per group at each time period. (**B**) Mean accuracies.

Notably, the classification accuracy was similar in both the contra-stimulated and ipsi-stimulated electrodes in the HC active group, suggesting that the higher accuracy in the ipsi-stimulated electrode in the CS active group is not likely to be driven by stimulation artifacts. Moreover, the classification accuracy in the ipsi-stimulated electrode in the CS active group is constant relative to the pre-stimulation state, again suggesting a physiological response to stimulation that peaks at the intra15 time period (Figure 5A,

Supplementary Tables S5 and S6). Overall, the classification accuracy for each electrode is dependent on the time of stimulation, i.e., whether stimulation is on or off.

4. Discussion

As neuromodulation becomes an accepted adjunct for chronic stroke recovery, the potential use of NIBS to assist in autonomously identifying movement phases from brain recordings has become a topic of great interest. The detection of movement intention alone using EEG recordings has been successfully performed in several studies using healthy and tetraplegic participants [26,47,48]. However, real movement classification in individuals with hemiparesis, the largest group of chronic stroke patients, is not well understood, including the dimensionality involved in modeling relevant parameters. Moreover, in implanted brain recording and stimulation systems, computation power is limited to the onboard processor, which prohibits the typical types of algorithms employed in these studies, e.g., deep learning and other neural network-based strategies. To surpass these constraints, using minimal signal preprocessing, we explored supervised ML modeling of EEG recordings in order to understand the range of accuracies and modeling times for potential use as control signal classifiers. The central problem addressed in this study is whether movement classification based on EEG recordings is plausible (i.e., above chance) in chronic stroke survivors undergoing NIBS during the performance of a functional task. Our overall finding is that EEG data recorded during different movement phases in individuals with chronic stroke can be accurately classified using ML. Additionally, tDCS improved classification accuracy primarily in the chronic stroke active stimulation group, with a slight preference for gamma frequency bands using the ensemble methods.

Comparing our results to similar studies that developed EEG classifiers for movement phases, SVMs performed modestly better in 46.4% of models than LDA and LR in classifying rest, simple arm movements, goal-oriented arm movements, and hand clenching during motor imagery [43]. In our study, the performance of LDA for hold vs. reach (at 74.5% accuracy during the intra15 time period) was within range of the accuracies reported by Yong and Menon [43] (75–81% accuracy) and higher than those of Rodrigo et al. [42] (64–68% accuracy). The voting classifiers trained in this study for hold vs. reach in the chronic stroke active stimulation cohort (at the intra15 time period, see Supplementary Table S3) achieved an accuracy of 73.8%, which is not as strong as the voting classifier created by Khrishna et al. [49] (86% accuracy). Similar to the dataset used by Mebarkia and Reffad [21], the dataset used by Khrishna et al. [49] classified motor imagery in the right arm, left arm, right foot, and left foot without any consideration for holding or preparation, which may explain the stronger performance. The AdaBoost classifier trained as part of our study achieved an accuracy of 75.3% using decision trees as the classifier base, which is comparable to the AdaBoost classifiers trained by Gao et al. [50], who used SVM and LDA bases to achieve accuracies of 74% and 72%, respectively.

A few studies have combined tDCS and BCI devices with mixed results. tDCS alters the brain's electric field, modulating ionic currents and neuronal membrane potentials to enhance synaptic plasticity [33,51]. These induced changes are hypothesized to improve information flow across neural networks, analogous to optimizing resonance frequencies in coupled oscillatory systems for greater synchronization and efficiency [51]. Matsumoto et al. [52]. used a motor imagery (MI) BCI in concert with multiple 1 mA 10 min tDCS sessions in six healthy participants. In their investigation, mu ERD improved with anodal and attenuated with cathodal stimulation [52]. Kasashima and colleagues [12] repeated this paradigm in participants with post-stroke hemiparesis, demonstrating similar results. In a study by Wei et al. [53], tDCS specifically modulated upper mu (10–14 Hz) and beta (14–26 Hz) frequencies in 32 healthy controls. Hong and colleagues [54] introduced diffusion and perfusion MRI following tDCS in combination with an MI-BCI. Tractography estimates showed significant changes on the ipsilesional side for participants receiving tDCS, although no difference in motor improvement was observed between active and sham groups [54]. Overall, tDCS did not appear to influence MI-BCI performance in a randomized, double-blinded controlled trial in 19 participants with stroke [10]. These results did not differ from a subsequent study in which functional MRI was used to derive a low-frequency fluctuation metric [32]. None of the studies outlined utilized real movements as a control.

Although we observed that the ensemble methods achieved the highest hold vs. reach classification using gamma power as a feature, the modulation of gamma power in individuals post-stroke is only sparsely reported. In a study by Tecchio et al. [55], increased gamma power (33.5–44 Hz) in the affected hemisphere of chronic stroke participants was correlated with motor improvement using magnetoencephalography recordings. Moreover, Pellegrino and colleagues [56] demonstrated that gamma reactivity to an auditory stimulus in chronic stroke participants was tightly correlated to the clinical outcome, as measured using the Barthel Index and Functional Independence Measure. Yet in a recent systematic analysis of randomized controlled trials examining the utility of BCIs in stroke motor recovery, gamma power was absent from the frequency bands investigated as a potential biomarker [7].

Limitations

This study was not without limitations. First, a small sample size limits the interpretability of our classification results. Notwithstanding, Bolognini et al. [57] performed a similar exploratory study evaluating the interaction between tDCS and constraint-induced movement therapy in chronic stroke and reported significant complementary effects using only 14 participants. Similarly, Biasucci et al. [58] performed a sham-controlled study evaluating the effect of BCI-coupled functional electric stimulation (FES) on motor outcomes and reported significant improvement when compared to sham, with only 27 participants and 14 in the experimental arm. These and other similar studies with limited participant numbers are reported in Supplementary Table S9. A limited sample size further predisposes to baseline differences between groups, as is demonstrated by the discrepancy in the classification accuracy between our pre-stimulation stroke cohorts. This group heterogeneity is further exacerbated by the sensitivity of machine learning models to cross-validation folds, which are randomly generated. In this situation, Watson and Holmes describe the use of a subgroup-specific statistical analysis plan for evaluation; however, this is impractical a priori in a single-subject chronic stroke BCI; furthermore, the use of modifications to machine learning models, as described in that study, may require computational resources and time that may not be available to BCI devices [59]. Other tDCS studies have employed a crossover design to mitigate this heterogeneity to some success [60,61], although this approach risks carryover effects from tDCS contaminating the sham phase, as was noted by Klomjai et al. [60]. We have attempted to address this using repeated measures, i.e., within-subject statistical design. Significant differences in the mean age between our chronic stroke and healthy cohorts were also observed, potentially confounding observed differences in the classification accuracy that may have been affected by age-related changes in the brain.

To attempt to mimic the processing and bit-rate constraints of a fully implantable BCI system [62], we avoided training more comprehensive deep learning models. Nevertheless, a robust classification pipeline using Convolutional Neural Networks was described by Lun et al. [63], who trained a five-layer model on the Physionet database [64,65]. Remarkably, using only 10 participants from that dataset, a global accuracy of 94% or above was demonstrated. Similar to our study, they limited the pre-processing of the EEG signal and still achieved high classification accuracies [63]. In our study, training was limited to twelve repetitions per task and four tasks per participant during a single one-hour experimental session. Nevertheless, our results are almost certainly confounded by some proportion of learning of the task, although we attempted to limit this by randomizing several features, including the time between movement phases and the location of reach target. Importantly, the effect of learning on our results should be mitigated by the controls that were incorporated, including healthy participants and sham stimulation. Also, significantly more task repetitions could potentially be limited by fatigue, which was not observed in our sample (i.e., no participants asked to take a break when offered).

We have not carried out a full examination of the preprocessing and feature extraction [66]. Finally, we did not include asymmetry index measures for the chronic stroke participants as some authors have [24]. Given the heterogeneity of each chronic stroke participant, the extent to which each of these factors contributes to the need to personalize the training models for each individual should be explored further in future studies. These findings may help lead to the development of a closed-loop device that can auto-detect movement phases and deliver therapeutic stimulation when most beneficial for chronic stroke rehabilitation.

Supplementary Materials: The following supporting information can be downloaded at: https://www.mdpi.com/article/10.3390/brainsci15010028/s1: Table S1: Participant demographics and characteristics; Table S2: For Figure 2. Disease state classification. Mean accuracies by stimulation state, algorithm and time period; Table S3: For Figure 3. Hold versus Reach classification. Mean accuracies by disease state, stimulation state, time period, and algorithm; Table S4 for Figure 4. Hold versus Reach classification by frequency band, time period, and algorithm.; Table S5 for Figure 5: Hold versus Reach classification. Mean accuracies by disease state, stimulation state, time period and electrode laterality relative to dominant hand; Supplementary Table S6: Statistical analyses; Supplementary Table S7: Multiple comparison of methods for Hold versus Reach for CS active only (Figure 4). Table S8: Mean training time for each algorithm (excluding the ensemble methods). Table S9: tDCS studies with small participant numbers limited subjects; Table S10: To summarize the outcomes from the literature with alternative approaches in machine learning; Table S11: Effect size calculations for main findings [10,14,27,50,57,58,63,66–78].

Author Contributions: N.C.R. performed the initial experiment and data collection. N.C.R. and R.E.S. performed the machine learning analyses. N.C.R., R.E.S., M.S.Z., B.F.S., P.G. and M.J.T. contributed to writing the manuscript and provided critical feedback. All authors have read and agreed to the published version of the manuscript.

Funding: Center of Biomedical Research Excellence (COBRE) in Stroke Recovery—Junior Investigator Research Project. Source: National Institutes of Health (5 P20 GM109040).

Institutional Review Board Statement: All participants were provided a thorough description of the study prior to participation. Approval was provided by the Medical University of South Carolina's Institutional Review Board(Pro#00087153; date: 16 July 2022).

Informed Consent Statement: Consent for publication was given by all participants.

Data Availability Statement: The data, as well as code, that support the findings of this study are available on request from the lead author.

Conflicts of Interest: The authors declare no conflicts of interest.

Abbreviations

ADA, adaptive boost; CS, chronic stroke; DT, decision tree; EEG, electroencephalogram; Global, soft-vote ensemble weighted based on the overall accuracy; Hard, hard-vote ensemble; KNN, k-nearest neighbors; LDA, linear discriminant analysis; LR, logistic regression; Me, soft-vote ensemble with weights chosen based on the mean of multiple base estimators; NB, Naïve Bayes; RF, random forest; Stim, stimulation; tDCS, transcranial direct current stimulation; Train, soft-vote ensemble weighted based on training performance; Uni, soft-vote ensemble weighted equally; VR, virtual reality; XG, XGboost.

References

1. Tsao, C.W.; Aday, A.W.; Almarzooq, Z.I.; Anderson, C.A.M.; Arora, P.; Avery, C.L.; Baker-Smith, C.M.; Beaton, A.Z.; Boehme, A.K.; Buxton, A.E.; et al. Heart Disease and Stroke Statistics-2023 Update: A Report From the American Heart Association. *Circulation* **2023**, *147*, e93–e621. [CrossRef]
2. Prust, M.L.; Forman, R.; Ovbiagele, B. Addressing disparities in the global epidemiology of stroke. *Nat. Rev. Neurol.* **2024**, *20*, 207–221. [CrossRef]
3. Chaudhary, U.; Birbaumer, N.; Ramos-Murguialday, A. Brain-computer interfaces for communication and rehabilitation. *Nat. Rev. Neurol.* **2016**, *12*, 513–525. [CrossRef] [PubMed]
4. Cervera, M.A.; Soekadar, S.R.; Ushiba, J.; Millán, J.D.R.; Liu, M.; Birbaumer, N.; Garipelli, G. Brain-computer interfaces for post-stroke motor rehabilitation: A meta-analysis. *Ann. Clin. Transl. Neurol.* **2018**, *5*, 651–663. [CrossRef] [PubMed]
5. Hughes, C.; Herrera, A.; Gaunt, R.; Collinger, J. Bidirectional brain-computer interfaces. *Handb. Clin. Neurol.* **2020**, *168*, 163–181. [CrossRef] [PubMed]
6. Young, M.J.; Lin, D.J.; Hochberg, L.R. Brain-Computer Interfaces in Neurorecovery and Neurorehabilitation. *Semin. Neurol.* **2021**, *41*, 206–216. [CrossRef]
7. Fu, J.; Chen, S.; Jia, J. Sensorimotor Rhythm-Based Brain-Computer Interfaces for Motor Tasks Used in Hand Upper Extremity Rehabilitation after Stroke: A Systematic Review. *Brain Sci.* **2022**, *13*, 56. [CrossRef]
8. Zhan, G.; Chen, S.; Ji, Y.; Xu, Y.; Song, Z.; Wang, J.; Niu, L.; Bin, J.; Kang, X.; Jia, J. EEG-Based Brain Network Analysis of Chronic Stroke Patients After BCI Rehabilitation Training. *Front. Hum. Neurosci.* **2022**, *16*, 909610. [CrossRef]
9. Mane, R.; Chouhan, T.; Guan, C. BCI for stroke rehabilitation: Motor and beyond. *J. Neural Eng.* **2020**, *17*, 041001. [CrossRef]
10. Chew, E.; Teo, W.P.; Tang, N.; Ang, K.K.; Ng, Y.S.; Zhou, J.H.; Teh, I.; Phua, K.S.; Zhao, L.; Guan, C. Using Transcranial Direct Current Stimulation to Augment the Effect of Motor Imagery-Assisted Brain-Computer Interface Training in Chronic Stroke Patients-Cortical Reorganization Considerations. *Front. Neurol.* **2020**, *11*, 948. [CrossRef] [PubMed]
11. Lima, J.P.S.; Silva, L.A.; Delisle-Rodriguez, D.; Cardoso, V.F.; Nakamura-Palacios, E.M.; Bastos-Filho, T.F. Unraveling Transformative Effects after tDCS and BCI Intervention in Chronic Post-Stroke Patient Rehabilitation-An Alternative Treatment Design Study. *Sensors* **2023**, *23*, 9302. [CrossRef] [PubMed]
12. Kasashima, Y.; Fujiwara, T.; Matsushika, Y.; Tsuji, T.; Hase, K.; Ushiyama, J.; Ushiba, J.; Liu, M. Modulation of event-related desynchronization during motor imagery with transcranial direct current stimulation (tDCS) in patients with chronic hemiparetic stroke. *Exp. Brain Res.* **2012**, *221*, 263–268. [CrossRef] [PubMed]
13. Pichiorri, F.; De Vico Fallani, F.; Cincotti, F.; Babiloni, F.; Molinari, M.; Kleih, S.C.; Neuper, C.; Kübler, A.; Mattia, D. Sensorimotor rhythm-based brain-computer interface training: The impact on motor cortical responsiveness. *J. Neural Eng.* **2011**, *8*, 025020. [CrossRef] [PubMed]
14. Mokienko, O.A.; Chervyakov, A.V.; Kulikova, S.N.; Bobrov, P.D.; Chernikova, L.A.; Frolov, A.A.; Piradov, M.A. Increased motor cortex excitability during motor imagery in brain-computer interface trained subjects. *Front. Comput. Neurosci.* **2013**, *7*, 168. [CrossRef]
15. Hänselmann, S.; Schneiders, M.; Weidner, N.; Rupp, R. Transcranial magnetic stimulation for individual identification of the best electrode position for a motor imagery-based brain-computer interface. *J. Neuroeng. Rehabil.* **2015**, *12*, 71. [CrossRef]
16. Gao, T.; Hu, Y.; Zhuang, J.; Bai, Y.; Lu, R. Repetitive Transcranial Magnetic Stimulation of the Brain Region Activated by Motor Imagery Involving a Paretic Wrist and Hand for Upper-Extremity Motor Improvement in Severe Stroke: A Preliminary Study. *Brain Sci.* **2022**, *13*, 69. [CrossRef]
17. Liu, X.; Zhang, W.; Li, W.; Zhang, S.; Lv, P.; Yin, Y. Effects of motor imagery based brain-computer interface on upper limb function and attention in stroke patients with hemiplegia: A randomized controlled trial. *BMC Neurol.* **2023**, *23*, 136. [CrossRef] [PubMed]
18. Al-Qazzaz, N.K.; Aldoori, A.A.; Ali, S.; Ahmad, S.A.; Mohammed, A.K.; Mohyee, M.I. EEG Signal Complexity Measurements to Enhance BCI-Based Stroke Patients' Rehabilitation. *Sensors* **2023**, *23*, 3889. [CrossRef] [PubMed]
19. Abiri, R.; Borhani, S.; Sellers, E.W.; Jiang, Y.; Zhao, X. A comprehensive review of EEG-based brain-computer interface paradigms. *J. Neural Eng.* **2019**, *16*, 011001. [CrossRef] [PubMed]
20. Mane, R.; Chew, E.; Phua, K.S.; Ang, K.K.; Robinson, N.; Vinod, A.P.; Guan, C. Prognostic and Monitory EEG-Biomarkers for BCI Upper-Limb Stroke Rehabilitation. *IEEE Trans. Neural Syst. Rehabil. Eng.* **2019**, *27*, 1654–1664. [CrossRef]

21. Mebarkia, K.; Reffad, A. Multi optimized SVM classifiers for motor imagery left and right hand movement identification. *Australas. Phys. Eng. Sci. Med.* **2019**, *42*, 949–958. [CrossRef] [PubMed]
22. Rabiller, G.; He, J.W.; Nishijima, Y.; Wong, A.; Liu, J. Perturbation of Brain Oscillations after Ischemic Stroke: A Potential Biomarker for Post-Stroke Function and Therapy. *Int. J. Mol. Sci.* **2015**, *16*, 25605–25640. [CrossRef]
23. Sato, Y.; Schmitt, O.; Ip, Z.; Rabiller, G.; Omodaka, S.; Tominaga, T.; Yazdan-Shahmorad, A.; Liu, J. Pathological changes of brain oscillations following ischemic stroke. *J. Cereb. Blood Flow. Metab.* **2022**, *42*, 1753–1776. [CrossRef] [PubMed]
24. Leonardi, G.; Ciurleo, R.; Cucinotta, F.; Fonti, B.; Borzelli, D.; Costa, L.; Tisano, A.; Portaro, S.; Alito, A. The role of brain oscillations in post-stroke motor recovery: An overview. *Front. Syst. Neurosci.* **2022**, *16*, 947421. [CrossRef]
25. Pfurtscheller, G.; Aranibar, A. Evaluation of event-related desynchronization (ERD) preceding and following voluntary self-paced movement. *Electroencephalogr. Clin. Neurophysiol.* **1979**, *46*, 138–146. [CrossRef] [PubMed]
26. Wolpaw, J.R.; McFarland, D.J. Control of a two-dimensional movement signal by a noninvasive brain-computer interface in humans. *Proc. Natl. Acad. Sci. USA* **2004**, *101*, 17849–17854. [CrossRef] [PubMed]
27. Ramos-Murguialday, A.; Broetz, D.; Rea, M.; Läer, L.; Yilmaz, O.; Brasil, F.L.; Liberati, G.; Curado, M.R.; Garcia-Cossio, E.; Vyziotis, A.; et al. Brain-machine interface in chronic stroke rehabilitation: A controlled study. *Ann. Neurol.* **2013**, *74*, 100–108. [CrossRef]
28. Bundy, D.T.; Souders, L.; Baranyai, K.; Leonard, L.; Schalk, G.; Coker, R.; Moran, D.W.; Huskey, T.; Leuthardt, E.C. Contralesional Brain-Computer Interface Control of a Powered Exoskeleton for Motor Recovery in Chronic Stroke Survivors. *Stroke* **2017**, *48*, 1908–1915. [CrossRef] [PubMed]
29. Dodd, K.C.; Nair, V.A.; Prabhakaran, V. Role of the Contralesional vs. Ipsilesional Hemisphere in Stroke Recovery. *Front. Hum. Neurosci.* **2017**, *11*, 469. [CrossRef]
30. Hasegawa, K.; Kasuga, S.; Takasaki, K.; Mizuno, K.; Liu, M.; Ushiba, J. Ipsilateral EEG mu rhythm reflects the excitability of uncrossed pathways projecting to shoulder muscles. *J. Neuroeng. Rehabil.* **2017**, *14*, 85. [CrossRef]
31. Rowland, N.C.; de Hemptinne, C.; Swann, N.C.; Qasim, S.; Miocinovic, S.; Ostrem, J.; Knight, R.T.; Starr, P.A. Task-related activity in sensorimotor cortex in Parkinson's disease and essential tremor: Changes in beta and gamma bands. *Front. Hum. Neurosci.* **2015**, *9*, 512. [CrossRef]
32. Hu, M.; Cheng, H.J.; Ji, F.; Chong, J.S.X.; Lu, Z.; Huang, W.; Ang, K.K.; Phua, K.S.; Chuang, K.H.; Jiang, X.; et al. Brain Functional Changes in Stroke Following Rehabilitation Using Brain-Computer Interface-Assisted Motor Imagery With and Without tDCS: A Pilot Study. *Front. Hum. Neurosci.* **2021**, *15*, 692304. [CrossRef] [PubMed]
33. Gandiga, P.C.; Hummel, F.C.; Cohen, L.G. Transcranial DC stimulation (tDCS): A tool for double-blind sham-controlled clinical studies in brain stimulation. *Clin. Neurophysiol.* **2006**, *117*, 845–850. [CrossRef] [PubMed]
34. Winkler, I.; Haufe, S.; Tangermann, M. Automatic Classification of Artifactual ICA-Components for Artifact Removal in EEG Signals. *Behav. Brain Funct.* **2011**, *7*, 30. [CrossRef] [PubMed]
35. McDermott, E.J.; Raggam, P.; Kirsch, S.; Belardinelli, P.; Ziemann, U.; Zrenner, C. Artifacts in EEG-Based BCI Therapies: Friend or Foe? *Sensors* **2021**, *22*, 96. [CrossRef]
36. Jiang, X.; Bian, G.-B.; Tian, Z. Removal of Artifacts from EEG Signals: A Review. *Sensors* **2019**, *19*, 987. [CrossRef]
37. Fatourechi, M.; Bashashati, A.; Ward, R.K.; Birch, G.E. EMG and EOG artifacts in brain computer interface systems: A survey. *Clin. Neurophysiol.* **2007**, *118*, 480–494. [CrossRef]
38. Mendenhall, W.; Sincich, T. *Statistics for Engineering and the Sciences*, 5th ed.; Pearson: London, UK, 2006.
39. Welch, P. The use of fast Fourier transform for the estimation of power spectra: A method based on time averaging over short, modified periodograms. *IEEE Trans. Audio Electroacoust.* **1967**, *15*, 70–73. [CrossRef]
40. Harris, C.R.; Millman, K.J.; van der Walt, S.J.; Gommers, R.; Virtanen, P.; Cournapeau, D.; Wieser, E.; Taylor, J.; Berg, S.; Smith, N.J.; et al. Array programming with NumPy. *Nature* **2020**, *585*, 357–362. [CrossRef] [PubMed]
41. Pedregosa, F.; Varoquaux, G.; Gramfort, A.; Michel, V.; Thirion, B.; Grisel, O.; Blondel, M.; Prettenhofer, P.; Weiss, R.; Dubourg, V.; et al. Scikit-learn: Machine Learning in Python. *J. Mach. Learn. Res.* **2012**, *12*, 2825–2830.
42. Rodrigo, M.; Montesano, L.; Minguez, J. Classification of resting, anticipation and movement states in self-initiated arm movements for EEG brain computer interfaces. In Proceedings of the 2011 Annual International Conference of the IEEE Engineering in Medicine and Biology Society, Boston, MA, USA, 30 August–3 September 2011; Volume 2011, pp. 6285–6288. [CrossRef]
43. Yong, X.; Menon, C. EEG classification of different imaginary movements within the same limb. *PLoS ONE* **2015**, *10*, e0121896. [CrossRef]
44. Chen, T.; Guestrin, C. XGBoost: A Scalable Tree Boosting System. In Proceedings of the 22nd ACM SIGKDD International Conference on Knowledge Discovery and Data Mining, San Francisco, CA, USA, 13–17 August 2016.
45. Kassambara, A. Pipe-Friendly Framework for Basic Statistical Tests [R Package Rstatix Version 0.6.0]. 2020. Available online: https://rpkgs.datanovia.com/rstatix/ (accessed on 1 November 2024).
46. Kassambara, A. Ggpubr: 'Ggplot2' Based Publication Ready Plots. 2022. Available online: https://rpkgs.datanovia.com/ggpubr/ (accessed on 1 November 2024).

47. Pfurtscheller, G.; Müller, G.R.; Pfurtscheller, J.; Gerner, H.J.; Rupp, R. 'Thought'--control of functional electrical stimulation to restore hand grasp in a patient with tetraplegia. *Neurosci. Lett.* **2003**, *351*, 33–36. [CrossRef]
48. Müller-Putz, G.R.; Scherer, R.; Pfurtscheller, G.; Rupp, R. EEG-based neuroprosthesis control: A step towards clinical practice. *Neurosci. Lett.* **2005**, *382*, 169–174. [CrossRef] [PubMed]
49. Khrishna, D.; Pasha, I.A.; Savithri, T. Multi-level voting method to classify motor imagery EEG signals. *ARPN J. Eng. Appl. Sci.* **2018**, *13*, 3815–3819.
50. Gao, L.; Cheng, W.; Zhang, J.; Wang, J. EEG classification for motor imagery and resting state in BCI applications using multi-class Adaboost extreme learning machine. *Rev. Sci. Instrum.* **2016**, *87*, 085110. [CrossRef] [PubMed]
51. Kuo, H.-I.; Bikson, M.; Datta, A.; Minhas, P.; Paulus, W.; Kuo, M.-F.; Nitsche, M.A. Comparing Cortical Plasticity Induced by Conventional and High-Definition 4 × 1 Ring tDCS: A Neurophysiological Study. *Brain Stimul.* **2013**, *6*, 644–648. [CrossRef] [PubMed]
52. Matsumoto, J.; Fujiwara, T.; Takahashi, O.; Liu, M.; Kimura, A.; Ushiba, J. Modulation of mu rhythm desynchronization during motor imagery by transcranial direct current stimulation. *J. Neuroeng. Rehabil.* **2010**, *7*, 27. [CrossRef]
53. Wei, P.; He, W.; Zhou, Y.; Wang, L. Performance of motor imagery brain-computer interface based on anodal transcranial direct current stimulation modulation. *IEEE Trans. Neural Syst. Rehabil. Eng.* **2013**, *21*, 404–415. [CrossRef] [PubMed]
54. Hong, X.; Lu, Z.K.; Teh, I.; Nasrallah, F.A.; Teo, W.P.; Ang, K.K.; Phua, K.S.; Guan, C.; Chew, E.; Chuang, K.H. Brain plasticity following MI-BCI training combined with tDCS in a randomized trial in chronic subcortical stroke subjects: A preliminary study. *Sci. Rep.* **2017**, *7*, 9222. [CrossRef] [PubMed]
55. Tecchio, F.; Pasqualetti, P.; Zappasodi, F.; Tombini, M.; Lupoi, D.; Vernieri, F.; Rossini, P.M. Outcome prediction in acute monohemispheric stroke via magnetoencephalography. *J. Neurol.* **2007**, *254*, 296–305. [CrossRef] [PubMed]
56. Pellegrino, G.; Arcara, G.; Cortese, A.M.; Weis, L.; Di Tomasso, S.; Marioni, G.; Masiero, S.; Piccione, F. Cortical gamma-synchrony measured with magnetoencephalography is a marker of clinical status and predicts clinical outcome in stroke survivors. *Neuroimage Clin.* **2019**, *24*, 102092. [CrossRef] [PubMed]
57. Bolognini, N.; Vallar, G.; Casati, C.; Latif, L.A.; El-Nazer, R.; Williams, J.; Banco, E.; Macea, D.D.; Tesio, L.; Chessa, C.; et al. Neurophysiological and behavioral effects of tDCS combined with constraint-induced movement therapy in poststroke patients. *Neurorehabil Neural Repair.* **2011**, *25*, 819–829. [CrossRef] [PubMed]
58. Biasiucci, A.; Leeb, R.; Iturrate, I.; Perdikis, S.; Al-Khodairy, A.; Corbet, T.; Schnider, A.; Schmidlin, T.; Zhang, H.; Bassolino, M.; et al. Brain-actuated functional electrical stimulation elicits lasting arm motor recovery after stroke. *Nat. Commun.* **2018**, *9*, 2421. [CrossRef] [PubMed]
59. Watson, J.A.; Holmes, C.C. Machine learning analysis plans for randomised controlled trials: Detecting treatment effect heterogeneity with strict control of type I error. *Trials* **2020**, *21*, 156. [CrossRef] [PubMed]
60. Klomjai, W.; Aneksan, B.; Pheungphrarattanatrai, A.; Chantanachai, T.; Choowong, N.; Bunleukhet, S.; Auvichayapat, P.; Nilanon, Y.; Hiengkaew, V. Effect of single-session dual-tDCS before physical therapy on lower-limb performance in sub-acute stroke patients: A randomized sham-controlled crossover study. *Ann. Phys. Rehabil. Med.* **2018**, *61*, 286–291. [CrossRef] [PubMed]
61. Hordacre, B.; Moezzi, B.; Ridding, M.C. Neuroplasticity and network connectivity of the motor cortex following stroke: A transcranial direct current stimulation study. *Hum. Brain Mapp.* **2018**, *39*, 3326–3339. [CrossRef] [PubMed]
62. Leuthardt, E.C.; Freudenberg, Z.; Bundy, D.; Roland, J. Microscale recording from human motor cortex: Implications for minimally invasive electrocorticographic brain-computer interfaces. *Neurosurg. Focus.* **2009**, *27*, E10. [CrossRef]
63. Lun, X.; Yu, Z.; Chen, T.; Wang, F.; Hou, Y. A Simplified CNN Classification Method for MI-EEG via the Electrode Pairs Signals. *Front. Hum. Neurosci.* **2020**, *14*, 338. [CrossRef] [PubMed]
64. Goldberger, A.L.; Amaral, L.A.; Glass, L.; Hausdorff, J.M.; Ivanov, P.C.; Mark, R.G.; Mietus, J.E.; Moody, G.B.; Peng, C.K.; Stanley, H.E. PhysioBank, PhysioToolkit, and PhysioNet: Components of a new research resource for complex physiologic signals. *Circulation* **2000**, *101*, E215–E220. [CrossRef] [PubMed]
65. Schalk, G.; McFarland, D.J.; Hinterberger, T.; Birbaumer, N.; Wolpaw, J.R. BCI2000: A general-purpose brain-computer interface (BCI) system. *IEEE Trans. Biomed. Eng.* **2004**, *51*, 1034–1043. [CrossRef]
66. Oken, B.S.; Orhan, U.; Roark, B.; Erdogmus, D.; Fowler, A.; Mooney, A.; Peters, B.; Miller, M.; Fried-Oken, M.B. Brain–Computer Interface With Language Model–Electroencephalography Fusion for Locked-In Syndrome. *Neurorehabilit. Neural Repair* **2013**, *28*, 387–394. [CrossRef] [PubMed]
67. Jia, T.; Mo, L.; Li, C.; Liu, A.; Li, Z.; Ji, L. 5 Hz rTMS improves motor-imagery based BCI classification performance. In Proceedings of the 2021 43rd Annual International Conference of the IEEE Engineering in Medicine & Biology Society (EMBC), Mexico City, Mexico, 1–5 November 2021; pp. 6116–6120. [CrossRef]
68. Epperson, J.D.; Meyers, E.C.; Pruitt, D.T.; Wright, J.M.; Hudson, R.A.; Adehunoluwa, E.A.; Nguyen-Duong, Y.N.; Rennaker, R.L.; Hays, S.A.; Kilgard, M.P. Characterization of an Algorithm for Autonomous, Closed-Loop Neuromodulation During Motor Rehabilitation. *Neurorehabilit. Neural Repair* **2024**, *38*, 493–505. [CrossRef] [PubMed]

69. Morone, G.; Capone, F.; Iosa, M.; Cruciani, A.; Paolucci, M.; Martino Cinnera, A.; Musumeci, G.; Brunelli, N.; Costa, C.; Paolucci, S.; et al. May Dual Transcranial Direct Current Stimulation Enhance the Efficacy of Robot-Assisted Therapy for Promoting Upper Limb Recovery in Chronic Stroke? *Neurorehabilit. Neural Repair* **2022**, *36*, 800–809. [CrossRef] [PubMed]
70. Watts, D.; Pulice, R.F.; Reilly, J.; Brunoni, A.R.; Kapczinski, F.; Passos, I.C. Predicting treatment response using EEG in major depressive disorder: A machine-learning meta-analysis. *Transl. Psychiatry* **2022**, *12*, 332. [CrossRef] [PubMed]
71. Hosseinifard, B.; Moradi, M.H.; Rostami, R. Classifying depression patients and normal subjects using machine learning techniques and nonlinear features from EEG signal. *Comput. Methods Programs Biomed.* **2013**, *109*, 339–345. [CrossRef] [PubMed]
72. Yuan, Q.; Zhou, W.; Li, S.; Cai, D. Epileptic EEG classification based on extreme learning machine and nonlinear features. *Epilepsy Res.* **2011**, *96*, 29–38. [CrossRef] [PubMed]
73. Yang, C.Y.; Chen, Y.Z. Support vector machine classification of patients with depression based on resting-state electroencephalography. *Asian Biomed. (Res. Rev. News)* **2024**, *18*, 212–223. [CrossRef] [PubMed]
74. Mumtaz, W.; Ali, S.S.A.; Yasin, M.A.M.; Malik, A.S. A machine learning framework involving EEG-based functional connectivity to diagnose major depressive disorder (MDD). *Med. Biol. Eng. Comput.* **2018**, *56*, 233–246. [CrossRef]
75. Maitín, A.M.; García-Tejedor, A.J.; Muñoz, J.P.R. Machine Learning Approaches for Detecting Parkinson's Disease from EEG Analysis: A Systematic Review. *Appl. Sci.* **2020**, *10*, 8662. [CrossRef]
76. Lee, M.; Hong, Y.; An, S.; Park, U.; Shin, J.; Lee, J.; Oh, M.S.; Lee, B.-C.; Yu, K.-H.; Lim, J.-S.; et al. Machine learning-based prediction of post-stroke cognitive status using electroencephalography-derived brain network attributes. *Front. Aging Neurosci.* **2023**, *15*, 1238274. [CrossRef] [PubMed]
77. Lin, W.Y.; Chen, C.H.; Tseng, Y.J.; Tsai, Y.T.; Chang, C.Y.; Wang, H.Y.; Chen, C.K. Predicting post-stroke activities of daily living through a machine learning-based approach on initiating rehabilitation. *Int. J. Med. Inf.* **2018**, *111*, 159–164. [CrossRef] [PubMed]
78. Ramirez Campos, M.S.; McCracken, H.S.; Uribe-Quevedo, A.; Grant, B.L.; Yielder, P.C.; Murphy, B.A. A Machine Learning Approach to Classifying EEG Data Collected with or without Haptic Feedback during a Simulated Drilling Task. *Brain Sci.* **2024**, *14*, 894. [CrossRef] [PubMed]

brain sciences

MDPI

Article

Dynamic Neural Network States During Social and Non-Social Cueing in Virtual Reality Working Memory Tasks: A Leading Eigenvector Dynamics Analysis Approach

Pinar Ozel

Electric and Electronic Engineering Department, Istanbul University-Cerrahpasa, Istanbul 34320, Turkey; pozell@gmail.com

Abstract: Background/Objectives: This research investigates brain connectivity patterns in reaction to social and non-social stimuli within a virtual reality environment, emphasizing their impact on cognitive functions, specifically working memory. Methods: Employing the LEiDA framework with EEG data from 47 participants, I examined dynamic brain network states elicited by social avatars compared to non-social stick cues during a VR memory task. Through the integration of LEiDA with deep learning and graph theory analyses, unique connectivity patterns associated with cue type were discerned, underscoring the substantial influence of social cues on cognitive processes. LEiDA, conventionally utilized with fMRI, was creatively employed in EEG to detect swift alterations in brain network states, offering insights into cognitive processing dynamics. Results: The findings indicate distinct neural states for social and non-social cues; notably, social cues correlated with a unique brain state characterized by increased connectivity within self-referential and memory-processing networks, implying greater cognitive engagement. Moreover, deep learning attained approximately 99% accuracy in differentiating cue contexts, highlighting the efficacy of prominent eigenvectors from LEiDA in EEG analysis. Analysis of graph theory also uncovered structural network disparities, signifying enhanced integration in contexts involving social cues. Conclusions: This multi-method approach elucidates the dynamic influence of social cues on brain connectivity and cognition, establishing a basis for VR-based cognitive rehabilitation and immersive learning, wherein social signals may significantly enhance cognitive function.

Keywords: dynamic brain connectivity; EEG; LEiDA; social cues; virtual reality; deep learning

1. Introduction

In recent years, the combination of neuroimaging and virtual reality (VR) has provided new opportunities for studying the brain dynamics of cognitive processes [1]. Working memory, an essential cognitive function in daily tasks, has been widely examined using conventional methods [2]. However, the emergence of VR technology offers a distinct opportunity to replicate real-life situations that involve both social and non-social cognitive processes [3]. Using electroencephalography (EEG) to measure neural activity during these tasks provides detailed and precise information regarding the brain states involved in these cognitive processes [4].

Research on the neural correlates of social and non-social cues has uncovered divergent patterns of cerebral activity. One study [5] investigates the impact of repetitive transcranial magnetic stimulation on functional connectivity in methamphetamine use disorder, offering insights into neural adaptations. The neural mechanisms underlying social contextual influences on adolescent risk-taking behavior have been examined, enhancing our comprehension of how intricate social environments affect neural responses. This is detailed in a study presented by [6].

VR is an appropriate platform for studying complicated cognitive tasks because it incorporates real-world features and allows for controlled modification of environmental stimuli [7]. The distinction between social and non-social cueing in VR can provide insight into how human cognition adjusts in various social situations, an area that has received less attention in traditional cognitive neuroscience [8]. Research has demonstrated that social interactions in VR can have a notable impact on cognitive load and the corresponding brain reactions [9]. This reveals the existence of separate neural networks that are activated during social activities compared with non-social tasks [10,11].

Studies have shown that VR interventions can improve social skills and emotional recognition in children and adolescents, including those with neurological disorders. One study [12] demonstrates the efficacy of VR in enhancing social and emotional learning. The influence of haptic feedback on social interactions in VR has been investigated, demonstrating its effect on user experience and the genuineness of social touch. Ref. [13] elucidates the influence of tactile feedback on social dynamics in virtual environments. The importance of social cues in mediated, bidirectional, multiparty interactions has been examined, highlighting their function in enhancing interpersonal communication. A study published in 2021 [14] examines the significance of different social cues in immersive environments.

VR has become a revolutionary instrument in multiple domains, offering immersive and interactive settings that replicate real-world situations. One of its most significant applications is in the medical field, where avatars in VR environments are employed for therapeutic and training objectives. VR interventions have demonstrated efficacy in treating mental health disorders, including social anxiety, post-traumatic stress disorder (PTSD), and depression, by enabling patients to interact with regulated virtual environments [15,16]. Moreover, avatars in VR are extensively utilized for cognitive rehabilitation, aiding patients in the re-acquisition of motor and cognitive skills after neurological injuries [17]. This research utilizes the distinctive features of VR to examine dynamic brain network states during interactions with social avatars.

Connectivity analysis grounded in graph theory provides critical insights into the structural organization and complexity of brain networks by depicting brain regions as nodes and significant functional connections as edges. This method is especially beneficial for examining dynamic brain states, as it facilitates the analysis of network characteristics, including the clustering coefficient and path length, which enhance comprehension of the neural mechanisms that underpin cognitive functions. Utilizing graph theory on EEG-derived coherence matrices allows for the examination of connectivity patterns across various cognitive states, especially in VR environments where social and non-social stimuli may elicit distinct network configurations [18–20].

Dynamic functional connectivity (DFC) refers to temporary and state-specific changes in connections between different brain networks [21]. It is considered an important factor in cognitive function. The high temporal resolution (TR) of EEG is particularly suitable for this type of analysis [22,23]. Some common DFC methods used with EEG are time–frequency resolution coherence analysis, phase synchronization analysis, sliding-window correlation analysis, graph-theory-based methods, and wavelet coherence. Time–frequency analysis was used to analyze the power and coherence of the EEG signals in different frequency bands. This method makes it possible to monitor coherence changes in signals over time and determine how synchronization between brain regions changes during specific tasks or sensory stimulation [24]. Phase synchronization analysis measures the extent to which the phases of EEG signals from different brain regions are in harmony. Phase synchronization is used to examine temporal connections and collaborations between neural networks [25]. The phase locking value (PLV) is a measure used to quantify the consistency of phase differences between signals over time [26]. Sliding-window correlation analysis continuously calculates the correlations between EEG signals using a specified window size. The window is shifted across the signal, revealing connectivity structures that change over time. This method is particularly suitable for tracking short-term changes in brain dynamics [27]. Moreover, EEG data can be analyzed using graph theory, in which brain regions are represented as nodes and the connectivity between them as edges. This approach helps us to understand the structural and functional properties of brain networks by evaluating their topological properties (e.g., the clustering coefficient and path length) [28]. Wavelet coherence measures the coherence between two signals at different time points and in different frequency bands, thereby enabling a more detailed dynamic connectivity analysis [29]. Progress in EEG data analysis has yielded novel techniques for evaluating functional connectivity [30]. A thorough review [31] examines contemporary methods for EEG seizure detection, emphasizing the challenges and prospective advancements in the domain. Furthermore, the amalgamation of EEG and ensemble learning has been suggested for precise grading and staging of diseases, highlighting the essential significance of time window length in assessing model stability and efficacy. This methodology is elucidated in a study [32]. The utilization of graph convolutional networks (GCNs) for processing graph-structured EEG data has demonstrated potential in patient-independent epileptic seizure detection, as outlined in [33].

Recent progress in DFC analysis approaches provides an exceptional opportunity to investigate these dynamics with very detailed spatial and temporal resolutions. The recently established Leading Eigenvector Dynamics Analysis (LEiDA) method has played a crucial role in capturing these changes by examining the patterns of leading eigenvectors derived from connectivity matrices [34]. This method offers a new way to observe changes [20] in neural network states linked to various cognitive demands.

The LEiDA method was developed to monitor the functional connectivity changes in the brain over time. This method is used specifically with functional magnetic resonance imaging (fMRI) data because fMRI measures changes in cerebral blood flow to visualize brain activity, which can reflect dynamic connectivity changes in the brain over long-term periods [35–37]. LEiDA analyzes the dynamics of leading eigenvectors in time-series data obtained from fMRI. These eigenvectors represent the strongest connectivity patterns in the brain at a given time point and show how these connections change over time. This method is particularly useful for understanding how brain networks organize and reorganize during complex cognitive processes [38,39].

The use of LEiDA with EEG is a newer field of research. EEG records brain waves with a high TR at the millisecond level, which provides the opportunity to observe much faster brain dynamics [40]. However, EEG has a lower spatial resolution than fMRI, which may limit some types of analysis [41]. The use of methods such as LEiDA combined with EEG data offers the potential to examine time-varying brain network dynamics in more detail. In particular, the high TR of EEG, when combined with the LEiDA method, can capture rapid changes in brain networks and allow more sensitive examination of cognitive processes. Additionally, Zhou et al. stated in their study that LEiDA has a superior TR compared to other DFC analysis methods, such as the sliding-window method [42]. Therefore, although the LEiDA method is mostly used with fMRI data, it can also be used with EEG data, and this combination can open new doors for a more comprehensive understanding of brain dynamics. Such an approach could provide great advantages, especially in studies on rapidly changing cognitive states and responses.

The term 'states' specifically denotes dynamic brain network states identified via LEiDA. These states signify persistent patterns of phase synchronization among various brain regions, as calculated from the dynamic phase-locking matrices (d) obtained from EEG data. Each state represents a distinct configuration of functional connectivity, reflecting the temporal dynamics of brain network organization throughout the task [34]. The identification of these states enables an examination of how the brain transitions between various connectivity patterns during interactions with social and non-social cues. In this study, the term PL states is used to refer to the states identified by LEiDA.

This research examines a significant deficiency in comprehending the dynamic neural mechanisms that underpin social cognition, especially during real-time interactions. Conventional investigations of brain connectivity typically depend on static analyses that average connectivity patterns over time, thereby neglecting the transient and rapidly changing states of brain networks during social interactions. This constraint is particularly problematic when examining intricate cognitive processes such as social cueing, which necessitate temporal precision to comprehend how the brain shifts between various functional states. This research employs the LEiDA method to address these challenges by providing a high-resolution, temporally sensitive framework for examining brain network dynamics.

This study incorporates an additional innovative element through the utilization of VR. In contrast to traditional experimental configurations, VR offers an immersive and ecologically valid environment for simulating authentic social interactions. This enables the examination of neural responses in an environment that closely mirrors natural social contexts, providing enhanced understanding of the mechanisms underlying social cognition. The utilization of VR to examine dynamic network states in reaction to social and non-social stimuli is both timely and innovative, as it connects laboratory research with the intricacies of real-world interactions.

LEiDA is particularly adept for this function as it directly examines the temporal dynamics of phase-locking matrices, discerning recurrent connectivity patterns with elevated temporal resolution. In contrast to sliding-window correlation (SWC), which is constrained by arbitrary window size and diminished sensitivity to swift transitions, LEiDA effectively captures nuanced, instantaneous alterations in brain connectivity. Likewise, although graph theory offers significant structural insights into brain networks, its metrics are frequently based on static representations and are incapable of analyzing temporal transitions. By integrating LEiDA with VR, I seek to offer a sophisticated comprehension of the brain's dynamic processing of social information in real time.

A primary impetus for this research is the opportunity to create targeted therapeutic interventions for disorders marked by impairments in social cognition, including autism spectrum disorder, schizophrenia, and social anxiety. Contemporary therapeutic approaches frequently neglect the dynamic characteristics of social interactions, prioritizing static or excessively simplistic models instead. By pinpointing particular brain network states linked to social and non-social interactions, my results may guide the development of VR-based interventions aimed at enhancing individuals' social functioning within a controlled yet realistic setting.

This research incorporates sophisticated analytical techniques, such as deep learning and graph theory, to enhance the LEiDA methodology. Deep learning facilitates the identification of complex patterns in high-resolution EEG data, revealing nuanced dynamics that conventional methods may overlook. Graph theory provides a structural perspective, illustrating how network attributes such as clustering coefficients and path lengths vary with different cognitive tasks. The amalgamation of these methodologies offers a thorough framework for comprehending the temporal and structural dimensions of brain network dynamics, facilitating a comprehensive exploration of cognitive processes.

This study enhances our comprehension of social cognition at the neural level by examining the transitions between dynamic brain network states during social and non-social cueing in VR. It also tackles the overarching challenge in neuroscience of examining brain networks in genuine, ecologically valid environments. This research addresses a significant gap in the literature and establishes a foundation for novel therapeutic strategies designed for practical social deficits, providing valuable applications for clinical populations [34,43].

2. Materials and Methods

In this study, three main themes were examined, namely LEiDA, graph theory, and deep learning. LEiDA analysis (Section 2.2) was performed to obtain PL states in the brain. In the graph theory part (Section 2.3), graph-based analysis was performed using different graphs. In the deep learning section (Section 2.4), classification results were evaluated using raw EEG, PLVs, and Eigenvectors. The methods applied are visualized in Figure 1.

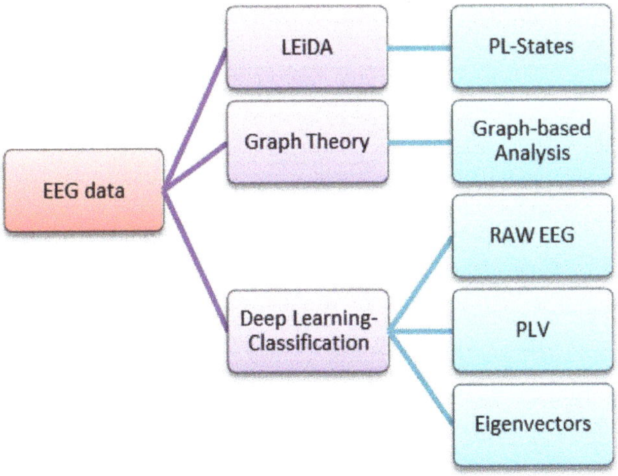

Figure 1. Applied methods (LEiDA, graph theory, and deep learning classification).

2.1. Dataset

2.1.1. Participant Demographics

This study utilized a dataset originally published by [44] comprising EEG recordings obtained during a VR working memory task involving the presentation of social and non-social cues to participants. The dataset was released to the public to facilitate research on cognitive processes within VR environments.

This research included 47 participants (25 females and 22 males) from a university community, all with normal or corrected-to-normal vision and no history of neurological or psychiatric disorders. Their average age was 24.3 years (SD = 4.7). Informed consent was obtained from all participants, and the study was approved by the institutional review board.

Although a formal power analysis to justify the sample size of 47 participants could not be performed due to time constraints, this limitation is addressed by referencing similar studies in the field that utilized a comparable number of participants. These references provide context and support for the adequacy of the sample size in this research [12,32].

2.1.2. VR Working Memory Task

Participants were immersed in a VR environment using a head-mounted display. The task assessed working memory through memorization of the location and characteristics of the objects displayed on a virtual table. Two types of attentional cues were utilized: social avatars (e.g., a virtual character making eye contact and pointing) and non-social stick cues (e.g., a stick arrow pointing). The task structure included multiple trials, each with cueing, memorization, and recall phases. The object details, participant responses, and reaction times were recorded for each trial. The criteria used to select the VR working memory task and its psychometric properties are detailed in reference [44]. This process is summarized in Figure 2 for this study.

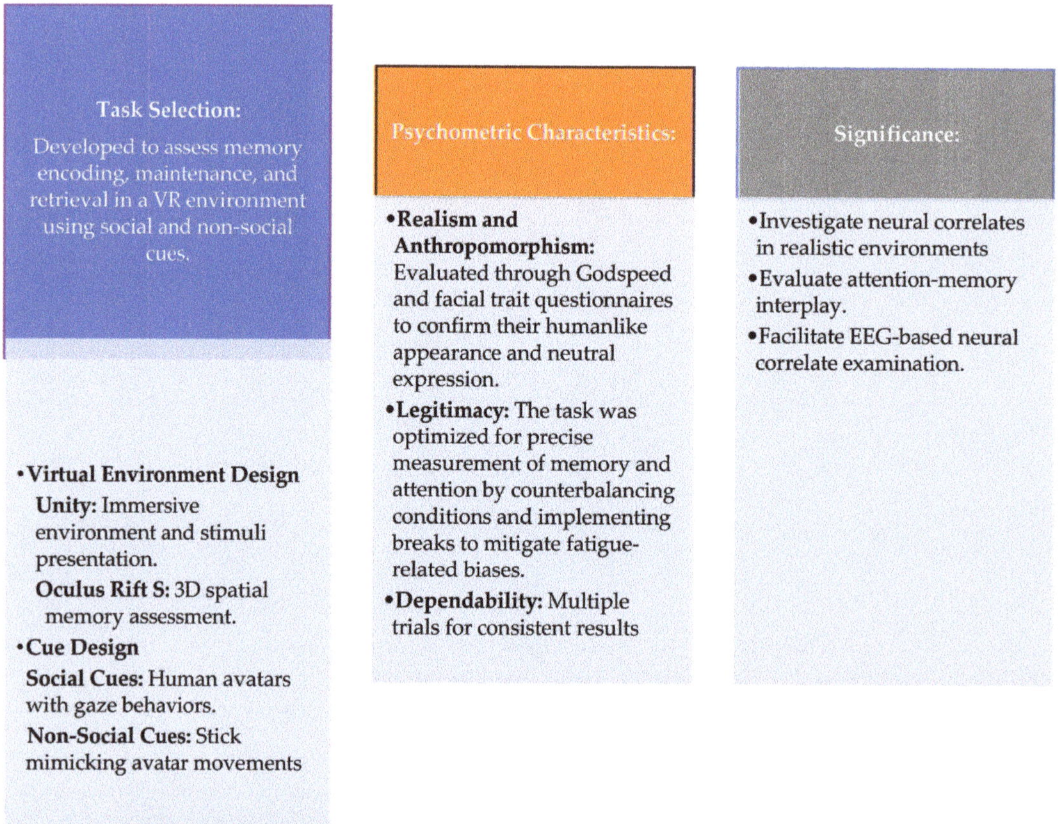

Figure 2. VR working memory task: selection and design schema.

2.1.3. Experiment Procedure

Using a head-mounted display, 3D avatars engaged participants by making eye contact before looking towards the intended object location, mimicking real eye movement behaviors. Target items—kitchen utensils such as bowls, plates, cups, and teapots—appeared on either side of a table, facilitating the study of cue influence in a realistic setting. Participants memorized multidimensional information about items, such as location and status, during the cue shift, encoding, and retrieval phases. A dynamic 3D stick served as a non-social control cue, paralleling the movement of social avatars (Figure 3).

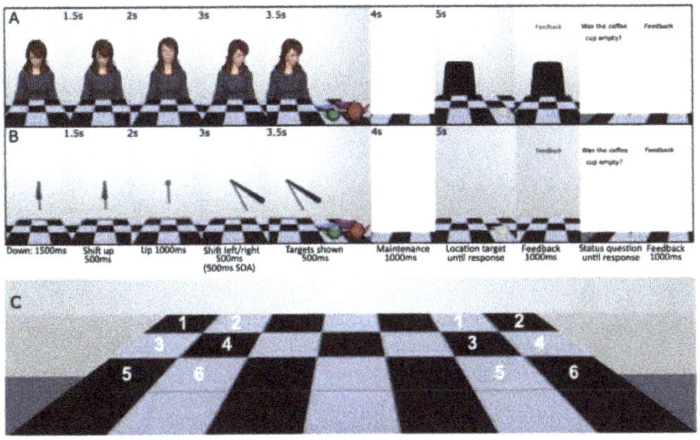

Figure 3. Depiction of the trial process (checkered pattern inspired by [45]). Utilizing the parameters of the conventional central cueing paradigm, the cue persisted on the screen for the duration of the trial (e.g., [46,47]). Panel (**A**) shows the social avatar cue, and Panel (**B**) shows the non-social stick cue. Timings, as depicted in the figure, were synchronized across cue types. The inter-trial interval was 1000 ms, during which a fixation cross was displayed. The experiment was a free-viewing study, allowing participants to move their eyes freely. Panel (**C**) shows the six possible left and right locations for the four encoding targets.

2.1.4. Apparatus and EEG Acquisition

The experiment was programmed in Unity and was conducted using a Lenovo Legion Y540-17IRH laptop (Beijing, China) with an Oculus Rift S PC-Powered VR Gaming Head-Mounted Display. EEG data were captured using a 64-channel ANT Neuro eego™ (Hengelo, The Netherlands) sports mobile EEG system following the international 10–20 electrode placement system. The data sampling rate was 500 Hz with online filtering between 0.1 and 100 Hz. Eye movements and blinks were monitored using additional electrodes. The preprocessing steps included band-pass filtering (1–40 Hz), artifact rejection, and independent component analysis (ICA) for correction of eye movement and muscle artifacts [44].

2.2. *LEiDA Analysis*

LEiDA, or Leading Eigenvector Dynamics Analysis [34], is a method used to detect phase-locked oscillatory patterns in large systems of coupled dynamical units, particularly in the context of brain activity recorded using fMRI. These patterns indicate meaningful functional subsystems related to cognitive and emotional processing. The analysis suggests that the macroscopic functional networks observed in resting brain activity reflect a repertoire of phase-locked solutions shaping hemodynamic fluctuations over ultra-slow timescales. In this study, it was applied to the EEG dataset. Figure 4 illustrates the methodological framework underlying the LEiDA technique.

At each time point, the principal eigenvector of dPL was calculated to identify the predominant connectivity pattern. The principal eigenvector, linked to the maximum eigenvalue, signifies the predominant direction of phase synchronization among brain regions at that particular moment. To discern recurring connectivity states, the leading eigenvectors were subsequently categorized using the k-means clustering algorithm, which classified them into discrete dynamic states based on their similarity.

Occupancy denotes the duration or proportion of time a brain network remains in a particular state during the VR task. This research quantifies the duration for which each participant's brain networks sustained specific dynamic connectivity states, identified through the LEiDA method, throughout the task. This indicates the involvement or predominance of specific functional brain networks in reaction to the cognitive requirements of the task, shaped by either social or non-social stimuli.

Temporal resolution (TR) denotes the frequency of measurements obtained during the analysis, reflecting the rate of data sampling over time intervals.

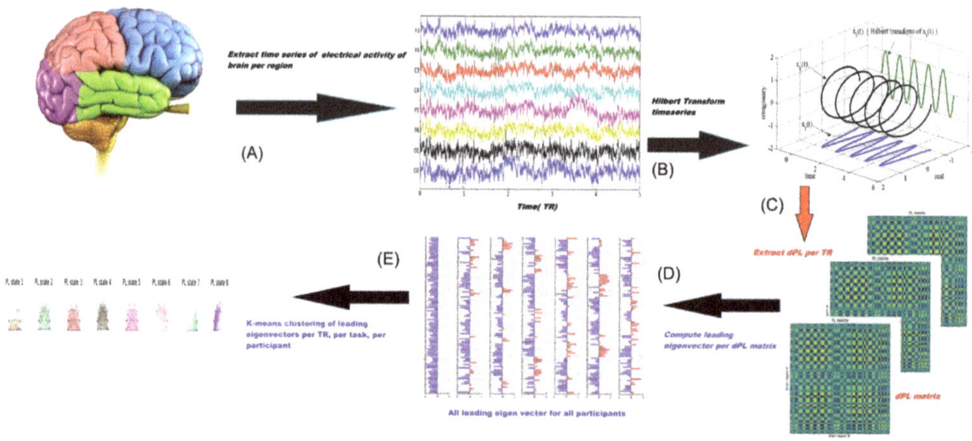

Figure 4. Extraction of EEG signal PL states. (**A**) For a given region, the EEG signal is first preprocessed. (**B**) Hilbert transformation is applied in order to acquire an analytic signal, whose phase can be represented over time and each TR (temporal resolution), which refers to the time interval between consecutive data samples, utilized for monitoring dynamic connectivity alterations. (**C**) The dPL(t) matrix quantifies the degree of phase synchronization between each pair of areas. The dominant eigenvector of the dPL(t) matrix, denoted as V(t), represents the primary direction of all phases. Each element in V(t) corresponds to the projection of the phase of each region onto V(t) (right). (**D**) The eigenvectors V(t) from all participants are combined and inputted into a k-means clustering algorithm, which separates the data points into a predetermined number of groups, k. (**E**) Every cluster centroid symbolizes a recurring PL state. dPL refers to dynamic phase-locking (Enhancing Clarity: Process Summary {1. Preprocessing →2. Hilbert Transformation →3. Dynamic Phase-Locking Matrix (dPL) →4. Leading Eigenvector Calculation →5. K-means Clustering →6. Identification of Recurrent Phase-Locking (PL) States}).

2.2.1. Functional Connectivity Matrix Calculation

For each time point or segment of EEG data, a functional connectivity matrix was calculated using PLV between all pairs of EEG channels. This results in a series of matrices that represent the connectivity patterns between brain regions over time. Initially, a comprehensive EEG matrix was acquired at each time point by computing dPL. This matrix gauges the alignment of phases between every pair of brain regions, with values ranging from 1 to −1 denoting signals changing in the same or opposite directions, respectively. Hilbert transformation was used to estimate the analytic phase of the averaged EEG signals. The Hilbert transform expresses any signal in polar coordinates, represented as x(t) = A(t)cos(θ(t)), where A(t) represents the instantaneous amplitude or envelope and θ(t) represents the instantaneous phase or phase angle. As shown in Figure 4A, the cosine of the phase angle effectively captured the fluctuations in the EEG signal. Given the phases of the EEG signals, the phase alignment, denoted as dPL (n, p, t), where 'n' and 'p' identify different brain regions and 't' represents the time point, is calculated using the cosine function as follows:

$$dPL(n.p.t) = cos(\theta(n,t) - \theta(p,t)) \quad (1)$$

The dPL matrix has dimensions N × N × T, where N denotes the brain regions and T denotes the recording frames per scan or time samples of the time series. Aligned EEG signals yield a PL value near 1 (cos (0°) = 1), whereas orthogonal signals yield a PL value of 0 (cos (90°) = 0).

2.2.2. Leading Eigenvector Extraction

The leading eigenvector represents the pattern of connectivity that accounts for the greatest variance in the data at that time point, effectively capturing the dominant state of the brain network connectivity. The next step involved computing the leading eigenvector of the EEG–dPL matrix. This eigenvector, denoted as V1(t), captures the primary alignment of EEG phases across brain regions. By focusing solely on the eigenvector associated with the largest-magnitude eigenvalue, data dimensionality is significantly reduced. The magnitude of the eigenvector elements reflects the strength of the association between the brain regions and their respective communities.

2.2.3. Clustering of Eigenvectors

A clustering algorithm was applied, namely k-means, to the collection of leading eigenvectors obtained from all time points. This step grouped similar eigenvectors together, identifying recurring patterns of connectivity or network states that emerge during a task.

2.2.4. Identification of Dynamic Network States and Statistical Analysis

Each cluster of the leading eigenvectors represents a distinct network state. By examining which time points are associated with which cluster, one can infer the temporal dynamics of brain connectivity, such as the frequency and order in which different network states occur.

Statistical analyses were performed to compare the prevalence and transition patterns of the identified network states across different conditions (e.g., social vs. non-social cueing), revealing insights into the neural mechanisms underlying cognitive processes.

Non-parametric permutation tests were utilized to compare the LEiDA-derived network states under social and non-social cueing conditions. In particular:

- Test procedure: For each LEiDA-derived dynamic phase-locking (dPL) state, the occupancy (percentage of time allocated to a specific state) was computed, and a two-sample *t*-test was conducted to compare the social and non-social conditions.
- Permutation framework: A permutation test with 10,000 iterations was conducted to assess statistical significance. In each iteration, the condition labels (social or non-social) were randomly shuffled, and the t-statistic was recalibrated to produce a null distribution.
- *p*-value calculation: The observed t-statistic was evaluated against the null distribution to ascertain the *p*-value, indicating the fraction of permuted test statistics that were as extreme or more extreme than the observed value.
- To mitigate the potential inflation of Type I errors resulting from multiple comparisons across the network states, FDR correction is implemented to obtain more robust significant *p*-values.

2.3. *Graph-Theory-Based Connectivity Analysis*

The graph-theory-based connectivity analysis involved constructing and analyzing functional brain networks derived from coherence matrices of EEG data. By thresholding coherence values and creating adjacency matrices, undirected weighted graphs representing significant connections between brain regions were built. The analysis of graph properties, such as degree distribution, clustering coefficient, and path length, provided insights into the functional organization of the brain networks and revealed differences in connectivity patterns between the two groups. This approach highlights the value of graph theory in understanding the complex dynamics of brain connectivity in cognitive neuroscience research. Utilizing graph theory metrics like centrality, modularity, and clustering coefficients enables the quantification of the structural and functional organization of brain networks. These metrics elucidate the dynamic interactions among neural regions over time, revealing the fundamental mechanisms of cognitive processes including memory, attention, and decision-making.

The steps involved in this method are detailed below:

- The raw EEG data from all subjects and channels underwent preprocessing, including band-pass filtering to remove noise and artifacts, followed by segmentation into task-specific epochs. This is the standard first process, which was explained in more detail in [44].
- For each subject and epoch, coherence matrices were computed to measure the synchronization between EEG signals across different brain regions.
- A threshold was then applied to these coherence matrices, retaining only significant coherence values (above a 0.2 threshold) to create adjacency matrices that represent connections between regions as undirected, weighted graphs.
- Using the network library in MATLAB 2023b, graphs were constructed with nodes as EEG channels and edges denoting meaningful coherence-based connections. This thresholding process focused the analysis on impactful relationships, reducing noise and allowing for the examination of robust connectivity patterns that reflect significant network structure.

The threshold value of 0.2 was selected based on empirical testing and the existing literature. This value is frequently employed in dynamic functional connectivity studies to optimize the inclusion of significant connections while reducing noise [48]. A threshold of 0.2 guarantees the retention of only the most robust and dependable PLV, which has demonstrated a stable representation of functional connectivity patterns.

Graph-theory-based connectivity analysis was conducted to examine the structure and organization of functional brain networks derived from EEG coherence matrices. Key graph properties were computed and compared between social and non-social cue conditions. This analysis provided insights into the organizational structure of brain networks and highlighted connectivity patterns that varied between the two conditions, allowing for a deeper understanding of functional brain network differences.

Graph Theory Metrics

To assess brain network organization under social and non-social cueing conditions, the following graph theory metrics are computed:

The clustering coefficient measures the tendency of nodes to create closely connected clusters, indicating local connectivity within the network. This metric was chosen to evaluate the local integration of cerebral regions. Increased clustering coefficients in the social cue condition may indicate enhanced local integration and more effective information processing.

Path length: The path length represents the average minimum distance between all pairs of nodes within the network, functioning as a measure of overall network efficiency. The diminished path lengths in the social cue condition indicate a more efficient network architecture, promoting rapid communication between brain regions.

Degree distribution: Degree distribution indicates the quantity of connections (edges) linked to each node. This metric was examined to ascertain if social cues affected the overall connectivity of brain networks. A higher average degree in the social cue condition may indicate improved overall integration among brain regions.

Modularity measures the degree to which a network can be divided into separate communities or modules, reflecting functional specialization. This metric would indicate whether social cues facilitate modular organization within brain networks.

Thresholding and analysis: Connectivity matrices were subjected to thresholding to preserve the top 20% of the most robust connections, as indicated by the existing literature. This facilitated the retention of critical edges while diminishing noise. Metrics of graph theory were computed for both social and non-social conditions to investigate alterations in brain network organization according to cue type.

2.4. Deep-Learning-Based Connectivity Analysis

To analyze the EEG data (using deep learning) from the VR working memory tasks, three types of input data were utilized: raw EEG data, phase-locking matrices, and leading eigenvectors derived from LEiDA. Each type of data was processed and classified using deep learning techniques to determine the efficacy of different feature sets in distinguishing between social and non-social cueing conditions.

Techniques for Model Validation:

i The dataset was divided into an 80–20 training and testing set to assess model performance on novel data. This division guarantees the segregation of data utilized for model training and performance assessment.

ii Cross-validation: Cross-validation was contemplated but ultimately disregarded due to computational constraints and the substantial sample size, which facilitated dependable generalization with the selected train–test methodology. To verify model stability, several training iterations utilizing various random seeds, ensuring uniform performance throughout these iterations, were conducted.

iii Early stopping: An early stopping criterion was employed during training to avert overfitting. The training process concluded when validation accuracy failed to enhance for five successive epochs.

iv Mitigation of Overfitting:
Regularization: Dropout layers with a rate of 0.5 were incorporated into the convolutional neural network (CNN) architecture to randomly deactivate neurons during training, thereby mitigating overfitting.
Batch normalization: Batch normalization layers were employed to enhance stability and accelerate training by normalizing inputs for each mini-batch.
Performance metrics: The model's efficacy was evaluated through metric of accuracy to confirm that the elevated accuracy was not due to overfitting to particular data patterns.

v Examination of Possible Overfitting:
The performance attained by the CNN model is recognized and has been underscored in the Discussion section as warranting careful interpretation Additional validation using independent datasets in subsequent studies is advocated to ascertain the model's generalizability.

The dataset comprised EEG-derived phase-locking matrices and leading eigenvectors gathered across multiple time points, yielding an adequate number of samples for training. This sample size allowed the CNN to discern intricate connectivity patterns while minimizing the risk of overfitting or underfitting.

Table 1 provides an overview of the CNN model architecture, detailing each layer's parameters, including the number of filters, kernel size, activation functions, and dropout rates.

Table 1. CNN model architecture and parameters.

Layer	Type	Number of Filters	Kernel Size	Activation Function	Pool Size	Dropout Rate
Conv1D	Convolution	64	3	ReLU	-	-
MaxPooling1D	Pooling	-	-	-	2	-
Dropout	Regularization	-	-	-	-	0.5
Flatten	Reshape	-	-	-	-	-
Dense	Fully Connected	100	-	ReLU	-	-
Output	Fully Connected	1	-	Sigmoid	-	-

2.4.1. Raw EEG

A 1D-CNN was employed to process the raw EEG signals. This decision was made to leverage the temporal structure inherent in EEG time-series data. The input size for raw EEG signals was [47 subject, 64 channels]. Raw EEG data underwent an inspection to eliminate channels exhibiting excessive noise or consistently subpar signal quality. The dataset preprocessing analysis was explained in more detail in [44].

Then, the data were divided into training and test sets, ensuring that the training set contained 80% of the data and the test set contained the remaining 20%. The model was compiled using the binary cross-entropy loss function, the Adam optimizer, and accuracy as the performance metric. It was trained for 10 epochs with a batch size of 10 and a validation split of 20%.

2.4.2. Phase-Locking Matrices

The PLV quantifies the synchronization of oscillatory activity between two brain regions, serving as a measure of functional connectivity. DFC can be evaluated by calculating PLV across consecutive time windows, thereby capturing the temporal progression of brain network states [26].

The phase-locking matrices were computed from the preprocessed EEG data using the Hilbert transform. The dPL matrices were then used as input for the deep learning model, following the same preprocessing, splitting, reshaping, and training procedures as described for the raw EEG data.

Despite being 2D connectivity matrices, 1D convolution was applied to each row (or column) of the matrix. This choice was made to simplify computational requirements while still capturing meaningful connectivity patterns. Each dPL was reshaped into a 1D vector (flattened row by row) and fed into the network.

2.4.3. Leading Eigenvectors from LEIDA

LEiDA was applied to the dPL matrices to extract the leading eigenvectors, which represent the most dominant patterns of connectivity in the brain at each time point. These eigenvectors were then combined across all subjects and time points to form the input data for the deep learning model.

As with the other data types, the eigenvectors were normalized, split into training and test sets, reshaped, and used to train the CNN model.

A deep learning technique was utilized to analyze EEG data, employing a CNN to classify data based on social and non-social cues. This method involved preprocessing EEG data, extracting phase-locking matrices, and deriving leading eigenvectors through LEiDA. The CNN model was trained to identify patterns associated with each cue type, providing a robust classification framework.

The step where the leading eigenvectors are obtained from all subjects for each time step produces a single dPL matrix, resulting in one leading eigenvector. Consequently, a total of #subjects × #samples 1D eigenvectors are generated. This output is then used as input to the CNN from LEiDA.

2.4.4. Managing Multiple Comparisons

To mitigate the probability of Type I errors resulting from multiple analyses (e.g., LEiDA, graph theory, and deep learning), suitable correction methods have been employed. In particular:

- LEIDA Evaluation:

FDR correction was utilized to adjust p-values for multiple comparisons in the statistical analysis of occupancy across dynamic brain states under social and non-social conditions.

The revised p-values are now clearly presented in the Results section for each network state derived from LEiDA.

- Analysis of Graph Theory:

In graph metrics (e.g., clustering coefficient, path length, and degree distribution), where multiple metrics were evaluated, Bonferroni correction was utilized to adjust for the quantity of metrics examined. This conservative methodology guarantees that only markedly significant differences are disclosed.

- Analysis of Deep Learning:

In deep learning analysis, assessing model performance, such as classification accuracy, renders adjustments for multiple comparisons unnecessary.

3. Results

3.1. Leida Analysis Results

Connectivity Matrix Heat Maps

See Figure 5 for connectivity matrix heat maps illustrating brain network synchronization patterns. Elevated PLVs in specific regions indicate that social cues may enhance neural coherence, activating brain regions associated with social cognition and memory processing. This observation corresponds with the current literature indicating that social stimuli augment functional connectivity within networks dedicated to social processing.

The exploration of dFC in a VR setting illuminated distinct behavioral responses in working memory when influenced by social versus non-social cues. By analyzing brain network interactions over time, significant variability and temporal characteristics in functional connectivity were identified, as depicted in Figure 5. The key findings of the presented analysis are as follows.

1. Consistency across models: The results consistently demonstrated specific brain states that distinguish responses to social cues from non-social cues across various partitioning schemes. This observation highlights the reliability of certain dynamic brain states in reflecting differences in cognitive processing related to cue type.
2. Identified states: A particular dynamic state was identified, the seventh, which showed significantly increased engagement in individuals exposed to social cues. This state is characterized by enhanced connectivity within networks that are typically involved in self-referential and memory-processing functions. The significance of this state is further elaborated in Figure 5, which shows the associated statistical results.
3. Role of specific regions: The analysis identified the precuneus as a crucial node across different dynamic states, underscoring its versatile role in facilitating connections and transitions between networks. Its varied involvement across states emphasizes its potential impact on how social and non-social cues are processed in working memory within a VR context (Figure 6).

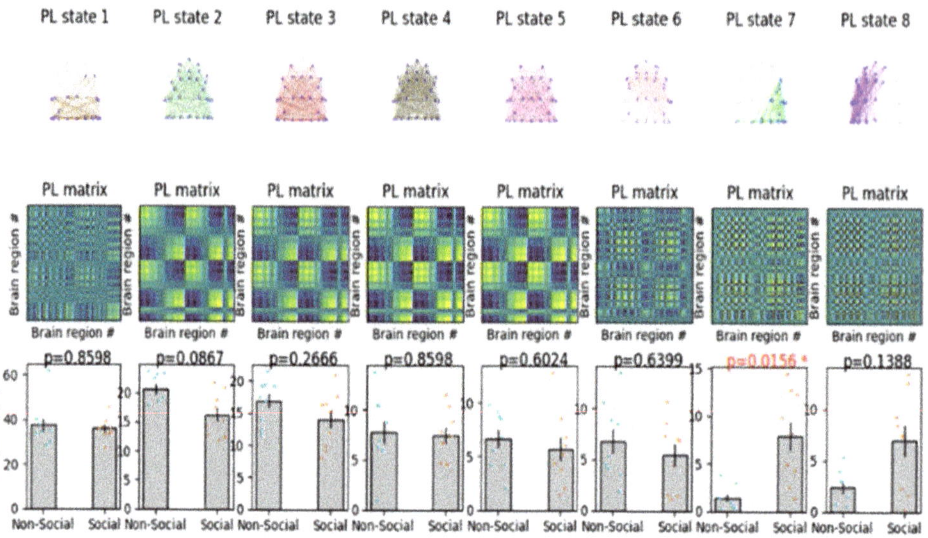

Figure 5. Repertoire of functional network states assessed with LEiDA and association to working memory. For a clustering solution of k = 8, PL State #7 is significantly correlated with enhanced working memory scores ($p = 0.0156$, (* refers to the significant p-value)), highlighted in a red color in the row of probabilities. The error bars represent the standard error of the mean across all 47 participants. These results underscore the role of DFC when clustered into 8 states in understanding the neural underpinnings of working memory, because the states and their connectivity after clustering results are the representation of dynamic function connectivity during the working memory tasks. Heat maps of the connectivity matrix display phase-locking values (PLVs) between EEG channels under social and non-social cue conditions. Warmer hues signify elevated PLVs, denoting enhanced functional connectivity among brain regions. Examining the variations in connectivity patterns between the two conditions may elucidate areas of increased synchronization in reaction to social cues, thereby corroborating the hypothesis of cue-specific brain network activation (the nodes represent the electrode locations).

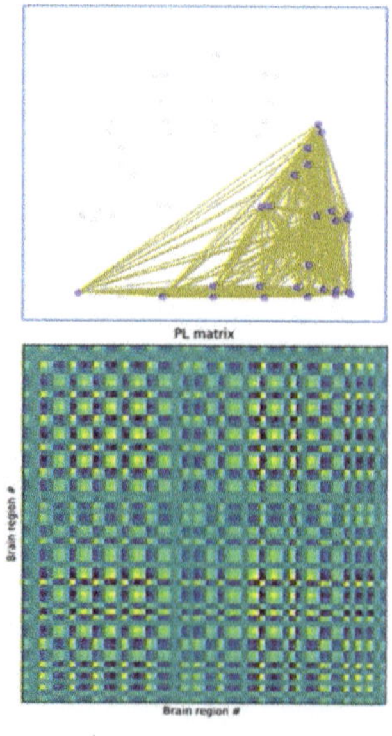

Figure 6. *Cont.*

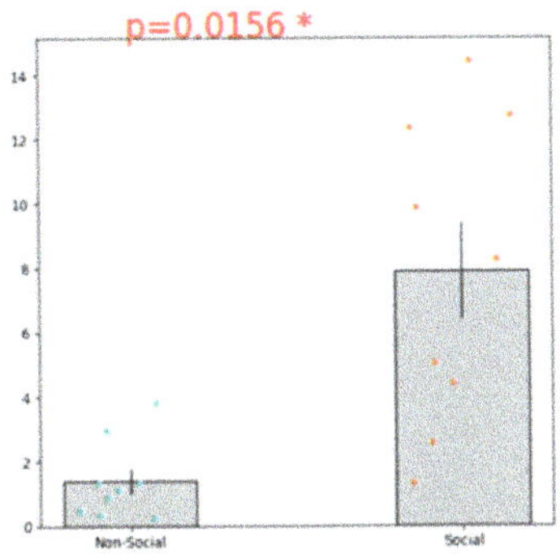

Figure 6. PL state 7 significantly differs for social compared to non-social working memory dynamic response. (**Top**) PL state is represented in the cortical space, where functionally connected brain regions (represented as spheres) are colored in blue. (**Middle**) PL states are also represented as the outer product of Vc, which is a 64 × 64 matrix representing the number of electrode regions. (**Bottom**) Significant (p-FDR < 0.05) differences in the percentage of occurrence between social compared to non-social working memory dynamic response. Dots represent individual data points; dark bars indicate the standard error of the mean. Analysis via non-parametric permutation-based *t*-test (N = 47 participants) (* refers to the significant *p*-value).

Although State 7 exhibited a statistically significant disparity in standard error across conditions, this result possesses limited predictive efficacy owing to the model's moderate classification accuracy. This outcome indicates that although variability in connectivity patterns is present, the observed difference may not entirely represent distinct brain states for social compared to non-social cues.

3.2. Graph Theory Analysis Results

Graph Representations

Graph representations of connectivity networks exhibit notable differences in clustering and degree distribution across conditions. The elevated clustering coefficient noted under social cues suggests a more cohesive network architecture, likely enhancing effective neural communication. This structural variation substantiates the hypothesis that social cues may influence brain connectivity, fostering a network architecture favorable to intricate social and cognitive processing. The results are presented in Figure 7.

Table 2 delineates the principal metrics contrasting brain connectivity in social and non-social cue conditions, encompassing coherence values, clustering coefficient, path length, and degree distribution. This table delineates notable disparities in network properties under varying conditions, with statistical significance marked where applicable.

The graph theory analysis of the EEG data yielded the following results:

Mean coherence values: The t-statistic and *p*-value indicate no significant difference in mean coherence values between the two groups.

Average clustering coefficient: Both groups have similar average clustering coefficients, with Group 2 having a slightly higher value, indicating a marginally higher tendency for nodes to form clusters.

Average path length: The NaN values indicate that the graphs for both groups are not fully connected, meaning there are disconnected components.

Degree distribution: The significant t-statistic and small *p*-value indicate that Group 2 has a higher average degree of connectivity among channels compared to Group 1, suggesting differences in network organization (Figure 7).

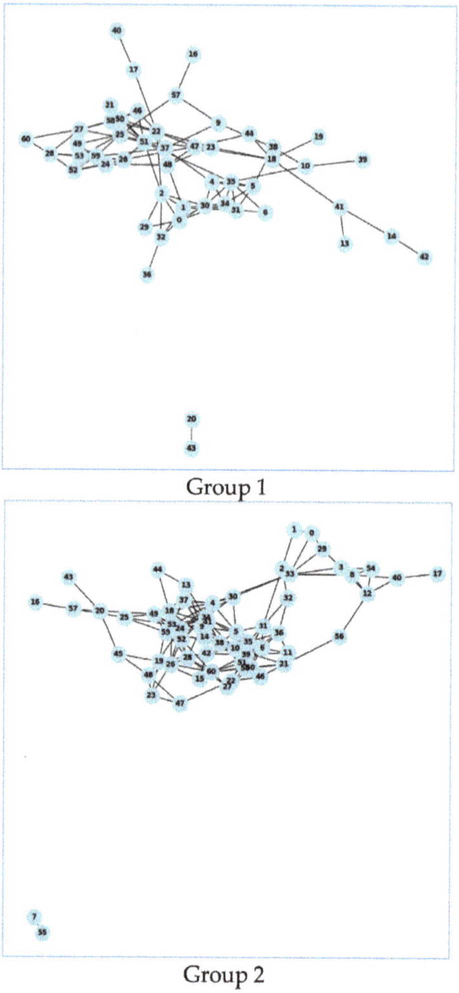

Figure 7. Graphical representations of brain connectivity networks under social and non-social cue conditions. Each node signifies a brain region, while edges indicate substantial coherence-based connections between regions. Essential network metrics, such as clustering coefficient and degree distribution, are presented to highlight the structural disparities in network organization across conditions. A more compact or clustered network architecture indicates improved integration within specific brain networks in reaction to social stimuli.

Table 2. Key results of brain connectivity analysis for social vs. non-social cue conditions.

Metric	Social Cue Condition	Non-Social Cue Condition	Statistical Significance (*p*-Value)
Mean Coherence Value	X (e.g., 0.45 ± 0.05)	Y (e.g., 0.42 ± 0.06)	$p = 0.336$
Average Clustering Coefficient	X (e.g., 0.375)	Y (e.g., 0.391)	Not significant
Average Path Length	NaN	NaN	-
Degree Distribution	X (e.g., 5.2 ± 1.1)	Y (e.g., 7.3 ± 1.3)	$p = 0.003$
PL State 7 Occurrence	Increased in Social Cues	Lower in Non-Social Cues	$p < 0.05$

3.3. Deep Learning Analysis Results

The analysis of the three types of input data yielded the following results:

- Raw EEG data: The CNN model trained on raw EEG data achieved an accuracy of 50–70%, indicating moderate discriminative power between the social and non-social cueing conditions. The performance evaluation of the CNN model revealed an accuracy range of 50% to 70% across various iterations. This range indicates fluctuations resulting from various factors, such as distinct random initializations of model parameters and minor discrepancies in the input data during each execution. By presenting an accuracy range, a realistic representation of the model's performance variability was offered, considering the inherent randomness in training methodologies and dataset sampling.
- Phase-locking matrices: Similar accuracy results (50–70%) were obtained when using the phase-locking matrices as input, suggesting that this feature set captures relevant information about the brain's connectivity patterns but does not significantly improve classification performance over raw EEG data.
- Leading eigenvectors from LEiDA: The most striking result was obtained with the leading eigenvectors derived from LEiDA, where the CNN model achieved an accuracy of around 99%. This high accuracy demonstrates that the leading eigenvectors are highly effective in distinguishing between the two cueing conditions, capturing the most salient features of the brain's dynamic connectivity patterns.

The CNN model demonstrated consistent performance across various runs with different random seeds, validating its robustness and stability. The final model achieved high accuracy on the test set, affirming the reliability of the results derived from the chosen train–test split. The early stopping criterion mitigated overfitting, as stable accuracy was noted after 10 epochs. The findings suggest that the dataset size and selected validation method established a robust framework for EEG classification.

4. Discussion

The results reveal that social cues not only trigger unique network states but also sustain these states for longer durations compared to non-social cues. This prolonged engagement with specific brain states suggests deeper cognitive processing or enhanced memory retention triggered by social interaction. The role of the precuneus highlighted in this research underscores its importance in integrating social information and memory processes. This finding suggests that social cues may engage more complex integrative cognitive mechanisms than previously understood, potentially enhancing the efficacy of memory encoding and retrieval by leveraging social context. This implies that the human brain may prioritize socially relevant information as a mechanism evolved to enhance cooperative interactions, which are critical in complex social environments.

The graph theory analysis reveals key differences in the functional brain networks between the two groups. While the mean coherence values and average clustering coefficients are not significantly different, the distribution of degrees shows a significant difference, with Group 2 exhibiting higher connectivity. This suggests that the functional organization of the brain networks in Group 2 is more densely connected compared to Group 1. The absence of fully connected graphs (as indicated by the NaN average path lengths; see Table 1) highlights the presence of disconnected components, which could be due to the inherent variability in EEG data or task-specific influences.

These findings underscore the importance of using graph theory analysis to uncover subtle differences in brain connectivity that may not be apparent through traditional coherence measures alone. The higher connectivity in Group 2 may reflect more efficient neural communication or greater integration of neural processes during the task.

The utilization of graph-theory-based connectivity analysis allowed me to detect subtle variations in the functional brain networks across different cueing conditions. The examination of network characteristics, including clustering coefficient and degree distribution, revealed the unique organizational patterns of brain connectivity in reaction to social compared to non-social stimuli. This indicates that social interactions may promote denser and more interconnected brain networks, potentially improving cognitive processing. These findings correspond with earlier research suggesting that densely interconnected networks may facilitate effective neural communication and cognitive involvement, especially in tasks related to social cognition.

The findings of this study underscore the significant role of social cues in enhancing cognitive processes through dynamic neural network modulation within a VR context. Employing LEiDA on EEG data, distinct patterns were discerned of brain connectivity that differentiate responses to social versus non-social cues. The novel application of LEiDA highlighted dynamic states correlating with varying performance in VR working memory tasks, enriching our understanding of the neural substrates involved in the cue-based modulation of brain networks.

The findings from this analysis highlight the effectiveness of using leading eigenvectors from LEiDA in deep learning models to classify EEG data. The superior performance of the LEiDA-derived features suggests that they encapsulate critical information about the brain's dynamic states that is not as readily apparent in the raw EEG data or phase-locking matrices. The moderate accuracy of the raw EEG data and phase-locking matrices indicates that while these features contain relevant information, they may also include noise or less discriminative features, which can hinder the classification performance. In contrast, LEiDA focuses on the most dominant connectivity patterns, providing a more robust and noise-resistant feature set for classification.

These results underscore the potential of LEiDA as a powerful tool for analyzing EEG data in cognitive neuroscience, particularly in studies involving complex tasks and conditions such as those involving social and non-social cues in a VR environment. The ability to achieve near-perfect classification accuracy with LEiDA-derived features paves the way for more nuanced and detailed investigations into the neural mechanisms underlying cognitive processes and their modulation by different types of cues.

The substantial disparity in the standard error of the mean for State 7 (see Figures 5 and 6), notwithstanding its statistical significance, resulted in poor predictive efficacy in differentiating social from non-social cues. This constraint may stem from the model's moderate classification accuracy, suggesting that supplementary features or more sophisticated modeling techniques may be necessary to discern finer distinctions in brain state connectivity.

Conversely, there is currently no directly comparable study in the literature, as no prior brain connectivity research has utilized the specific dataset employed in this study. Ref. [49] provides a comprehensive review of various approaches used to assess brain functional connectivity with EEG data, covering pairwise and multivariate connectivity metrics across time, frequency, and information-theoretic domains, and discussing their respective strengths and limitations. This work informed the broader context for the presented study, offering a comparative foundation for the methodologies. Consequently, this research is the first to apply connectivity analysis using EEG data with the LEiDA method in the literature. Furthermore, Ref. [50] highlights the diversity of EEG connectivity techniques and proposes a checklist to assess the quality of these methods, emphasizing the importance of standardization in EEG connectivity studies. Such standardization enables researchers to align study designs and findings with established practices in the field. Evaluating this presented work in conjunction with these prior works can help guide future research and support the attainment of robust and reliable results in EEG connectivity analysis.

This study utilized PLV to evaluate DFC from EEG data, reflecting the temporal progression of brain network states in reaction to social and non-social stimuli. PLV quantifies the synchronization of oscillatory activity among brain regions, acting as an indicator of functional connectivity. This method is consistent with prior studies that employed PLV to examine task-related alterations in neural synchronization [26].

This analysis demonstrated that social cues prompted elevated clustering coefficients in brain networks, signifying improved local connectivity. This discovery indicates increased cognitive involvement and social processing in response to social stimuli, aligning with research that has noted comparable patterns of local network integration in social settings [20].

Furthermore, discrepancies in path length between social and non-social conditions were noted, indicating variations in global network efficiency. Reduced path lengths during social cue processing indicate enhanced information transfer within the brain, consistent with previous studies on functional brain network dynamics during social interactions [18,20].

The utilization of graph theory metrics, including the clustering coefficient and path length, offered an in-depth insight into the brain's functional organization across different conditions. This methodology has been successfully employed in prior studies to examine brain connectivity patterns [51].

These findings enhance the expanding literature on the neural mechanisms of social cognition. This study illustrates that social cues can influence both local and global brain network characteristics, providing insights into the brain's dynamic adaptation to social information. These findings align with current research highlighting the significance of network flexibility in social cognitive processes [52].

To guarantee the reliability of my results, stringent corrections for multiple comparisons were implemented. The application of FDR correction in the examination of LEiDA-derived network states guaranteed that notable differences in occupancy were not attributable to random variation. Likewise, the Bonferroni correction utilized in graph theory metrics reduced the likelihood of Type I errors while upholding a conservative stance on statistical significance.

A principal limitation of EEG is its inferior spatial resolution relative to fMRI, which complicates the precise localization of neural sources associated with dynamic brain states. This limitation stems from the inverse problem associated with EEG, wherein neural sources must be deduced from signals captured on the scalp. Although LEiDA was initially developed for fMRI data, which provide accurate spatial localization, its use in EEG may lead to reduced spatial specificity.

To mitigate these limitations, various measures were instituted in this study. Preprocessing methods, such as independent component analysis (ICA), were employed to remove noise and artifacts, thus enhancing signal quality. Furthermore, coherence-based connectivity analysis was utilized to enhance network definitions. The application of high-density EEG (64 channels) enhanced spatial resolution, enabling a more precise representation of extensive brain networks.

Notwithstanding these measures. It is recognized that the diminished spatial resolution of EEG may influence the interpretation of particular findings, especially concerning the localization of specific network nodes, such as the precuneus. The principal eigenvector obtained from dPL matrices signifies a global connectivity pattern rather than specific spatial origins, requiring careful interpretation of network localization.

This study's findings indicate that social cues in VR environments improve cognitive engagement and task performance by fostering more integrated brain network states. These findings possess significant implications for cognitive rehabilitation and immersive education. The observed enhancements in behavior and neural patterns linked to social cues may guide the creation of VR-based interventions aimed at social cognition and working memory.

These findings corroborate earlier research (e.g., Refs. [9,11]—no real-life dataset was used for comparison in these studies), which has established the effectiveness of VR in replicating authentic social interactions. It is recognized that the regulated characteristics of the VR environment and the restricted scope of tasks in this study may not adequately represent the intricacies of genuine social interactions and learning environments. Individual differences, task complexity, and environmental variability in real-world contexts may affect the generalizability of these findings.

The application of LEiDA in this research signifies a notable progression in the examination of dynamic brain network states during real-time social interactions. By concentrating on the principal eigenvectors of phase-locking matrices, LEiDA elucidates the predominant connectivity patterns at each temporal juncture, facilitating the recognition of ephemeral network states that are challenging to discern through conventional techniques. This method offers essential insights into the neural mechanisms of social cognition, facilitating future research on dynamic functional connectivity in real-world contexts.

This study integrates VR with EEG-based dynamic connectivity analysis, effectively linking ecologically valid experimental designs with high-resolution neural data collection. The identification of unique dynamic brain network states during social interactions in VR establishes a framework for investigating social cognition in more realistic contexts, facilitating future research and therapeutic applications.

5. Conclusions

This research substantially enhances our comprehension of the dynamic neural mechanisms that govern working memory, especially in VR settings. Through the application of LEiDA to EEG data analysis, the modulation of brain connectivity by social cues was illustrated, uncovering distinct network states and highlighting the potential of VR for investigating intricate cognitive phenomena. The integration of LEiDA with deep-learning- and graph-theory-based connectivity analysis has augmented our cognitive neuroscience toolkit, improving the precision of brain state classification and offering structural insights into network organization under diverse cognitive demands.

The integration of DFC analysis, deep learning, and graph theory enabled me to elucidate both temporal and structural brain dynamics, providing a comprehensive understanding of cognitive function modulation in VR. This multi-method approach demonstrates the distinct impact of social cues on brain connectivity and underscores the importance of analyzing transient states to enhance the understanding of cognitive resilience, adaptability, and performance, particularly in dynamic environments. This dynamic viewpoint transcends conventional static metrics, facilitating the identification of biomarkers and therapeutic targets for cognitive resilience and neurorehabilitation.

This study elucidates dynamic interactions, thereby advancing our comprehension of social processing in the brain and paving the way for tailored interventions to modulate specific neural dynamics. This holistic method indicates potential uses in cognitive rehabilitation, immersive educational settings, and therapeutic strategies adapted to changing neural patterns linked to psychiatric and neurological disorders.

These findings underscore the potential of VR as a medium for investigating dynamic brain network states in ecologically valid environments. Future research should investigate the evolution of these states over extended interaction durations and determine if analogous patterns are present in clinical populations, such as individuals with autism spectrum disorder or social anxiety.

Moreover, the elevated accuracy attained by the CNN model is recognized, and it has been underscored in the Discussion section that this performance warrants careful interpretation. Additional validation using independent datasets in subsequent studies is advocated to ascertain the model's generalizability.

Future research should incorporate a comprehensive outlier analysis to enhance the validation of the findings. Such analyses could enhance the robustness and generalizability of the results, particularly in studies involving dynamic neural connectivity. Systematically addressing potential data deviations enhances the reliability of interpretations and facilitates comparisons with other research in the field.

Funding: This research received no external funding.

Institutional Review Board Statement: The study was conducted according to the guidelines of the Declaration of Helsinki and approved by the Institutional Review Board of Istanbul University (protocol code IRB-2022-47, approved on 10 January 2022).

Informed Consent Statement: Informed consent was obtained from all subjects involved in the study.

Data Availability Statement: The dataset supporting the reported results is publicly available at Gregory et al., 2022, including EEG recordings from 47 participants collected during a VR working memory task. Additional data and materials can be requested from the corresponding author upon reasonable request.

Acknowledgments: The author would like to thank the participants and the technical staff for their support during data collection and preprocessing. Special thanks to Gregory et al. for making the dataset available to the research community.

Conflicts of Interest: The author declares no conflict of interest.

References

1. Sahu, M.; Gupta, R.; Ambasta, R.K.; Kumar, P. IoT-driven augmented reality and VR systems in neurological sciences. *Internet Things* **2024**, *25*, 101098. [CrossRef]
2. Cowan, N. Working memory underpins cognitive development, learning, and education. *Educ. Psychol. Rev.* **2014**, *26*, 197–223. [CrossRef] [PubMed]
3. South, L.; Yildirim, C.; Pavel, A.; Borkin, M.A. Barriers to Photosensitive Accessibility in VR. In Proceedings of the CHI'24: 2024 CHI Conference on Human Factors in Computing Systems, Honolulu, HI, USA, 11–16 May 2024; Volume 58, pp. 1–13.
4. Taherysayah, F.; Malathouni, C.; Liang, H.-N.; Westermann, C. VR and electroencephalography in architectural design: A systematic review of empirical studies. *J. Build. Eng.* **2024**, *85*, 108611. [CrossRef]
5. Li, Y.; Yang, B.; Ma, J.; Li, Y.; Zeng, H.; Zhang, J. Assessment of rTMS treatment effects for methamphetamine addiction based on EEG functional connectivity. *Cogn. Neurodyn.* **2024**, *18*, 2373–2386. [CrossRef] [PubMed]
6. Kwon, S.J.; Turpyn, C.C.; Duell, N.; Telzer, E.H. Neural underpinnings of social contextual influences on adolescent risk-taking. *Curr. Addict. Rep.* **2020**, *7*, 413–420. [CrossRef] [PubMed]
7. Wu, J.; Chen, X.; Zhao, M.; Xue, C. Cognitive characteristics in wayfinding tasks in commercial and residential districts during daytime and nighttime: A comprehensive neuroergonomic study. *Adv. Eng. Inform.* **2024**, *61*, 102534. [CrossRef]
8. Kwon, J.-H.; Won, J.; Cho, G.-H. Investigating the dynamics of collective behavior among pedestrians crossing roads: A multi-user VR approach. *Accid. Anal. Prev.* **2024**, *199*, 107477. [CrossRef] [PubMed]
9. Elkin, R.L.; Beaubien, J.M.; Damaghi, N.; Chang, T.P.; Kessler, D.O. Dynamic cognitive load assessment in VR. *Simul. Gaming* **2024**, *55*, 4. [CrossRef]
10. Trudel, N.; Lockwood, P.L.; Rushworth, M.F.S.; Wittmann, M.K. Neural activity tracking identity and confidence in social information. *eLife* **2023**, *12*, e71315. [CrossRef] [PubMed]
11. Zhou, S.; Yang, H.; Yang, H.; Liu, T. Bidirectional understanding and cooperation: Interbrain neural synchronization during social navigation. *Soc. Cogn. Affect. Neurosci.* **2023**, *18*, nsad031. [CrossRef] [PubMed]
12. Zhang, F.; Zhang, Y.; Li, G.; Luo, H. Using Virtual Reality Interventions to Promote Social and Emotional Learning for Children and Adolescents: A Systematic Review and Meta-Analysis. *Children* **2024**, *11*, 41. [CrossRef] [PubMed]
13. Jacucci, G.; Bellucci, A.; Ahmed, I.; Harjunen, V.; Spape, M.; Ravaja, N. Haptics in social interaction with agents and avatars in virtual reality: A systematic review. *Virtual Real.* **2024**, *28*, 170. [CrossRef]

14. Sharan, N.N.; Toet, A.; Mioch, T.; Niamut, O.; van Erp, J.B.F. The Relative Importance of Social Cues in Immersive Mediated Communication. In *Human Interaction, Emerging Technologies and Future Systems V; Lecture Notes in Networks and Systems; Proceedings of the IHIET 2021, Paris, France, 27–29 August 2021*; Ahram, T., Taiar, R., Eds.; Springer: Cham, Switzerland, 2021; Volume 319, p. 319. [CrossRef]
15. Freeman, D.; Reeve, S.; Robinson, A.; Ehlers, A.; Clark, D.; Spanlang, B.; Slater, M. Virtual reality in the assessment, understanding, and treatment of mental health disorders. *Psychol. Med.* **2017**, *47*, 2393–2400. [CrossRef] [PubMed]
16. Riva, G.; Mancuso, V.; Cavedoni, S.; Stramba-Badiale, C. Virtual reality in neurorehabilitation: A review of its effects on multiple cognitive domains. *Expert Rev. Med. Devices* **2020**, *17*, 1035–1061. [CrossRef] [PubMed]
17. Catania, V.; Rundo, F.; Panerai, S.; Ferri, R. Virtual Reality for the Rehabilitation of Acquired Cognitive Disorders: A Narrative Review. *Bioengineering* **2024**, *11*, 35. [CrossRef] [PubMed]
18. Bordier, C.; Nicolini, C.; Bifone, A. Graph Analysis and Modularity of Brain Functional Connectivity Networks: Searching for the Optimal Threshold. *Front. Neurosci.* **2017**, *11*, 441. [CrossRef]
19. Wang, J.; Zuo, X.; He, Y. Graph-Based Network Analysis of Resting-State Functional MRI. *Front. Syst. Neurosci.* **2010**, *4*, 16. [CrossRef]
20. Farahani, F.V.; Karwowski, W.; Lighthall, N.R. Application of Graph Theory for Identifying Connectivity Patterns in Human Brain Networks: A Systematic Review. *Front. Neurosci.* **2019**, *13*, 585. [CrossRef]
21. Nielsen, S.F.V.; Schmidt, M.N.; Madsen, K.H.; Mørup, M. Predictive assessment of models for dynamic functional connectivity. *Neuroimage* **2018**, *171*, 116–134. [CrossRef] [PubMed]
22. Shou, G.; Yuan, H.; Li, C.; Chen, Y.; Chen, Y.; Ding, L. Whole-brain electrophysiological functional connectivity dynamics in resting-state EEG. *J. Neural Eng.* **2020**, *17*, 026016. [CrossRef] [PubMed]
23. Hasan, A.; Pandey, D.; Khan, A. Application of EEG Time-Varying Networks in the Evaluation of Dynamic Functional Brain Networks. *Augment. Hum. Res.* **2021**, *6*, 8. [CrossRef]
24. Khaleghi, N.; Hashemi, S.; Peivandi, M.; Ardabili, S.Z.; Behjati, M.; Sheykhivand, S.; Danishvar, S. EEG-based functional connectivity analysis of brain abnormalities: A systematic review study. *Inform. Med. Unlocked* **2024**, *47*, 101476. [CrossRef]
25. Chen, J.; Cui, Y.; Wang, H.; He, E.; Alhudhaif, A. Deep learning approach for detection of unfavorable driving state based on multiple phase synchronization between multi-channel EEG signals. *Inf. Sci.* **2024**, *658*, 120070. [CrossRef]
26. Ruchilekha; Srivastava, V.; Singh, M.K. Emotion Recognition Using Phase-Locking-Value Based Functional Brain Connections Within-Hemisphere and Cross-Hemisphere. In *Intelligent Human Computer Interaction*; Choi, B.J., Singh, D., Tiwary, U.S., Chung, W.Y., Eds.; Springer: Cham, Switzerland, 2024; Volume 14531. [CrossRef]
27. Nadasdy, Z.; Fogarty, A.S.; Fisher, R.S.; Primiani, C.T.; Graber, K. Technical validation of the Zeto wireless, dry electrode EEG system. *TechRxiv* **2024**, 1–8. [CrossRef]
28. Sun, Q.; Liu, Y.; Li, S. Weighted directed graph-based automatic seizure detection with effective brain connectivity for EEG signals. *Signal Image Video Process* **2024**, *18*, 899–909. [CrossRef]
29. Wang, T.; Xia, M.; Wang, J.; Zhilenkov, A.; Wang, J.; Xi, X.; Li, L. Delay estimation for cortical-muscular interaction with wavelet coherence time lag. *J. Neurosci. Methods* **2024**, *405*, 110098. [CrossRef]
30. Ozel, P.; Karaca, A.; Olamat, A.; Akan, A.; Ozçoban, M.A.; Tan, O. Intrinsic synchronization analysis of brain activity in obsessive–compulsive disorders. *Int. J. Neural Syst.* **2020**, *30*, 2050046. [CrossRef]
31. Ein Shoka, A.A.; Dessouky, M.M.; El-Sayed, A.; Hemdan, E.E.D. EEG seizure detection: Concepts, techniques, challenges, and future trends. *Multimed. Tools Appl.* **2023**, *82*, 42021–42051. [CrossRef] [PubMed]
32. Luo, X.; Zhou, B.; Fang, J.; Cherif-Riahi, Y.; Li, G.; Shen, X. Integrating EEG and ensemble learning for accurate grading and quantification of generalized anxiety disorder: A novel diagnostic approach. *Diagnostics* **2024**, *14*, 1122. [CrossRef]
33. Alqirshi, R.; Belhaouari, S. EEG-Based Patient Independent Epileptic Seizure Detection Using GCN-BRF. In *Deep Learning Theory and Applications. DeLTA 2024*; Communications in Computer and Information Science; Springer: Cham, Switzerland, 2024; Volume 2172. [CrossRef]
34. Cabral, J.; Vidaurre, D.; Marques, P.; Magalhães, R.; Moreira, P.S.; Soares, J.M.; Deco, G.; Sousa, N.; Kringelbach, M.L. Cognitive performance in healthy older adults relates to spontaneous switching between states of functional connectivity. *Sci. Rep.* **2017**, *7*, 5135. [CrossRef]
35. Wu, S.; Peng, H.; Deng, H.; Guo, Z.; Jiang, Z.; Mu, Q. Insomnia disorder characterized by probabilistic metastable substates using blood-oxygenation-level-dependent (BOLD) phase signals. *Sleep Breath.* **2024**, *28*, 1409–1414. [CrossRef]
36. De Alteriis, G.; MacNicol, E.; Hancock, F.; Ciaramella, A.; Cash, D.; Expert, P.; Turkheimer, F.E. LEiDA: A lossless approach for dynamic functional connectivity; application to fMRI data of a model of ageing. *Imaging Neurosci.* **2024**, *2*, 1–22. [CrossRef]
37. Zhang, S.; Ge, M.; Cheng, H.; Chen, S.; Li, Y.; Wang, K. Classification of cognitive ability of healthy older individuals using resting-state functional connectivity magnetic resonance imaging and an extreme learning machine. *BMC Med. Imaging* **2024**, *24*, 72. [CrossRef] [PubMed]

38. França, L.G.S.; Ciarrusta, J.; Gale-Grant, O.; Fenn-Moltu, S.; Fitzgibbon, S.; Chew, A.; Falconer, S.; Dimitrova, R.; Cordero-Grande, L.; Price, A.N.; et al. Neonatal brain dynamic functional connectivity in term and preterm infants and its association with early childhood neurodevelopment. *Nat. Commun.* **2024**, *15*, 16. [CrossRef]
39. van der Horn, H.J.; Ling, J.M.; Wick, T.V.; Dodd, A.B.; Robertson-Benta, C.R.; McQuaid, J.R.; Zotev, V.; Vakhtin, A.A.; Ryman, S.G.; Cabral, J.; et al. Dynamic Functional Connectivity in Pediatric Mild Traumatic Brain Injury. *Neuroimage* **2024**, *285*, 120470. [CrossRef]
40. Yang, Y.; Luo, S.; Wang, W.; Gao, X.; Yao, X.; Wu, T. From bench to bedside: Overview of magnetoencephalography in basic principle, signal processing, source localization and clinical applications. *Neuroimage Clin.* **2024**, *42*, 103608. [CrossRef]
41. Zhang, W.; Jiang, M.; Teo, K.A.C.; Bhuvanakantham, R.; Fong, L.; Sim, W.K.J.; Guo, Z.; Foo, C.H.V.; Chua, R.H.J.; Padmanabhan, P.; et al. Revealing the spatiotemporal brain dynamics of covert speech compared with overt speech: A simultaneous EEG-fMRI study. *Neuroimage* **2024**, *293*, 120629. [CrossRef]
42. Zhou, B.; Wu, X.; Tang, L.; Li, C. Dynamics of the Brain Functional Network Associated with Subjective Cognitive Decline and Its Relationship to Apolipoprotein E ε4 Alleles. *Front. Aging Neurosci.* **2022**, *14*, 806032. [CrossRef]
43. Li, J.; Li, S.; Zeng, S.; Wang, X.; Liu, M.; Xu, G.; Ma, X. Static and Temporal Dynamic Alterations of Local Functional Connectivity in Chronic Insomnia. *Brain Imaging Behav.* **2024**. *advance online publication*. [CrossRef] [PubMed]
44. Gregory, S.E.A.; Wang, H.; Kessler, K. A Dataset of EEG Recordings from 47 Participants Collected during a VR Working Memory Task Where Attention Was Cued by a Social Avatar and Non-Social Stick Cue. *Data Brief* **2022**, *41*, 107827. [CrossRef] [PubMed]
45. Harkin, B.; Rutherford, H.; Kessler, K. Impaired Executive Functioning in Subclinical Compulsive Checking with Ecologically Valid Stimuli in a Working Memory Task. *Front. Psychol.* **2011**, *2*, 78. [CrossRef] [PubMed]
46. Friesen, C.K.; Kingstone, A. The Eyes Have It! Reflexive Orienting Is Triggered by Nonpredictive Gaze. *Psychon. Bull. Rev.* **1998**, *5*, 490–495. [CrossRef]
47. Driver, J.; Davis, G.; Ricciardelli, P.; Kidd, P.; Maxwell, E.; Baron-Cohen, S. Gaze Perception Triggers Reflexive Visuospatial Orienting. *Vis. Cogn.* **1999**, *6*, 509–540. [CrossRef]
48. Blaney, G.; Sassaroli, A.; Fantini, S. Algorithm for Determination of Thresholds of Significant Coherence in Time-Frequency Analysis. *Biomed. Signal Process. Control* **2020**, *56*, 101704. [CrossRef] [PubMed]
49. Chiarion, G.; Sparacino, L.; Antonacci, Y.; Faes, L.; Mesin, L. Connectivity Analysis in EEG Data: A Tutorial Review of the State of the Art and Emerging Trends. *Bioengineering* **2023**, *10*, 372. [CrossRef] [PubMed]
50. Miljevic, A.; Bailey, N.W.; Vila-Rodriguez, F.; Herring, S.E.; Fitzgerald, P.B. EEG-Connectivity: A Fundamental Guide and Checklist for Optimal Study Design and Evaluation. *arXiv* **2021**, *preprint*. arXiv:2108.13611.
51. Mijalkov, M.; Kakaei, E.; Pereira, J.B.; Westman, E.; Volpe, G. BRAPH: A Graph Theory Software for the Analysis of Brain Connectivity. *PLoS ONE* **2017**, *12*, e0178798. [CrossRef]
52. Marzetti, L.; Basti, A.; Chella, F.; D'Andrea, A.; Syrjälä, J.; Pizzella, V. Brain Functional Connectivity through Phase Coupling of Neuronal Oscillations: A Perspective from Magnetoencephalography. *Front. Neurosci.* **2019**, *13*, 964. [CrossRef]

brain sciences

Article

The Modulatory Effects of Transcranial Alternating Current Stimulation on Brain Oscillatory Patterns in the Beta Band in Healthy Older Adults

Kenya Morales Fajardo [1,2], Xuanteng Yan [2,3], George Lungoci [2,4], Monserrat Casado Sánchez [2,4], Georgios D. Mitsis [3] and Marie-Hélène Boudrias [1,2,*]

1 School of Physical and Occupational Therapy, McGill University, Montréal, QC H3G 1Y5, Canada; kenya.moralesfajardo@mail.mcgill.ca
2 Center for Interdisciplinary Research in Rehabilitation of Greater Montreal (CRIR), Montréal, QC H3S 1M9, Canada; xuanteng.yan@mail.mcgill.ca (X.Y.); george.lungoci@mail.mcgill.ca (G.L.); monserrat.casado@mail.mcgill.ca (M.C.S.)
3 Department of Bioengineering, McGill University, Montréal, QC H3A 0E9, Canada; georgios.mitsis@mcgill.ca
4 Integrated Program in Neuroscience, McGill University, Montréal, QC H3A 1A1, Canada
* Correspondence: mh.boudrias@mcgill.ca; Tel.: +1-(514) 398-5457

Abstract: Background: In the last few years, transcranial alternating current stimulation (tACS) has attracted attention as a promising approach to interact with ongoing oscillatory cortical activity and, consequently, to enhance cognitive and motor processes. While tACS findings are limited by high variability in young adults' responses, its effects on brain oscillations in older adults remain largely unexplored. In fact, the modulatory effects of tACS on cortical oscillations in healthy aging participants have not yet been investigated extensively, particularly during movement. This study aimed to examine the after-effects of 20 Hz and 70 Hz High-Definition tACS on beta oscillations both during rest and movement. **Methods:** We recorded resting state EEG signals and during a handgrip task in 15 healthy older participants. We applied 10 min of 20 Hz HD-tACS, 70 Hz HD-tACS or Sham stimulation for 10 min. We extracted resting-state beta power and movement-related beta desynchronization (MRBD) values to compare between stimulation frequencies and across time. **Results:** We found that 20 Hz HD-tACS induced a significant reduction in beta power for electrodes C3 and CP3, while 70 Hz did not have any significant effects. With regards to MRBD, 20 Hz HD-tACS led to more negative values, while 70 Hz HD-tACS resulted in more positive ones for electrodes C3 and FC3. **Conclusions:** These findings suggest that HD-tACS can modulate beta brain oscillations with frequency specificity. They also highlight the focal impact of HD-tACS, which elicits effects on the cortical region situated directly beneath the stimulation electrode.

Keywords: tACS; aging; MRBD; EEG; movement; beta oscillations; non-invasive brain stimulation

1. Introduction

Advanced age often comes with a decline in sensorimotor control and functioning that affects the ability to perform activities of daily living. Indeed, it has been shown that movements become slower and/or less accurate and more cognition-dependent as we age [1]. Motor declines overall significantly impact motor independence, which is essential for older adults' quality of life and interactions with their environment. Additionally, behavioral evidence indicates that aging is frequently linked to slower movements, reduced capacity for learning new motor skills, and diminished ability to adjust a movement plan after initiation [2]. In fact, Mild Parkinsonian Symptoms (**MPSs**), such as rigidity, bradykinesia, and tremor, are commonly diagnosed during clinical examination of older adults who do not have a diagnosed neurological disease [3].

Advances in neuroimaging techniques have made contributions to a better understanding of the aging brain. For instance, aging impacts brain structure, leading to a decrease in gray and white matter volume, along with an increase in cerebrospinal fluid in ventricles, fissures, and sulci [4,5]. These aging-related processes affect almost the entire cortex and underlying white matter, with a steeper decline in the primary motor cortex (**M1**) and frontal subcortical white matter [6]. For instance, aging is also associated with a complex pattern of atrophy [7], demyelination [8], free tissue water, and iron reduction within somatosensory and motor areas [6]. Moreover, age-related atrophy of motor cortical regions and the corpus callosum has been shown to coincide with motor declines such as balance gait deficits and coordination deficits [9].

Sensorimotor cortex oscillations measured by electroencephalography (**EEG**) in the beta band (13–30 Hz) are a predominant feature of movement production and have been shown to be generated by local field potentials within the motor cortex [10]. Beta oscillations exhibit a robust pattern of movement-related changes, such as pre-movement beta Event-Related Desynchronization (**ERD**), Movement-Related Beta Desynchronization (**MRBD**), and post-movement beta Event-Related Synchronization (**ERS**) or beta rebound [11,12]. In terms of how these oscillations relate to motor performance, an association between MRBD and the accuracy with which subjects performed a bimanual task has been demonstrated, where subjects with more negative MRBD values exhibited worse task performance [13]. Greater MRBDs were also shown to correlate with a longer movement duration to complete a finger-tapping sequence [14]. In older adults, a greater (i.e. more negative) MRBD in both motor and premotor areas has been observed in subjects performing cued finger button presses [15,16] and handgrip tasks [13]. Additionally, older age has been associated with greater baseline beta power (15–29 Hz) at rest [17], suggesting that the alterations in brain structure and biochemistry during aging could be the reason behind the observed altered neural activation patterns. Given the association between movement production and beta band features, there is a high interest in modulating these oscillations non-invasively to improve motor ability and performance in older adults.

Transcranial alternating current stimulation (**tACS**) is a non-invasive brain stimulation (**NIBS**) technique that can alter oscillatory brain rhythms through synchronization of neural networks in a frequency-dependent manner [18]. This method is believed to entrain endogenous brain oscillations through the synchronization of two oscillatory systems that occurs when a driving external oscillatory force coordinates with another oscillating system [19–22]. The effects of tACS depend on key parameters: stimulation location, intensity, and frequency [23]. In terms of location, the acquisition of motor skills is linked to a number of cortical and subcortical brain regions, but among these, M1 is thought to play a central role [24–27], making it a popular target for neurostimulation. Regarding intensity, it is generally set between 1 mA and 2 mA because it is well tolerated and it has been shown to modulate cortex excitability and alter cognitive function [28]. Additionally, using higher intensities raise concerns about safety and side effects [29]. Regarding frequency, motor cortex activity during movement predominantly oscillates at 20 Hz (beta band) [30] and 70 Hz (gamma band) [31]. Beta band activity within the motor system has been linked to an antikinetic role, as it is associated with slower voluntary movements in both healthy individuals [21,32,33] and those with motor disorders [34]. A reason for this may be that, in the cortex–basal ganglia circuit, beta activity is associated with promoting tonic rather than voluntary movement [35,36]. Also, motor impairments in Parkinson's disease (**PD**) have also been linked to elevated beta band activity in the motor cortex and subthalamic nucleus [37]. In contrast, gamma band activity is thought to be prokinetic, as it increases in the basal ganglia–cortical motor circuit during voluntary movement [38]. Behaviorally, the use of 20 Hz tACS has been shown to slow voluntary movement, while 70 Hz tACS enhances motor learning along with an increase in beta power [32]. However, these results involved participants in the younger range (32.7 $\pm$ 6.8 years) and the influence of these tACS frequencies on aging-related brain neural activity has not yet been studied.

Other NIBS techniques, such as transcranial direct current stimulation (**tDCS**), have been shown to have greater effects on motor performance when applied during a motor task, a technique often called *online stimulation*, compared to before the motor task [39].

There is also evidence that applying tDCS during practice triggers effects that outlast the stimulation period and facilitate neuroplasticity [28]. Previous studies have reported mixed results regarding the effects of NIBS on young adults, and these results cannot be easily transferred to older adults. The stimulation sites and frequencies that modulate brain oscillatory activity in young adults may not result in the same effect in older adults, and functional reorganization of the aging brain may be an explanation [40].

In recent years, standard double-electrode tACS has shown limitations in controlling the stimulation focus and intensity. The use of different electrode montages, such as High-Definition tACS (**HD-tACS**), has allowed more precise stimulation control. Prior research has indicated that HD-tACS yields a more pronounced focalization of its effects through multiple smaller electrodes, possibly resulting from reduced distribution of the electrical field compared to conventional tACS [41,42]. Notably, online HD-tACS, applied during a motor task, induces phase- and frequency-dependent effects on cortical excitability [43,44].

After-effects on brain oscillations are a common outcome following tACS [45]. For instance, 10 Hz tACS stimulation of the parieto-occipital area resulted in an enhancement of the EEG-recorded alpha amplitude during the stimulation and this effect was seen to last at least 30 min after a 10 min stimulation period [46,47]. Other NIBS techniques, such as tDCS, have induced long-lasting excitability elevations in the human motor cortex [48], and in animals, a stimulation period of 5 to 30 min causes an effect lasting for hours after the end of stimulation [49].

This study aims to explore the after-effects of 70 Hz and 20 Hz HD-tACS on beta brain oscillatory patterns in healthy older adults. Based on the previously mentioned effect of tACS on motor performance on younger cohorts and how their beta brain oscillations differ from older ones, we hypothesized that 70 Hz HD-tACS would decrease resting-state beta power and promote a more positive MRBD (lower desynchronization). Conversely, we hypothesized that 20 Hz HD-tACS would increase beta power at rest and induce more negative MRBD (higher desynchronization).

2. Materials and Methods

2.1. Participants

In this single-blinded, sham-controlled study, 15 healthy individuals (7 males and 8 females) over 65 years old (age criteria as suggested by the Organization for Economic Co-operation and Development [50]) were recruited via advertisements. All participants signed a written informed consent form and were compensated for their participation. Inclusion criteria included having right-hand dominance as assessed through The Edinburgh Handedness Inventory [51] and scoring higher than 3 in the Mini-Cog Test [52]. We excluded subjects who had a personal history of neurological and psychiatric disorders, had any contraindications related to HD-tACS assessed through our NIBS safety questionnaire, and had received tDCS or tACS in the previous three months. Participants completed the following motor performance upper limb screening assessment tests: Box and Block Test (**BBT**) [53], Purdue Pegboard Test (**PPT**) [54], and Handgrip Strength (**HGS**) [55].

2.2. Experimental Design

The paradigm flow is shown in Figure 1. There were 3 experimental sessions in addition to the eligibility visit. A 64-channel EEG system (Brain Products, Gilching, Germany) was used to collect data, and stimulation was delivered using an EEG-compatible HD-tACS device (Soterix Medical, Woodbridge, NJ, USA). Baseline EEG at rest was recorded for 5 min, during which participants were seated in front of a screen displaying a centered white cross. They were asked to relax, look at the cross, and stay as still as possible. During the handgrip task, participants held a grip force response dynamometer with their right hand. They were required to squeeze a hand-clench dynamometer (BIOPAC, Goleta, CA, USA), which produced a linear force measurement output based on the pressure applied with the hand. A blue bar moved up and down according to the gripping force produced by the participants, who were asked to reach a red bar higher up as fast and accurately as possible. The force required to reach the target was 15% of their maximum voluntary contraction (**MVC**), which they had to hold for 4 s with an interval of 8–10 s resting between each handgrip (Figure 2). The handgrips were repeated 50 times (10 min). The hand dynamometer was connected to a BIOPAC system that converted the input to electrical signals. The signals were then transferred to a recording computer that displayed the force that was being applied by the participant in real time. Participants practiced until they understood the goal of the task (10 trials).

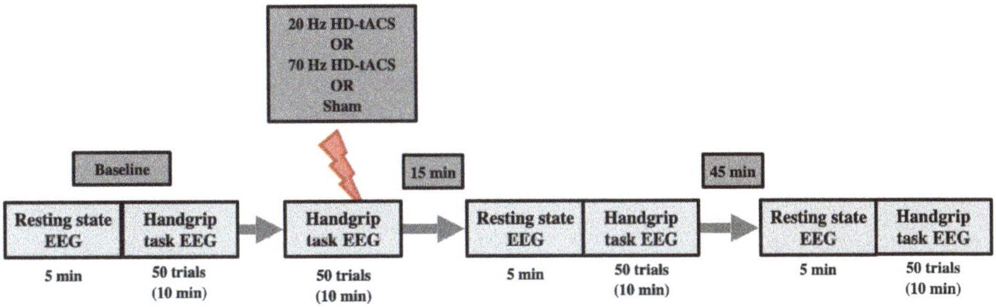

Figure 1. Schematic of the experimental timeline. Participants attended three different experimental sessions (each separated at least by a week), two active HD-tACS sessions (20 Hz and 70 Hz), and one Sham (control) session. Sessions were counterbalanced across participants. Each session started with baseline EEG recording for 5 min at resting state followed by 50 trials of the handgrip task. After that, active or Sham HD-tACS stimulation was applied while performing another 50 trials of the handgrip task. Resting-state EEG and handgrip-task EEG were performed again 15 min and 45 min post-tACS/Sham.

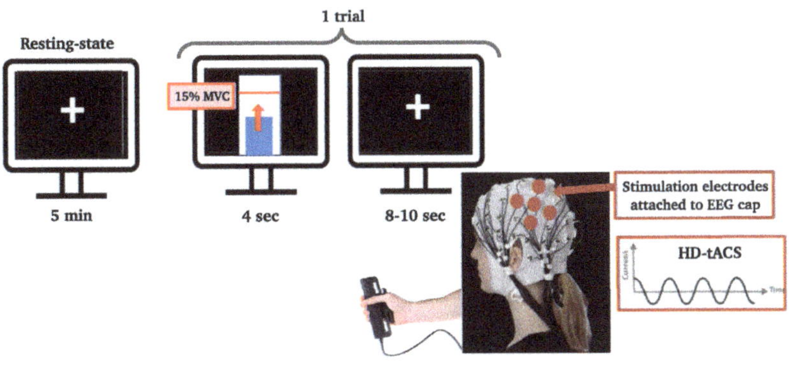

Figure 2. Computer screens showing the paradigm to the participant. During resting state, the participant looked at a black screen with a white cross in the center for 5 min. During the handgrip task, one trial consisted of reaching a threshold (red line inside the white bar), which was set to 15% of their maximum voluntary contraction, with a dynamometer using their right hand and staying on that threshold for 4 s. Each trial was followed by an inter-trial resting interval of 8 to 10 s. The stimulation electrodes that delivered HD-tACS were positioned on 5 recording electrodes over left M1 (anode: C3; cathodes: FC5, FC1, C3, CP5, and CP1).

After the baseline recordings, participants received either 20 Hz, 70 Hz HD-tACS, or Sham stimulation for 10 min while repeating the 50 handgrip task as EEG signals were simultaneously recorded. The order in which participants received the type of stimulation was randomized. After the stimulation ended, participants were asked to fill out a questionnaire to monitor the following possible adverse effects of HD-tACS: headache, neck pain, scalp pain, tingling, itching, burning sensation, skin redness, sleepiness, trouble concentrating, and acute mood change [56]. Participants were also asked if they thought the stimulation was active or sham to assess for protocol blindness. EEG recordings (rest and handgrips) were repeated 15 and 45 min after stimulation ended.

2.3. Data Acquisition and Pre-Processing

EEG signals were amplified and sampled at 2500 Hz. All electrodes were referenced to FCz. Electrode impedances were kept below 20 kΩ. HD stimulation was delivered by a current regulator (Soterix Medical, Germany). The EEG cap covered the individual's entire scalp, but the stimulation was delivered at pre-selected electrodes over the left sensorimotor cortex, with a target on M1 [28]. The anode was located in electrode C3 and the cathodes in FC5, FC1, CP5, and CP1. The stimulation lasted 10 min and it was delivered in the form of a sinusoid waveform with a peak-to-peak value of 1 mA and frequencies of 20 Hz and 70 Hz.

Offline EEG data were pre-processed using the Brainstorm MATLAB toolbox (version September 2024) [57]. Electrodes with an atypical power spectrum density were rejected from analysis. The rest of the EEG data were filtered (0.5–100 Hz bandpass, 60 Hz notch), resampled at 250 Hz, and then re-referenced to an average reference. Noisy segments (e.g., muscle, head, and jaw movement artifacts) were rejected by visual inspection. Independent components analysis (ICA) was used to identify and remove eye movement, muscle, and heart artifacts. Criteria for rejection included components' topography and time history [58].

2.4. Data Analysis

2.4.1. Behavioral Scores

Desrosiers et al. [59] developed predictive equations for PPT scores based on normative data resulting from their study. The normative data portion of the study involved 360 healthy participants over the age of 60 years. Student's *t*-tests were used to determine if the predicted scores were significantly different from the scores obtained by our participants.

2.4.2. Resting-State EEG

Signals from the resting state recording were epoched in 5 s segments ($n = 51 \pm 5.95$/recording). EEG signals were convoluted using a Morlet wavelet transformation with a frequency range from 1 Hz to 55 Hz with 1 Hz steps (time resolution = 3 s; central frequency = 1 Hz) [60]. We further analyzed signals from electrodes FC3, FC1, C5, C3, C1, CP5, CP3, and CP1, since we were interested in the HD effects over M1 and surrounding areas. We excluded FC5, since it was a faulty electrode. The beta frequency was extracted (15–29 Hz) and averaged in this range to calculate the beta power at rest. We selected this range to avoid including any power from the contiguous alpha and gamma bands.

2.4.3. Motor Task EEG

Signals were epoched from 1 s before to 8 s after the appearance of the blue bar that triggered the initiation of the handgrips. The first 5 trials were rejected for each subject and each recording. Additionally, trials were once again visually inspected and if still contaminated with artifacts they were manually rejected ($n = 35 \pm 6.18$ trials/recording). EEG signals were then examined in the time–frequency domain using a Morlet wavelet transformation with a frequency range from 1 Hz to 55 Hz with 1 Hz steps (time resolution = 3 s; central frequency = 1 Hz) [60]. Time–frequency maps were averaged within the beta band (15–29 Hz). *MRBD* was calculated as follows:

$$MRBD = \frac{P(t) - B}{B} \times 100\% \quad (1)$$

where $P(t)$ is the absolute power at time t and B is the mean power of the baseline, which was defined as 0.9–0.1 s before the start of each trial. MRBD% was averaged over 0.5–3.5 s after the appearance of the visual cue during sustained contraction at 15% MVC [61]. The analyzed signals were focused on the same electrodes as the resting-state analysis.

The EEG signals recorded during active HD-tACS were contaminated with very large artifacts; therefore, they were not included in the analysis.

2.5. Statistical Analysis

Repeated-measures analysis (**rmANOVA**) tests were conducted for beta features during rest and movement, with an a-level of 0.05 and factors Stimulation (20 Hz, 70 Hz, and Sham) and Time (baseline, post-15 min, and post-45 min). Post-hoc Bonferroni-corrected [62] *t*-tests were also used to test for differences across time (before applying stimulation vs. 15- and 45-min post-stimulation).

3. Results

3.1. Behavioral Assessment

The results from the behavioral assessments are reported in Table 1. The predicted scores for each of the participants based on their age were calculated according to Desrosiers et al. [59] (Table 2). There was no significant difference between the predicted scores of this model and the scores of our participants.

Table 1. Subject characteristics and behavioral scores.

	Mean ± SD	Min–Mix
Age (years)	69.7 ± 4.2	65–78
Handedness (/100)	94.6 ± 6.8	80–100
MiniCog (points)	4.7 ± 0.5	4–5
BBT Right Hand (blocks)	56.4 ± 5.5	46–65
BBT Left Hand (blocks)	57.2 ± 6.9	43–65
PPT Right Hand (pins)	13.4 ± 2.3	9–17
PPT Left Hand (pins)	12.0 ± 2.4	6–15
PPT Both Hands (pins)	10.4 ± 2.0	6–14
PPT Assembly (pins)	26.7 ± 5.1	17–34
HGS Right Hand (kg)	32.7 ± 11.3	17.6–55.6
HGS Left Hand (kg)	30.7 ± 9.5	15.6–54.3

BBT = Box and Blocks Test, PPT = Purdue Pegboard Test, HGS = Handgrip Strength.

Table 2. Participants' PPT scores (predicted from Desrosiers' [59] model vs. real scores).

PPT Subtests	Predicted	Real	*p*-Value	Cohen's d
Right Hand	12.8 ± 0.8	13.4 ± 2.3	0.136	0.326
Left Hand	12.0 ± 0.7	12.0 ± 2.4	0.439	0.048
Both Hands	9.8 ± 0.6	10.4 ± 2.0	0.128	0.358
Assembly	26.7 ± 2.2	26.7 ± 5.1	0.493	0.005

PPT = Purdue Pegboard Test.

3.2. Participant Blinding

After the end of the stimulation, participants were asked if they believed they had received active stimulation and, if they believed they did, at which frequency. For all the sessions, 33% of participants correctly identified Sham stimulation, and 47% and 33% correctly identified the 20 Hz and 70 Hz active stimulation, respectively. Cochran's Q test did not reveal significant differences between conditions ($p = 0.716$).

3.3. Effects of HD-tACS on Resting-State Beta Power

First, to test that our baseline measurements were comparable across stimulation conditions we ran Friedman tests (Table A1). The lack of significant differences supports the validity of our comparative analyses, as the starting points for each condition were not significantly different. The analysis of resting-state beta power across electrodes revealed varied effects of stimulation and time (Table 3. At electrode C3, there was a significant main effect of Stimulation ($F = 0.087$, $p = 0.016$), whereas no significant effects were observed for Time or the Time × Stimulation interaction. Similarly, electrode CP3 showed a significant main effect of Time ($F = 3.958$, $p = 0.030$) but no significant effects for Stimulation or their interaction. All other electrodes (FC3, FC1, C5, C1, CP5, and CP1) demonstrated no significant main effects or interactions (all $p > 0.05$).

Table 3. Summary of rmANOVA results for resting-state beta power at each electrode.

Electrode	Main Effect of Time (F, p)	Main Effect of Stimulation (F, p)	Time × Stimulation Interaction (F, p)
FC3	0.751, 0.480	0.303, 0.740	0.207, 0.933
FC1	0.800, 0.458	1.654, 0.209	0.492, 0.741
C5	0.258, 0.773	0.666, 0.521	1.535, 0.204
C3	0.795, 0.461	0.087, 0.016	0.429, 0.786
C1	0.229, 0.796	0.112, 0.893	1.458, 0.227
CP5	0.751, 0.481	0.465, 0.632	1.221, 0.312
CP3	3.958, 0.030	0.137, 0.872	0.849, 0.500
CP1	2.206, 0.128	0.371, 0.693	0.586, 0.673

Using post hoc *t*-tests of C3 and CP3 (Bonferroni-corrected) (Figure 3), spontaneous beta power significantly decreased only 45 min post-20 Hz HD-tACS ($p < 0.001$ in both electrodes) but not after 15 min ($p > 0.05$).

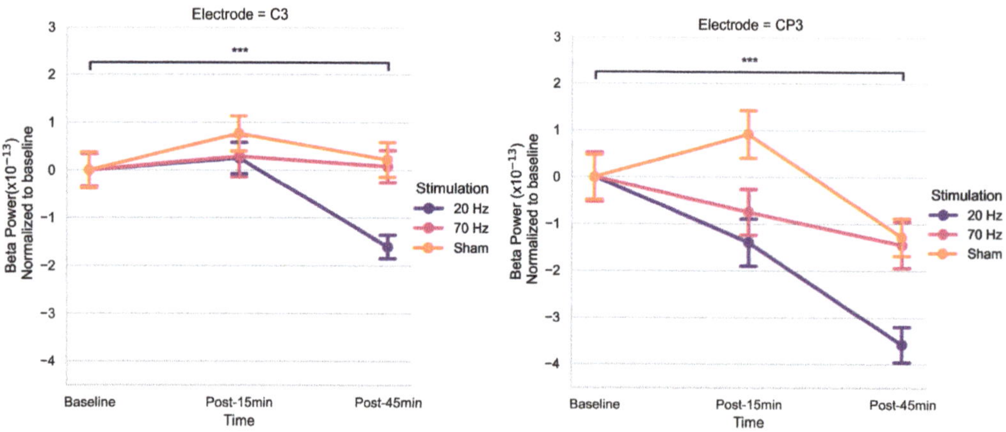

Figure 3. Effects of HD-tACS on average beta power at electrode C3 and CP3 normalized to baseline. Error bars represent standard error. Color indicates stimulation condition: orange = Sham, purple = 20 Hz, and pink = 70 Hz. (*** = $p < 0.001$).

3.4. Effects of HD-tACS on MRBD

Baseline measurements were comparable across stimulation conditions as shown in Table A2. The lack of significant differences supports the validity of our comparative analyses, as the starting points for each condition were not significantly different. The results of the rmANOVA on MRBD across electrodes are summarized in Table 4. At electrode FC3, a significant interaction between Time and Stimulation was observed ($F = 4.144$, $p = 0.005$), while no main effects of Time or Stimulation were detected. Similarly, at electrode C3, a significant Time × Stimulation interaction was found ($F = 2.694$, $p = 0.040$), with no main effects. At all other electrodes, no significant main effects or interactions were observed ($p > 0.05$).

The timecourses for MRBD at electrodes FC3 and C3 at the three time points during the three stimulation sessions can be seen in Figure 4.

Using post hoc t-tests for electrodes FC3 and C3 (Bonferroni-corrected) (Figure 5), a significant increase in MRBD percentage (more negative) was found 15 min post-20 Hz HD-tACS for both (FC3: $p = 0.039$; C3: $p = 0.011$). Applying 70 Hz HD-tACS elicited a decrease in MRBD percentage for both electrodes (FC3: $p < 0.001$; C3: $p = 0.039$), which persisted post-45 min only in FC3 ($p = 0.036$). There were no significant differences in MRBD values during Sham stimulation for either of them ($p > 0.05$).

Table 4. Results of rmANOVA on MRBD at each electrode.

Electrode	Main Effect of Time (F, p)	Main Effect of Stimulation (F, p)	Time × Stimulation Interaction (F, p)
FC3	0.035, 0.965	1.276, 0.294	4.144, 0.005
FC1	0.577, 0.567	1.049, 0.363	2.367, 0.063
C5	0.492, 0.616	0.866, 0.431	1.896, 0.123
C3	0.616, 0.546	0.964, 0.393	2.694, 0.040
C1	0.079, 0.923	1.759, 0.190	1.271, 0.292
CP5	0.189, 0.828	0.827, 0.447	1.561, 0.197
CP3	1.449, 0.251	1.877, 0.171	2.156, 0.085
CP1	0.238, 0.789	0.340, 0.714	0.686, 0.604

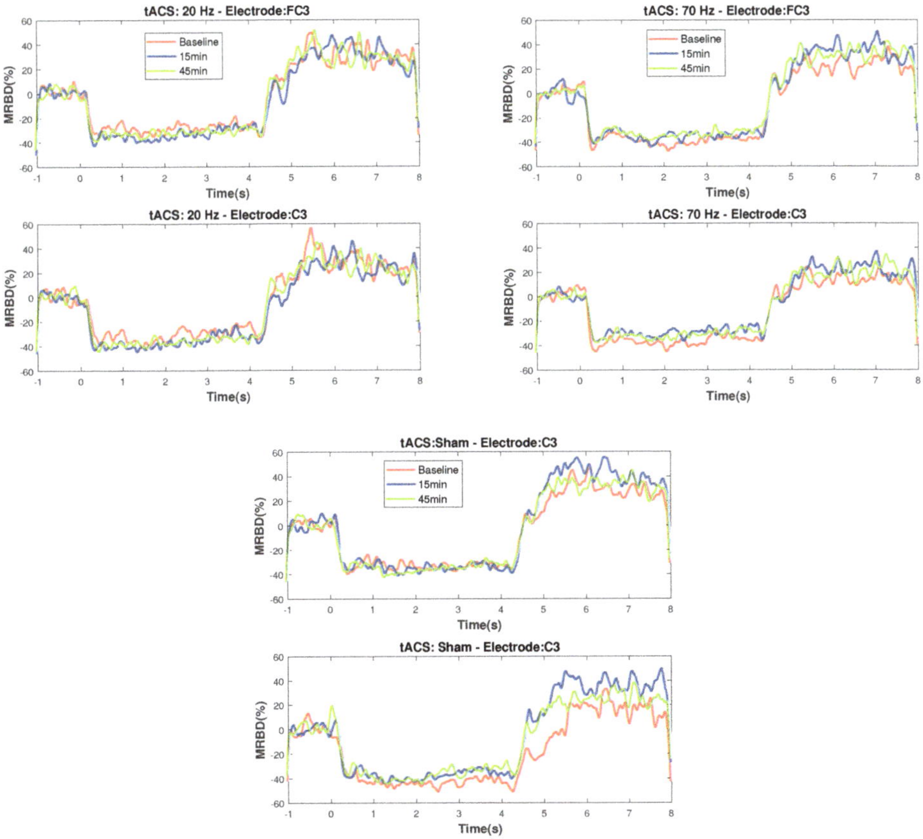

Figure 4. Temporal evolution of MRBD on electrodes that showed significant changes (C3 and FC3) across the different NIBS (20 Hz HD-tACS, 70 Hz HD-tACS, and Sham). Time zero is motor task onset and Time 4 is motor task offset. Applying 20 Hz HD-tACS induced a more negative MRBD only after 15 min and 70 Hz HD-tACS induced a more positive MRBD after 15 min in both electrodes and after 45 min only on FC3. No changes were significant during Sham stimulation.

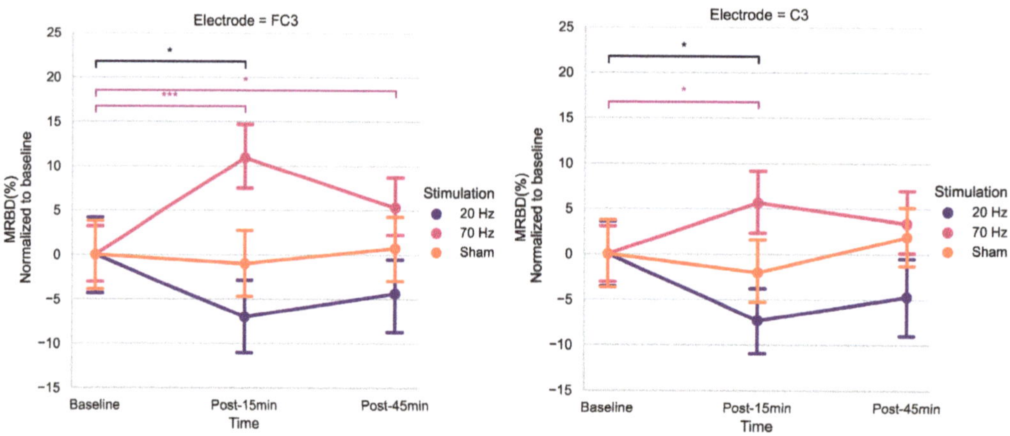

Figure 5. Effects of HD-tACS on average MRBD percentage at electrode FC3 and C3 normalized to baseline. Error bars represent standard error. Color indicates stimulation condition: orange = Sham, purple = 20 Hz, and pink = 70 Hz. (* = $p < 0.05$; *** = $p < 0.001$)).

4. Discussion

This study aimed to quantitatively examine the after-effects of HD-tACS on electrophysiological features in healthy older adults. Our rmANOVA anaysis for beta power revealed a noteworthy main effect of *Stimulation* for C3, as well as a significant effect of *Time* for CP3. Conversely, no other electrodes displayed any statistically significant main effects. Upon post hoc analysis of t-tests, a notable reduction in resting-state beta power was observed post-45 min for 20 Hz HD-tACS for both electrodes. In contrast, in the case of 70 Hz HD-tACS and Sham, no significant changes were observed in either electrode.

When analyzing MRBD effects, there was a significant *Time* × *Stimulation* interaction for FC3 and C3. The results of the post hoc t-tests showed higher MRBD values (more negative) 15 min after 20 Hz HD-tACS in both electrodes. Applying 70 Hz HD-tACS resulted in significant reductions in MRBD values after 15 min in FC3 and C3 and after 45 min only in FC3. Notably, no significant changes were observed during Sham stimulation. In the subsequent subsections, the implications of these results on beta oscillatory patterns are discussed in greater detail.

4.1. Population Behavioral Scores and Baseline Features

To ensure the representativeness of our participants concerning motor performance, we employed the Desrosiers et al. [59] model to predict the PPT score for each participant based on their age when performing the motor tests during the eligibility session. The PPT quantifies fingertip dexterity and gross movement of the hand, fingers, and arm. The predicted and actual scores of the PPT were similar in all the subtests, which suggests that the motor performance of our participants fell within the range of normal scores for the older population.

4.2. Modulation of Resting-State Beta Power

After administering 20 Hz HD-tACS, a delayed reduction in beta power was observed for electrodes C3 and CP3. This modulation appeared only 45 min post-stimulation, with no change detected at 15 min, suggesting that the impact on beta power is gradual rather than immediate. This finding aligns with previous studies showing no after-effects on beta power within 5 to 20 min after 20 Hz tACS [63]. However, prior research demonstrated that tACS could have effects on endogenous EEG power in the range of the stimulation frequency up to 70 min after the stimulation [64], which guided our decision to monitor changes up to 45 min post-stimulation.

The delayed response may reflect plasticity mechanisms, long-lasting modifications in neural connections and activity, with dynamic brain changes emerging over time. These results are consistent with reports of late plasticity changes in corticospinal excitability following 20 Hz tACS [65]. A relationship between beta oscillations and corticospinal excitability has been observed, with increased spontaneous beta oscillatory activity linked to smaller motor-evoked potential (**MEP**) amplitudes [66]. Furthermore, beta tACS has been shown to elevate cortical excitability in the M1 during stimulation, as evidenced by increased MEP amplitudes [67,68]. Heise et al. [41] also reported that more focal HD stimulation is significantly more effective in modulating MEPs post-stimulation. Additionally, delayed effects induced by tACS may be attributed to spike-timing-dependent plasticity (**STDP**), a process where the precise timing of neural activity determines whether synaptic connections are strengthened or weakened [69]. According to STDP principles, synapses in circuits resonating at frequencies similar to repetitive inputs are strengthened during stimulation. After stimulation ends, these synaptic modifications persist, leading to enhanced neural activity at the circuits' resonant frequencies. This aligns with findings that beta tACS can sustain elevated beta oscillations and cortical excitability for at least an hour post-stimulation [64,65,70]. Currently, evidence on STDP and its specific manifestations in aging populations is limited. However, studies have shown that synaptic plasticity mechanisms, including STDP, may be less efficient in older adults, potentially due to age-related changes in synaptic connectivity, neurotransmitter levels, and cortical excitability [71]. This could influence the extent and timing of tACS-induced plasticity effects. While our findings do not directly investigate these mechanisms, they provide a basis for future research to explore how aging might modulate tACS effects through STDP-related processes.

Our results disagree with recent research showing that 20 Hz HD-tACS increased beta power following stimulation of the visual [72] and parietal cortexes [73]. Also, a study reported no significant effects when applying tACS at 20 Hz, also in M1 during rest, on the beta power in younger participants using the standard double-electrode tACS montage [74]. We can identify three possible reasons why our results differ from the aforementioned studies. First, the difference in age of our participants is a possible reason, as it has been shown before that older groups demonstrated a decrease in tACS-induced neuroplasticity compared to a younger cohort [75]. We chose an older population, since these HD-tACS frequencies have not been studied before in aging, and we specifically examined its effects on beta oscillations. Secondly, in our study, HD-tACS was applied while participants were performing a handgrip task (online) as opposed to HD-tACS during rest (offline). Differences have been observed across studies comparing tACS-induced changes in online and offline protocols [46]. Additionally, a recent study comparing online and offline HD-tDCS showed that only the online stimulation reduced the power of the alpha rhythm during motor skill execution [76]. Thirdly, we used a different electrode montage and it has been shown that an HD-tACS electrode montage delivers a more focal current to M1 than the standard double-electrode tACS montage [77,78]. It is important to note that beta-tACS has been shown to have mixed results on other outcomes, such as corticospinal excitability and motor function [79]. Therefore, more research should focus on applying HD-tACS at this frequency, while keeping similar parameters to the ones used in studies showing significant effects.

For 70 Hz, there were no significant changes in resting beta power, which does not align with our initial hypothesis. This null finding underscores a critical limitation in current HD-tACS research: the application of fixed-frequency stimulation across diverse neural populations. Sugata et al. [32] found frequency-specific increases in beta power, while our results are more consistent with Mastakouri et al. [80], who demonstrated that gamma-tACS effects are heterogeneous and subject-specific. The recent literature increasingly emphasizes the importance of individualized stimulation protocols. Multiple studies [69,81–84] have shown that tACS delivered at a subject-specific frequency can produce more robust effects on cortical oscillations compared to fixed-frequency approaches. Moreover, age-related neuroplasticity differences may further complicate our findings. Guerra et al. [85] demonstrated that gamma stimulation mechanisms differ between younger and older adults, with older participants showing enhanced motor skill consolidation despite potentially reduced long-term potentiation (LTP)-like plasticity.

One promising solution emerges in the form of closed-loop brain–computer interfaces (BCIs). As highlighted by Xu et al., these systems provide a dynamic approach to neural modulation by continuously monitoring outcomes and adaptively modifying stimulation parameters [86]. By identifying specific neural biomarkers in real time, such as oscillatory patterns in gamma or beta frequency ranges, closed-loop systems can adjust stimulation accordingly. This approach addresses the current limitations of fixed-frequency stimulation by enabling real-time adaptation to an individual's unique neural dynamics. These nuanced findings suggest that future research should achieve the following:

- Develop individualized stimulation protocols based on baseline neural characteristics;
- Consider age-related neuroplasticity differences;
- Employ a more comprehensive assessment of stimulation effects;
- Explore closed-loop systems for real-time neural modulation.

Our null findings for 70 Hz stimulation should not be interpreted as a complete absence of effect, but rather as an invitation to develop more sophisticated, personalized neuromodulation approaches [87].

4.3. Modulation of MRBD

We found that after 15 min of 20 Hz HD-tACS, MRBD values became more negative in FC3 and C3. Conversely, after 15 min of 70 Hz HD-tACS, MRBD values became more positive in both FC3 and C3, and the effect persisted after 45 min only in FC3. These results align closely with our initial hypothesis. To the best of our knowledge, there are no other studies investigating the specific effects of HD-tACS on MRBD values in older adults. A recent study demonstrated that tACS at 10 Hz enhanced MRBD during a motor imagery task compared to pseudo-stimulation, indicating the capability of tACS to modify movement-related brain oscillations [88]. Additionally, other forms of NIBS, such as tDCS, have been shown to induce more negative MRBDs during motor imagery after 15 min of stimulation.

Xifra-Porxas et al. [13] previously mentioned that the motor performance decline observed in healthy aging may not be due to an impairment in the capacity to modulate beta oscillations. In fact, they observed a larger modulation in older compared to younger adults. On the other hand, beta oscillations at rest are greater in older adults [17], which suggests that increased desynchronization is needed to reach a threshold to initiate a movement. This would mean that modulating this desynchronization could later translate into a change in motor performance. While our study primarily focused on brain oscillations as a potential mechanism of HD-tACS-induced effects, we also evaluated motor task performance through measures of accuracy and reaction time. However, no significant changes were observed, potentially due to the simplicity of the motor task employed. Future research aiming to validate this hypothesis should utilize more complex motor tasks, such as the sequential visual isometric pinch task (**SVIPT**), which better resembles the level of difficulty encountered in everyday life skills [89]. Tasks like the SVIPT have been shown to challenge motor learning and performance to a degree comparable to real-world activities [90], making them ideal for studying the interplay between MRBD, beta modulation, and motor performance.

Finally, the fact that electrode C3, the anode of our HD-tACS montage, showed significant changes in both resting-state beta power and MRBD values suggests a focalization of the current right on the electrode that delivers the current. This focalization possibly results from a reduced distribution of the electrical field compared to the conventional tACS montage, which utilizes two distant patch electrodes [41].

Taken together, the results of this study are partly aligned with our initial hypothesis. In terms of beta power, only 20 Hz HD-tACS showed a decrease in beta power, but only 45 min after the end of stimulation, while 70 Hz HD-tACS did not show any significant changes, contrary to our hypothesis that it would decrease beta power. Regarding MRBD, both 20 Hz and 70 Hz showed the expected results: the former resulted in a more negative MRBD and the latter in a more positive. However, these changes in beta oscillations may not necessarily translate to improvements in motor performance. We do not report effects on motor tasks, since the complexity of our task is not enough to show impacts. Future research should focus on assessing the impact of HD-tACS on motor performance using tasks that are more complex and reflective of real-world daily activities, particularly for older adults.

There are certain limitations in our study that may explain some of variability in our HD-tACS outcomes and previously reported outcomes, such as interindividual differences, including skull thickness and the actual amount of current that reaches the cortex [91,92]. Because of these differences, individualized stimulation frequencies and current amplitudes, validated by studies such as Yamaguchi et al. [93], emphasize the parameter-dependent nature of tACS effects. In addition, our limited stimulation duration (10 min) contrasts with longer-lasting effects seen in extended gamma-tACS in mice, suggesting prolonged sessions or multiple-day approaches akin to tDCS studies for in-depth exploration [49,94]. Finally, the analysis of ongoing brain signals during concurrent stimulation is of primary interest and would also shed light on the immediate effects and mechanisms of HD-tACS. Recent advances in artifact removal algorithms will enable this type of analysis in our data [95].

5. Conclusions

To our knowledge, this is the first HD-tACS study that looks at the beta oscillation effects on healthy older adults. Future research should focus on replicating protocols that have been shown to have an effect on the desired outcomes in a bigger cohort to establish a more robust effect. In summary, our study reveals that HD-tACS has a modulating effect on beta oscillations during movement. Notably, different HD-tACS frequencies led to specific alterations in MRBD values, indicating frequency-specific effects on movement-related brain oscillations. The focal impact observed at electrode C3, the site of HD-tACS anodal stimulation, underscores the technique's precision over brain regions related to motor control. Future studies should explore personalized protocols tailored to individual neural characteristics, potentially employing advanced tACS methods such as HD-tACS, phase-shifted tACS, amplitude-modulated tACS, temporally interfering, and intersectional short-pulse techniques. Addressing these intricacies will enhance our understanding of tACS efficiency, guiding its optimized application in clinical settings. This research not only contributes to the ongoing discourse on brain stimulation but also holds promise for therapeutic interventions, making HD-tACS a promising avenue for exploration in diverse clinical populations such as stroke and PD patients. Further investigations in these domains will unveil the full potential of HD-tACS as a targeted therapeutic tool. Exploring HD-tACS effects in healthy older adults is crucial in providing insights into designing targeted interventions, addressing the complex interplay between neural oscillations, brain stimulation, and motor performance in the aging population.

Author Contributions: Conceptualization, G.D.M. and M.-H.B.; methodology, K.M.F., X.Y., G.D.M. and M.-H.B.; software, K.M.F. and X.Y.; validation, K.M.F., X.Y., G.D.M. and M.-H.B.; formal analysis, K.M.F.; investigation, K.M.F., X.Y., G.L., M.C.S. and M.-H.B.; resources, G.D.M. and M.-H.B.; data collection, K.M.F., X.Y., G.L. and M.C.S.; data curation, K.M.F., X.Y., G.D.M. and M.-H.B.; writing—original draft preparation, K.M.F.; writing—review and editing, G.D.M. and M.-H.B.; visualization, K.M.F.; supervision, G.D.M. and M.-H.B.; project administration, K.M.F. and M.-H.B.; funding acquisition, K.M.F. and M.-H.B. All authors have read and agreed to the published version of the manuscript.

Funding: This work was supported by the Canadian Foundation for Innovation grant (No. 34277) [M.H.B.], the Natural Sciences and Engineering Research Council of Canada (NSERC) Discovery Grants (No. RGPIN-2017-05270) [M.H.B.], and a scholarship from the Mexican National Council of

Humanities, Science and Technology (CONAHCYT) granted to K.M.F (No. 795453). M.H.B. holds a Fonds de recherche du Québec–Santé (FRQ-S) Research Scholar Award.

Institutional Review Board Statement: This study was conducted in accordance with the Declaration of Helsinki, and approved by the McGill University Ethical Advisory Committee #A06-B35-20B (20-05-055) date of approval 8 November 2020.

Informed Consent Statement: Informed consent was obtained from all subjects involved in this study.

Data Availability Statement: The raw data supporting the conclusions of this article will be made available by the authors on request due to ethical reasons, since we did not mention to our Ethics Board that we would make the data available. The code is publicly available in the following repository: https://github.com/kenyamelissamf/tACS_Aging (accessed on 20 November 2024).

Acknowledgments: The authors thank the participants for their cooperation and commitment to data collection. We would also like to thank Fabien Dal Maso for support in data pre-processing. We are grateful to Franziska Keller for providing support with the EEG system. We thank Rémi Dagenais for providing parts of the code and for useful discussions.

Conflicts of Interest: The authors declare no conflicts of interest.

Abbreviations

The following abbreviations are used in this manuscript:

BBT	Box and Block Test
CMC	Corticomuscular Coherence
ECR	Extensor Carpi Radialis
EEG	Electroencephalography
EMG	Electromyography
FDI	First Dorsal Interosseous
GABA	Gamma-aminobutyric Acid
HD-tACS	High-Definition Transcranial Alternating Current Stimulation
HGS	Handgrip Strength Test
iAPF	Individual Alpha Peak Frequency
ICA	Independent Component Analysis
ISP	Intersectional Short Pulse
LTP	Long-term Potentiation
M1	Primary Motor Cortex
MEG	Magnetoencephalography
MEP	Motor-Evoked Potential
MPS	Mild Parkinsonian Sign
MRBD	Movement-Related Beta Desynchronization
MVC	Maximum Voluntary Contraction
NREM	Non-rapid Eye Movement
NIBS	Non-Invasive Brain Stimulation
PD	Parkinson's Disease
PPT	Purdue Pegboard Test
rmANOVA	Repeated-Measures Analysis of Variances
SICI	Short-Interval Intracortical Inhibition
SVIPT	Sequential Visual Isometric Pinch Task
tACS	Transcranial Alternating Current Stimulation
tDCS	Transcranial Direct Current Stimulation
TI	Temporally Interfering
TMS	Transcranial Magnetic Stimulation

Appendix A

Table A1. Baseline comparison of beta power across stimulation conditions.

Electrode	Chi-Square	*p*-Value	Median Beta Power ($\times 10^{-13}$)		
			Sham	20 Hz	70 Hz
FC3	0.400	0.819	1.103	1.235	1.248
FC1	0.933	0.627	0.670	0.803	0.849
C5	2.133	0.344	0.946	1.146	0.804
C3	0.133	0.936	0.699	0.761	0.702
C1	1.733	0.420	0.939	1.095	1.067
CP5	0.533	0.766	0.620	0.716	0.715
CP3	0.400	0.819	1.110	1.500	1.049
CP1	0.133	0.936	0.717	0.783	0.794

Note: Friedman tests showed no significant differences between conditions at baseline (all p-values > 0.344).

Table A2. Baseline comparison of MRBD across stimulation conditions.

Electrode	Chi-Square	*p*-Value	Median MRBD (%)		
			Sham	20 Hz	70 Hz
FC3	4.933	0.085	−21.459	−7.981	−25.445
FC1	4.133	0.127	−18.151	−7.596	−23.386
C5	3.733	0.155	−10.502	2.976	−12.650
C3	4.933	0.085	−20.433	−5.745	−21.876
C1	5.733	0.057	−28.285	−21.597	−33.539
CP5	2.800	0.247	−20.020	−5.455	−20.274
CP3	5.200	0.074	−24.733	−18.127	−28.332
CP1	0.933	0.627	−19.511	−11.305	−22.868

Note: Friedman tests showed no significant differences between conditions at baseline (all p-values indicated above).

References

1. Maes, C.; Gooijers, J.; Orban de Xivry, J.J.; Swinnen, S.P.; Boisgontier, M.P. Two hands, one brain, and aging. *Neurosci. Biobehav. Rev.* **2017**, *75*, 234–256. [CrossRef] [PubMed]
2. Carmona, J.; Michan, S. Biology of Healthy Aging and Longevity. *Rev. Investig. Clin. Organo Del Hosp. Enfermedades Nutr.* **2016**, *68*, 7–16 .
3. Louis, E.D.; Tang, M.X.; Schupf, N.; Mayeux, R. Functional Correlates and Prevalence of Mild Parkinsonian Signs in a Community Population of Older People. *Arch. Neurol.* **2005**, *62*, 297–302. [CrossRef] [PubMed]
4. Giorgio, A.; Santelli, L.; Tomassini, V.; Bosnell, R.; Smith, S.; De Stefano, N.; Johansen-Berg, H. Age-related changes in grey and white matter structure throughout adulthood. *NeuroImage* **2010**, *51*, 943–951. [CrossRef]
5. Sullivan, E.V.; Pfefferbaum, A. Neuroradiological characterization of normal adult ageing. *Br. J. Radiol.* **2014**, *80*, S99–S108. [CrossRef]
6. Taubert, M.; Roggenhofer, E.; Melie-Garcia, L.; Muller, S.; Lehmann, N.; Preisig, M.; Vollenweider, P.; Marques-Vidal, P.; Lutti, A.; Kherif, F.; et al. Converging patterns of aging-associated brain volume loss and tissue microstructure differences. *Neurobiol. Aging* **2020**, *88*, 108–118. [CrossRef]
7. Good, C.D.; Johnsrude, I.S.; Ashburner, J.; Henson, R.N.; Friston, K.J.; Frackowiak, R.S. A Voxel-Based Morphometric Study of Ageing in 465 Normal Adult Human Brains. *NeuroImage* **2001**, *14*, 21–36. [CrossRef]
8. Bartzokis, G. Age-related myelin breakdown: A developmental model of cognitive decline and Alzheimer's disease. *Neurobiol. Aging* **2004**, *25*, 5–18. [CrossRef]
9. Seidler, R.D.; Bernard, J.A.; Burutolu, T.B.; Fling, B.W.; Gordon, M.T.; Gwin, J.T.; Kwak, Y.; Lipps, D.B. Motor control and aging: Links to age-related brain structural, functional, and biochemical effects. *Neurosci. Biobehav. Rev.* **2010**, *34*, 721–733. [CrossRef]
10. Kilavik, B.E.; Ponce-Alvarez, A.; Trachel, R.; Confais, J.; Takerkart, S.; Riehle, A. Context-Related Frequency Modulations of Macaque Motor Cortical LFP Beta Oscillations. *Cereb. Cortex* **2011**, *22*, 2148–2159. [CrossRef]
11. Cheyne, D.; Ferrari, P. MEG studies of motor cortex gamma oscillations: Evidence for a gamma "fingerprint" in the brain? *Front. Hum. Neurosci.* **2013**, *7*, 575. [CrossRef] [PubMed]
12. Kilavik, B.E.; Zaepffel, M.; Brovelli, A.; MacKay, W.A.; Riehle, A. The ups and downs of beta oscillations in sensorimotor cortex. *Exp. Neurol.* **2013**, *245*, 15–26 . [CrossRef] [PubMed]

13. Xifra-Porxas, A.; Niso, G.; Larivière, S.; Kassinopoulos, M.; Baillet, S.; Mitsis, G.D.; Boudrias, M.H. Older adults exhibit a more pronounced modulation of beta oscillations when performing sustained and dynamic handgrips. *NeuroImage* **2019**, *201*, 116037. [CrossRef] [PubMed]
14. Heinrichs-Graham, E.; Wilson, T.W. Is an absolute level of cortical beta suppression required for proper movement? Magnetoencephalographic evidence from healthy aging. *NeuroImage* **2016**, *134*, 514–521. [CrossRef]
15. Bardouille, T.; Bailey, L. Evidence for age-related changes in sensorimotor neuromagnetic responses during cued button pressing in a large open-access dataset. *NeuroImage* **2019**, *193*, 25–34. [CrossRef]
16. Heinrichs-Graham, E.; McDermott, T.J.; Mills, M.S.; Wiesman, A.I.; Wang, Y.P.; Stephen, J.M.; Calhoun, V.D.; Wilson, T.W. The lifespan trajectory of neural oscillatory activity in the motor system. *Dev. Cogn. Neurosci.* **2018**, *30*, 159–168. [CrossRef]
17. Rossiter, H.E.; Davis, E.M.; Clark, E.V.; Boudrias, M.H.; Ward, N.S. Beta oscillations reflect changes in motor cortex inhibition in healthy ageing. *NeuroImage* **2014**, *91*, 360–365. [CrossRef]
18. Reato, D.; Rahman, A.; Bikson, M.; Parra, L.C. Effects of weak transcranial alternating current stimulation on brain activity—A review of known mechanisms from animal studies. *Front. Hum. Neurosci.* **2013**, *7*, 687. [CrossRef]
19. Antal, A.; Paulus, W. Transcranial alternating current stimulation (tACS). *Front. Hum. Neurosci.* **2013**, *7*, 317. [CrossRef]
20. Herrmann, C.S.; Rach, S.; Neuling, T.; Strüber, D. Transcranial alternating current stimulation: A review of the underlying mechanisms and modulation of cognitive processes. *Front. Hum. Neurosci.* **2013**, *7*, 279. [CrossRef]
21. Pogosyan, A.; Gaynor, L.D.; Eusebio, A.; Brown, P. Boosting Cortical Activity at Beta-Band Frequencies Slows Movement in Humans. *Curr. Biol.* **2009**, *19*, 1637–1641. [CrossRef] [PubMed]
22. Lakatos, P.; Gross, J.; Thut, G. A New Unifying Account of the Roles of Neuronal Entrainment. *Curr. Biol.* **2019**, *29*, R890–R905. [CrossRef] [PubMed]
23. Antal, A.; Herrmann, C.S. Transcranial Alternating Current and Random Noise Stimulation: Possible Mechanisms. *Neural Plast.* **2016**, *2016*, 3616807. [CrossRef]
24. Diedrichsen, J.; Kornysheva, K. Motor skill learning between selection and execution. *Trends Cogn. Sci.* **2015**, *19*, 227–233. [CrossRef]
25. Krakauer, J.W.; Hadjiosif, A.M.; Xu, J.; Wong, A.L.; Haith, A.M. Motor Learning. In *Comprehensive Physiology*; John Wiley & Sons, Ltd.: Hoboken, NJ, USA, 2019; pp. 613–663. [CrossRef]
26. Sanes, J.N.; Donoghue, J.P. Plasticity and Primary Motor Cortex. *Annu. Rev. Neurosci.* **2000**, *23*, 393–415. [CrossRef]
27. Yarrow, K.; Brown, P.; Krakauer, J. Inside the brain of an elite athlete: The neural processes that support high achievement in sports. *Nat. Rev. Neurosci.* **2009**, *10*, 585–596. [CrossRef]
28. Woods, A.; Antal, A.; Bikson, M.; Boggio, P.; Brunoni, A.; Celnik, P.; Cohen, L.; Fregni, F.; Herrmann, C.; Kappenman, E.; et al. A technical guide to tDCS, and related non-invasive brain stimulation tools. *Clin. Neurophysiol.* **2016**, *127*, 1031–1048. [CrossRef]
29. Workman, C.; Kamholz, J.; Rudroff, T. The Tolerability and Efficacy of 4 mA Transcranial Direct Current Stimulation on Leg Muscle Fatigability. *Brain Sci.* **2020**, *10*, 244. [CrossRef]
30. Chakarov, V.; Naranjo, J.R.; Schulte-Mönting, J.; Omlor, W.; Huethe, F.; Kristeva, R. Beta-Range EEG-EMG Coherence With Isometric Compensation for Increasing Modulated Low-Level Forces. *J. Neurophysiol.* **2009**, *102*, 1115–1120. [CrossRef]
31. Muthukumaraswamy, S.D. Functional Properties of Human Primary Motor Cortex Gamma Oscillations. *J. Neurophysiol.* **2010**, *104*, 2873–2885. [CrossRef] [PubMed]
32. Sugata, H.; Yagi, K.; Yazawa, S.; Nagase, Y.; Tsuruta, K.; Ikeda, T.; Matsushita, K.; Hara, M.; Kawakami, K.; Kawakami, K. Modulation of Motor Learning Capacity by Transcranial Alternating Current Stimulation. *Neuroscience* **2018**, *391*, 131–139. [CrossRef] [PubMed]
33. Gilbertson, T.; Lalo, E.; Doyle, L.; Di Lazzaro, V.; Cioni, B.; Brown, P. Existing Motor State Is Favored at the Expense of New Movement during 13–35 Hz Oscillatory Synchrony in the Human Corticospinal System. *J. Neurosci.* **2005**, *25*, 7771–7779. [CrossRef] [PubMed]
34. Hammond, C.; Bergman, H.; Brown, P. Pathological synchronization in Parkinson's disease: Networks, models and treatments. *Trends Neurosci.* **2007**, *30*, 357–364 .[CrossRef] [PubMed]
35. Brown, P. Abnormal oscillatory synchronisation in the motor system leads to impaired movement. *Curr. Opin. Neurobiol.* **2007**, *17*, 656–664 . [CrossRef] [PubMed]
36. Engel, A.K.; Fries, P. Beta-band oscillations—Signalling the status quo? *Curr. Opin. Neurobiol.* **2010**, *20*, 156–165. [CrossRef]
37. Mallet, N.; Pogosyan, A.; Sharott, A.; Csicsvari, J.; Bolam, J.P.; Brown, P.; Magill, P.J. Disrupted Dopamine Transmission and the Emergence of Exaggerated Beta Oscillations in Subthalamic Nucleus and Cerebral Cortex. *J. Neurosci.* **2008**, *28*, 4795–4806. [CrossRef]
38. Brown, P. Oscillatory nature of human basal ganglia activity: Relationship to the pathophysiology of Parkinson's disease. *Mov. Disord.* **2003**, *18*, 357–363. [CrossRef]
39. Sriraman, A.; Oishi, T.; Madhavan, S. Timing-dependent priming effects of tDCS on ankle motor skill learning. *Brain Res.* **2014**, *1581*, 23–29. [CrossRef]

40. Perceval, G.; Flöel, A.; Meinzer, M. Can transcranial direct current stimulation counteract age-associated functional impairment? *Neurosci. Biobehav. Rev.* **2016**, *65*, 157–172. [CrossRef]
41. Heise, K.F.; Kortzorg, N.; Saturnino, G.B.; Fujiyama, H.; Cuypers, K.; Thielscher, A.; Swinnen, S.P. Evaluation of a Modified High-Definition Electrode Montage for Transcranial Alternating Current Stimulation (tACS) of Pre-Central Areas. *Brain Stimul.* **2016**, *9*, 700–704. [CrossRef]
42. Douglas, Z.H.; Maniscalco, B.; Hallett, M.; Wassermann, E.M.; He, B.J. Modulating Conscious Movement Intention by Noninvasive Brain Stimulation and the Underlying Neural Mechanisms. *J. Neurosci.* **2015**, *35*, 7239–7255. [CrossRef] [PubMed]
43. Heise, K.F.; Monteiro, T.S.; Inge, L.; Dante, M.; Swinnen, S.P. Distinct online and offline effects of alpha and beta transcranial alternating current stimulation (tACS) on continuous bimanual performance and task-set switching. *Sci. Rep. (Nat. Publ. Group)* **2019**, *9*, 3144. [CrossRef] [PubMed]
44. Pozdniakov, I.; Nunez, V.A.; Giulia, G.; Rossi, S.; Matteo, F. Online and offline effects of transcranial alternating current stimulation of the primary motor cortex. *Sci. Rep. (Nat. Publ. Group)* **2021**, *11*, 3854. [CrossRef] [PubMed]
45. Veniero, D.; Vossen, A.; Gross, J.; Thut, G. Lasting EEG/MEG Aftereffects of Rhythmic Transcranial Brain Stimulation: Level of Control Over Oscillatory Network Activity. *Front. Cell. Neurosci.* **2015**, *9*, 477. [CrossRef]
46. Helfrich, R.F.; Schneider, T.R.; Rach, S.; Trautmann-Lengsfeld, S.A.; Engel, A.K.; Herrmann, C.S. Entrainment of Brain Oscillations by Transcranial Alternating Current Stimulation. *Curr. Biol.* **2014**, *24*, 333–339. [CrossRef]
47. Neuling, T.; Rach, S.; Herrmann, C.S. Orchestrating neuronal networks: Sustained after-effects of transcranial alternating current stimulation depend upon brain states. *Front. Hum. Neurosci.* **2013**, *7*, 161. [CrossRef]
48. Nitsche, M.A.; Paulus, W. Excitability changes induced in the human motor cortex by weak transcranial direct current stimulation. *J. Physiol.* **2000**, *527*, 633–639. [CrossRef]
49. Nitsche, M.A.; Paulus, W. Sustained excitability elevations induced by transcranial DC motor cortex stimulation in humans. *Neurol. J. Am. Heart Assoc.* **2001**, *57*, 1899–1901. [CrossRef]
50. OECD. *Elderly Population*; OECD: Paris, France, 2014. [CrossRef]
51. Oldfield, R. The assessment and analysis of handedness: The Edinburgh inventory. *Neuropsychologia* **1971**, *9*, 97–113. [CrossRef]
52. Borson, S.; Scanlan, J.; Brush, M.; Vitaliano, P.; Dokmak, A. The Mini-Cog: A cognitive 'vital signs' measure for dementia screening in multi-lingual elderly. *Int. J. Geriatr. Psychiatry* **2000**, *15*, 1021–1027. [CrossRef]
53. Mathiowetz, V.; Volland, G.; Kashman, N.; Weber, K. Adult Norms for the Box and Block Test of Manual Dexterity. *Am. J. Occup. Ther.* **1985**, *39*, 386–391. [CrossRef] [PubMed]
54. Lindstrom-Hazel, D.; VanderVlies Veenstra, N. Examining the Purdue Pegboard Test for Occupational Therapy Practice. *Open J. Occup. Ther.* **2015**, *3*, 5. [CrossRef]
55. Bohannon, R.W.; Peolsson, A.; Massy-Westropp, N.; Desrosiers, J.; Bear-Lehman, J. Reference values for adult grip strength measured with a Jamar dynamometer: A descriptive meta-analysis. *Physiotherapy* **2006**, *92*, 11–15. [CrossRef]
56. Matsumoto, H.; Ugawa, Y. Adverse events of tDCS and tACS: A review. *Clin. Neurophysiol. Pract.* **2017**, *2*, 19–25. [CrossRef]
57. Tadel, F.; Baillet, S.; Mosher, J.C.; Pantazis, D.; Leahy, R.M. Brainstorm: A User-Friendly Application for MEG/EEG Analysis. *Comput. Intell. Neurosci.* **2011**, *2011*, 879716. [CrossRef]
58. Delorme, A.; Makeig, S. EEGLAB: An open source toolbox for analysis of single-trial EEG dynamics including independent component analysis. *J. Neurosci. Methods* **2004**, *134*, 9–21. [CrossRef]
59. Desrosiers, J.; Hébert, R.; Bravo, G.; Dutil, E. The Purdue Pegboard Test: Normative data for people aged 60 and over. *Disabil. Rehabil.* **1995**, *17*, 217–224. [CrossRef]
60. Tallon-Baudry, C.; Bertrand, O.; Delpuech, C.; Pernier, J. Stimulus Specificity of Phase-Locked and Non-Phase-Locked 40 Hz Visual Responses in Human. *J. Neurosci.* **1996**, *16*, 4240–4249. [CrossRef]
61. Dal Maso, F.; Desormeau, B.; Boudrias, M.H.; Roig, M. Acute cardiovascular exercise promotes functional changes in cortico-motor networks during the early stages of motor memory consolidation. *NeuroImage* **2018**, *174*, 380–392. [CrossRef]
62. Holm, S. A Simple Sequentially Rejective Multiple Test Procedure. *Scand. J. Stat.* **1979**, *6*, 65–70.
63. Rumpf, J.J.; Barbu, A.; Fricke, C.; Wegscheider, M.; Classen, J. Posttraining Alpha Transcranial Alternating Current Stimulation Impairs Motor Consolidation in Elderly People. *Neural Plast.* **2019**, *2019*, 2689790. [CrossRef] [PubMed]
64. Kasten, F.H.; Dowsett, J.; Herrmann, C.S. Sustained Aftereffect of gamma-tACS Lasts Up to 70 min after Stimulation. *Front. Hum. Neurosci.* **2016**, *10*, 245. [CrossRef] [PubMed]
65. Wischnewski, M.; Engelhardt, M.; Salehinejad, M.A.; Schutter, D.J.L.G.; Kuo, M.F.; Nitsche, M.A. NMDA Receptor-Mediated Motor Cortex Plasticity After 20 Hz Transcranial Alternating Current Stimulation. *Cereb. Cortex* **2018**, *29*, 2924–2931. [CrossRef] [PubMed]
66. Mäki, H.; Ilmoniemi, R.J. EEG oscillations and magnetically evoked motor potentials reflect motor system excitability in overlapping neuronal populations. *Clin. Neurophysiol.* **2010**, *121*, 492–501. [CrossRef] [PubMed]
67. Feurra, M.; Bianco, G.; Santarnecchi, E.; Del Testa, M.; Rossi, A.; Rossi, S. Frequency-Dependent Tuning of the Human Motor System Induced by Transcranial Oscillatory Potentials. *J. Neurosci.* **2011**, *31*, 12165–12170. [CrossRef]
68. Feurra, M.; Pasqualetti, P.; Bianco, G.; Santarnecchi, E.; Rossi, A.; Rossi, S. State-Dependent Effects of Transcranial Oscillatory Currents on the Motor System: What You Think Matters. *J. Neurosci.* **2013**, *33*, 17483–17489. [CrossRef]

69. Zaehle, T.; Rach, S.; Herrmann, C.S. Transcranial Alternating Current Stimulation Enhances Individual Alpha Activity in Human EEG. *PLoS ONE* **2010**, *5*, e13766. [CrossRef]
70. Vossen, A.; Gross, J.; Thut, G. Alpha Power Increase After Transcranial Alternating Current Stimulation at Alpha Frequency (alpha-tACS) Reflects Plastic Changes Rather Than Entrainment. *Brain Stimul.* **2015**, *8*, 499–508. [CrossRef]
71. Burke, S.N.; Barnes, C.A. Neural plasticity in the ageing brain. *Nat. Reviews. Neurosci.* **2006**, *7*, 30–40. [CrossRef]
72. Nakazono, H.; Ogata, K.; Takeda, A.; Yamada, E.; Kimura, T.; Tobimatsu, S. Transcranial alternating current stimulation of gamma but not beta frequency sharpens multiple visual functions. *Brain Stimul.* **2020**, *13*, 343–352. [CrossRef]
73. Battaglini, L.; Ghiani, A.; Casco, C.; Ronconi, L. Parietal tACS at beta frequency improves vision in a crowding regime. *NeuroImage* **2020**, *208*, 116451. [CrossRef] [PubMed]
74. Louis-Philippe, L.; Murray, A.; Manon, D.; Pacheco-Barrios, K.; Felipe, F.; Tremblay, S.; Saint-Amour, D.; Jean-François, L.; Hugo, T. No aftereffects of high current density 10 Hz and 20 Hz tACS on sensorimotor alpha and beta oscillations. *Sci. Rep. (Nat. Publ. Group)* **2021**, *11*, 21416. [CrossRef]
75. Fresnoza, S.; Christova, M.; Feil, T. The effects of transcranial alternating current stimulation (tACS) at individual alpha peak frequency (iAPF) on motor cortex excitability in young and elderly adults. *Exp. Brain Res.* **2018**, *236*, 2573–2588. [CrossRef]
76. Guo, F.; Niu, M.; Hanson, N.J.; Guo, J.; Zhou, K.; Zhao, T.; Ren, Y. Enhancing motor skill learning through multiple sessions of online high-definition transcranial direct current stimulation in healthy adults: Insights from EEG power spectrum. *Cereb. Cortex* **2024**, *34*, bhae395. [CrossRef]
77. Datta, A.; Elwassif, M.; Battaglia, F.; Bikson, M. Transcranial current stimulation focality using disc and ring electrode configurations: FEM analysis. *J. Neural Eng.* **2008**, *5*, 163. [CrossRef]
78. Edwards, D.; Cortes, M.; Datta, A.; Minhas, P.; Wassermann, E.M.; Bikson, M. Physiological and modeling evidence for focal transcranial electrical brain stimulation in humans: A basis for high-definition tDCS. *NeuroImage* **2013**, *74*, 266–275. [CrossRef]
79. Rostami, M.; Lee, A.; Frazer, A.K.; Akalu, Y.; Siddique, U.; Pearce, A.J.; Tallent, J.; Kidgell, D.J. Determining the corticospinal, intracortical and motor function responses to transcranial alternating current stimulation of the motor cortex in healthy adults: A systematic review and meta-analysis. *Brain Res.* **2024**, *1822*, 148650. [CrossRef]
80. Mastakouri, A.A.; Schölkopf, B.; Grosse-Wentrup, M. Beta Power May Meditate the Effect of Gamma-TACS on Motor Performance. In Proceedings of the 2019 41st Annual International Conference of the IEEE Engineering in Medicine and Biology Society (EMBC), Berlin, Germany, 23–27 July 2019; pp. 5902–5908. [CrossRef]
81. Schilberg, L.; Engelen, T.; ten Oever, S.; Schuhmann, T.; de Gelder, B.; de Graaf, T.A.; Sack, A.T. Phase of beta-frequency tACS over primary motor cortex modulates corticospinal excitability. *Cortex* **2018**, *103*, 142–152. [CrossRef]
82. Janssens, S.E.W.; Sack, A.T.; Ten Oever, S.; de Graaf, T.A. Calibrating rhythmic stimulation parameters to individual electroencephalography markers: The consistency of individual alpha frequency in practical lab settings. *Eur. J. Neurosci.* **2022**, *55*, 3418–3437. [CrossRef]
83. Kemmerer, S.; Sack, A.; de Graaf, T.; ten Oever, S.; De Weerd, P.; Schuhmann, T. Frequency-specific transcranial neuromodulation of alpha power alters visuospatial attention performance. *Brain Res.* **2022**, *1782*, 147834. [CrossRef]
84. Ayanampudi, V.; Kumar, V.; Krishnan, A.; Walker, M.P.; Ivry, R.B.; Knight, R.T.; Gurumoorthy, R. Personalized transcranial alternating current stimulation improves sleep quality: Initial findings. *Front. Hum. Neurosci.* **2023**, *16*, 1066453. [CrossRef] [PubMed]
85. Guerra, A.; Asci, F.; Zampogna, A.; D'Onofrio, V.; Berardelli, A.; Suppa, A. The effect of gamma oscillations in boosting primary motor cortex plasticity is greater in young than older adults. *Clin. Neurophysiol.* **2021**, *132*, 1358–1366. [CrossRef] [PubMed]
86. Xu, S.; Liu, Y.; Lee, H.; Li, W. Neural interfaces: Bridging the brain to the world beyond healthcare. *Exploration* **2024**, *4*, 20230146. [CrossRef] [PubMed]
87. Krause, V.; Wach, C.; Suedmeyer, M.; Ferrea, S.; Schnitzler, A.; Pollok, B. Cortico-muscular coupling and motor performance are modulated by 20 Hz transcranial alternating current stimulation (tACS) in Parkinson's disease. *Front. Hum. Neurosci.* **2014**, *7*, 928. [CrossRef]
88. Xie, J.; Peng, M.; Lu, J.; Xiao, C.; Zong, X.; Wang, M.; Gao, D.; Qin, Y.; Liu, T. Enhancement of Event-Related Desynchronization in Motor Imagery Based on Transcranial Electrical Stimulation. *Front. Hum. Neurosci.* **2021**, *15*, 635351. [CrossRef]
89. Kuo, H.I.; Bikson, M.; Datta, A.; Minhas, P.; Paulus, W.; Kuo, M.F.; Nitsche, M.A. Comparing cortical plasticity induced by conventional and high-definition 4 × 1 ring tDCS: A neurophysiological study. *Brain Stimul.* **2013**, *6*, 644–648. [CrossRef]
90. Reis, J.; Schambra, H.M.; Cohen, L.G.; Buch, E.R.; Fritsch, B.; Zarahn, E.; Celnik, P.A.; Krakauer, J.W. Noninvasive cortical stimulation enhances motor skill acquisition over multiple days through an effect on consolidation. *Proc. Natl. Acad. Sci. USA* **2009**, *106*, 1590–1595. [CrossRef]
91. Li, L.M.; Uehara, K.; Hanakawa, T. The contribution of interindividual factors to variability of response in transcranial direct current stimulation studies. *Front. Cell. Neurosci.* **2015**, *9*, 181. [CrossRef]
92. Pellegrini, M.; Zoghi, M.; Jaberzadeh, S. Biological and anatomical factors influencing interindividual variability to noninvasive brain stimulation of the primary motor cortex: A systematic review and meta-analysis. *Rev. Neurosci.* **2018**, *29*, 199–222. [CrossRef]
93. Yamaguchi, T.; Svane, C.; Forman, C.; Beck, M.; Geertsen, S.; Lundbye-Jensen, J.; JB, N. Transcranial Alternating Current Stimulation of the Primary Motor Cortex after Skill Acquisition Improves Motor Memory Retention in Humans: A Double-Blinded Sham-Controlled Study. *Cereb. Cortex Commun.* **2020**, *1*, tgaa047. [CrossRef]

94. Wu, L.; Cao, T.; Li, S.; Yuan, Y.; Zhang, W.; Huang, L.; Cai, C.; Fan, L.; Li, L.; Wang, J.; et al. Long-term gamma transcranial alternating current stimulation improves the memory function of mice with Alzheimer's disease. *Front. Aging Neurosci.* **2022**, *14*, 980636. [CrossRef] [PubMed]
95. Yan, X.; Boudrias, M.H.; Mitsis, G.D. Removal of Transcranial Alternating Current Stimulation EEG Artifacts Using Blind Source Separation and Wavelets. *IEEE Trans. Biomed. Eng.* **2022**, *69*, 3183–3192. [CrossRef] [PubMed]

brain sciences

MDPI

Article

The Effect of Processing Techniques on the Classification Accuracy of Brain-Computer Interface Systems

András Adolf [1,2,3,*], Csaba Márton Köllőd [2,3], Gergely Márton [2,3], Ward Fadel [1,2,3] and István Ulbert [2,3,4]

1 Roska Tamás Doctoral School of Sciences and Technology, Práter utca 50/a, 1083 Budapest, Hungary; fadel.ward@ttk.hu
2 Faculty of Information Technology and Bionics, Pázmány Péter Catholic University, Práter utca 50/a, 1083 Budapest, Hungary; kollod.csaba@itk.ppke.hu (C.M.K.); marton.gergely@ttk.hu (G.M.); ulbert.istvan@ttk.hu (I.U.)
3 Institute of Cognitive Neuroscience and Psychology, HUN-REN Research Centre for Natural Sciences, Magyar Tudósok Körútja 2, 1117 Budapest, Hungary
4 Department of Neurosurgery and Neurointervention, Faculty of Medicine, Semmelweis University, Amerikai út 57, 1145 Budapest, Hungary
* Correspondence: adolf.andras@itk.ppke.hu

Abstract: Background/Objectives: Accurately classifying Electroencephalography (EEG) signals is essential for the effective operation of Brain-Computer Interfaces (BCI), which is needed for reliable neurorehabilitation applications. However, many factors in the processing pipeline can influence classification performance. The objective of this study is to assess the effects of different processing steps on classification accuracy in EEG-based BCI systems. **Methods**: This study explores the impact of various processing techniques and stages, including the FASTER algorithm for artifact rejection (AR), frequency filtering, transfer learning, and cropped training. The Physionet dataset, consisting of four motor imagery classes, was used as input due to its relatively large number of subjects. The raw EEG was tested with EEGNet and Shallow ConvNet. To examine the impact of adding a spatial dimension to the input data, we also used the Multi-branch Conv3D Net and developed two new models, Conv2D Net and Conv3D Net. **Results**: Our analysis showed that classification accuracy can be affected by many factors at every stage. Applying the AR method, for instance, can either enhance or degrade classification performance, depending on the subject and the specific network architecture. Transfer learning was effective in improving the performance of all networks for both raw and artifact-rejected data. However, the improvement in classification accuracy for artifact-rejected data was less pronounced compared to unfiltered data, resulting in reduced precision. For instance, the best classifier achieved 46.1% accuracy on unfiltered data, which increased to 63.5% with transfer learning. In the filtered case, accuracy rose from 45.5% to only 55.9% when transfer learning was applied. An unexpected outcome regarding frequency filtering was observed: networks demonstrated better classification performance when focusing on lower-frequency components. Higher frequency ranges were more discriminative for EEGNet and Shallow ConvNet, but only when cropped training was applied. **Conclusions**: The findings of this study highlight the complex interaction between processing techniques and neural network performance, emphasizing the necessity for customized processing approaches tailored to specific subjects and network architectures.

Keywords: artifact rejection; brain-computer interface; electroencephalography; motor imagery; faster; CNN

1. Introduction

Brain-computer Interfaces (BCI) represent a rapidly evolving research field. As depicted in Figure 1, the general structure of these systems consists of physiological measurements of the brain, digitizing and processing the signal, and finally, giving commands based on the classification of the received data. These devices could provide significant help to people with different serious disabilities; however, reliable classification accuracy is essential for use in real life [1,2]. Non-invasive EEG measurements can be conducted more easily than invasive ones because they do not require surgery preparation. On the other hand, the classification accuracy is poorer due to the distortion from the various tissues between the source and the recording electrode and the lower resolution.

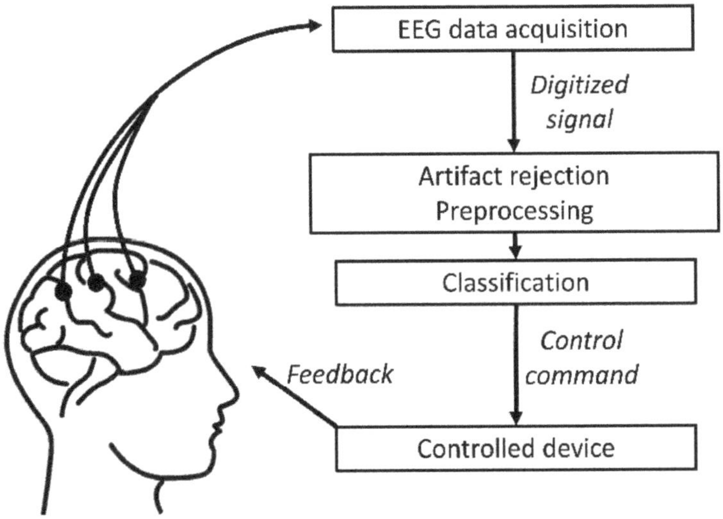

Figure 1. The general structure of a BCI system. Acquired data are digitized and go through preprocessing, such as artifact rejection. Afterward, a classifier decides what the intention of the user is and gives a control command based on the decision.

An important aspect to consider when working with EEG signals is the presence of artifacts. Artifacts refer to unwanted signals that contaminate the EEG recordings, potentially introducing significant distortions that can affect the analysis of underlying neurological phenomena [3]. It is also important to note that artifact rejection (AR) algorithms may interact with various classification methods, the interaction of which has not been well identified to date. We have limited knowledge about the extent to which artifact rejection improves (or deteriorates) the classification accuracy of various machine learning methods. Moreover, small variations in feature extraction methods may also play an important role in the classification accuracy of a BCI system. It is also unknown how small computational changes in the feature extraction methods affect the overall classification accuracy of the whole system. Finally, subjects interacting with a BCI system may not use similar strategies to achieve control over the given mental imagery task. Also, their brain anatomy and function may differ. Thus, their EEG parameters may also be different with respect to the same task; therefore, individual variations may well have a considerable effect on the classification accuracy of a BCI system. In this work, we examined how changes in the processing pipeline modify the classification accuracy of a BCI system, including individual subject variations. As the specific features that neural networks learn from are not well understood, it is possible that training the classifier interacts with the removal of artifacts and small changes in the processing pipeline. Furthermore, it cannot be guaranteed that the AR method filters out all artifactual components. It is also interesting to consider that networks trained on polluted data may also focus on the artifacts.

1.1. EEG Artifacts

Considering the field of EEG, two distinct categories of artifacts are recognized: physiological/biological and non-physiological. The latter can be caused by the measurement instrument, meaning faulty electrodes, powerline- and environment noise, high impedance of electrodes, cable, or body movement artifacts. These can be mostly avoided by a precise recording system and strict recording procedures [4].

Physiological artifacts may arise from various sources, including cardiac activity, pulse, respiratory patterns, and glossokinetic effects. The two most significant contributors to physiological artifacts are Electrooculograms (EOG) and Electromyograms (EMG). EOGs can come from ocular movements such as eye blinking, eye movement, and eye flatter, while EMGs can result from chewing, clenching, swallowing, sniffing, and talking [3,5].

It is important to remove artifacts from the EEG signal prior to processing, as they can potentially interfere with the interpretation of the original data.

1.2. Artifact Rejection for Imaginary Movement EEG Classification

To remove artifact contamination, a filtering algorithm needs to be employed. In the academic literature, several methods have been proposed for this purpose, of which the Independent Component Analysis (ICA) is one of the most frequently utilized mathematical methods [6–8]. In addition to ICA, other methods such as Wavelet Transformation [9], Canonical Correlation Analysis [10], Empirical-Mode Decomposition [11], and further Hybrid approaches [4] are also commonly employed. One of the widely used algorithms for AR is the FASTER algorithm [12], also utilized in this work, which, besides using the ICA method, also performs filtering and interpolation over global and epoch-vise bad channels.

1.3. Comparison of Artifact Rejected and Raw EEG Data Classification

To our knowledge, only a limited number of studies have examined the effects of artifact rejection (AR) methods on BCI performance, prompting our focus on the FASTER algorithm in this paper. The following studies provide relevant insights.

Zhong and Qi [13] investigated the impact of AR on P300-based BCI performance, utilizing Independent Component Analysis (ICA) for artifact removal and Bayesian Linear Discriminant Analysis (BLDA) for classification. They found AR methods can improve accuracy and information transfer rate, though their study was limited by manual filtering and a small sample of eight participants. Similarly, Kim et al. [14] assessed various AR methods (ICA, adaptive filtering, and Artifact Subspace Reconstruction) on P300/N200 classification with Support Vector Machine (SVM), finding ICA-denoised data achieved the highest accuracy at 62.87%.

Mohammadi and Mosavi [15] compared two ICA-based EOG AR methods and observed better accuracy using ICA with wavelet decomposition on the BCI Competition IV 2a dataset [16]. Winkler et al. [8] developed an ICA-based AR method for BCI use and noted minimal impact on accuracy across ERP [17] and MI-BCI [18] datasets, though individual differences were observed. Iqbal et al. [19] achieved an 80.5% accuracy using an EEGNet and Temporal Convolution Network (TCN)-based classifier with an EOG removal system, improving consistency across subjects.

Other studies explored diverse approaches to AR. Bou Assi et al. [20] combined ICA with K-means clustering for AR, observing improved accuracy from 66% to 88.1% using the Physionet database. Thompson et al. [21] found that automated AR led to a decline in P300-based BCI performance, with certain methods (SOBI, JADER, EFICA) minimizing, but not eliminating, accuracy loss. Frølich et al. [22] examined how different artifacts (e.g., blinks, muscle movements) affected motor imagery BCI and found muscle artifacts notably impacted performance, especially with broader electrode coverage.

Several studies focused on EOG-based AR. Mannan et al. [23] identified that using a lowpass filter on the EOG signal improved classification when paired with the CSP classifier on BCI Competition IV 2a data. Daly et al. [24] presented the FORCe algorithm for

AR and observed it significantly improved accuracy over raw EEG, especially compared to the FASTER algorithm, in cerebral palsy patients. Merinov et al. [25] found no significant accuracy differences between four AR methods on the BCI Competition IV 2a dataset, though classification performance varied significantly between subjects.

Lastly, studies by Stigt et al. [26] and Chen et al. [27] analyzed the impact of artifact removal on EEG data. Stigt et al. observed that AR did not improve normal versus abnormal EEG classification accuracy but did slightly expedite training. Chen et al. examined specific artifact types, finding that removing EMG, powerline interference, electrode artifacts, and EOG improved accuracy by 5.5%, 4.0%, 3.1%, and 1.7%, respectively. Islam et al. [28] and Anjum et al. [29] both applied probability mapping for AR, demonstrating increased classification performance, though Anjum's study was limited by the low electrode count of the Emotiv Pro headset used.

1.4. Frequency Dependence of the EEG Signal Classification

Ali Al-Saegh et al. [30] reviewed 36 studies on motor imagery EEG classification and found a strong consensus on using the 8–25 Hz frequency range associated with mu and beta bands for effective feature extraction in motor-related brain activity. Lower frequencies (0–5 Hz) were rarely used, appearing in only seven studies, indicating a preference for higher frequencies. In contrast, R. Salazar Valas and Roberto A. Vazquez [31] examined the effect of cutoff frequencies on classification accuracy with the BCI Competition IV 2a dataset and found that even the 0–10 Hz range could yield meaningful classification results, challenging the typical emphasis on higher frequencies.

1.5. Aim of the Study

Our article stands out by providing a comprehensive evaluation of the FASTER artifact rejection algorithm in conjunction with other preprocessing and processing methods, such as frequency filtering, transfer learning, and cropped training, across multiple neural network architectures, including EEGNet, Shallow ConvNet, and our custom classifiers. Unlike prior studies, which typically focus on a single preprocessing technique or classifier, we examine the combined effects of processing steps on classification accuracy. This approach highlights the complex interactions between artifact rejection, filtering, and neural network performance, offering unique insights into how these factors collectively influence BCI performance across different subjects and network configurations.

2. Materials and Methods

2.1. The Physionet Database-EEG Motor Movement/Imagery Dataset

We performed our research on the EEG motor movement/imagery dataset recorded by Schalk et al. as part of the Physionet Database [32]. Data were recorded with a 64-channeled 10–10 EEG system with 160 Hz sampling frequency and by using the BCI2000 framework without hardware filters [33]. It is one of the largest EEG datasets of motor imaginary tasks, consisting of recordings from 109 subjects, 14 files for each. In our work, we excluded subjects 88, 92, and 100 due to the sampling frequency and data structure mismatch. We also omitted subject 89, where labels were found to be incorrect. These problems were also reported previously [34,35].

The dataset contains five classes of motor imagery tasks, namely baseline activity and imagined activities of right-hand, left-hand, both-hand, and both-leg movements. Although the database includes EEG signals of tasks where movements were actually realized, we only used data from imagined movements to train and test the systems, as tetraplegic people are the target patients for BCI research for whom only imagined movements are available. During experiments, we used four-way classification for the following classes: right-hand, left-hand, both hands, and both legs. There are 90 motor imagery trials for each subject.

2.2. The FASTER Algorithm

The Fully Automated Statistical Thresholding for the EEG artifact Rejection method (FASTER algorithm) was designed by H. Nolan et al. [12]. In this chapter, the steps of this method are explained, with particular attention paid to the parts we have modified to the original algorithm.

We employed a method comprising four sequential steps, with a deviation from the original algorithm, by excluding the final artifact detection step across subjects. We chose to omit this step to avoid excluding any subjects based on artifact contamination. The algorithm utilized statistical criteria to identify channels and components that exhibited deviations exceeding three times the standard deviation of the computed parameters.

In preparation for algorithm application, frequency filtering was implemented, employing a 5th-order Butterworth filter. While the original article specified a frequency range from 1 to 95 Hz with a notch filter at 50 Hz, our study deviated by utilizing distinct frequency ranges tailored to our specific experimental parameters and objectives. The exact frequencies are detailed in Section 2.5.

The initial step of the algorithm involved the identification of globally artefactual channels. Channels exceeding predetermined thresholds for variance, the mean of the channel's correlation coefficients with other channels, or Hurst exponent parameters were flagged as faulty. Subsequently, the algorithm proceeded to eliminate epochs containing artifacts. The examined parameters for this step included amplitude range in epoch, deviation from each channel's average value, and variance in each epoch. The third step involved the utilization of ICA to segregate time-dependent data into statistically independent waveforms. During this process, epochs and channels labeled as defective were disregarded. A transformation was performed by using the fast-ICA algorithm. Components displaying excessive correlation with the signal of electrodes proximal to the ocular region were omitted in the resultant space, as well as components that failed to meet the Z-score criterion for kurtosis, power gradient, Hurst exponent, and median gradient parameters. In the last step, defective channels were determined on an epoch-by-epoch basis, while subsequently, both globally and individually impaired channels were replaced through the spherical spline interpolation technique. Ultimately, the data were referenced to the average of all scalp electrodes.

2.3. Feature Extraction and Classification

We tested different models to assess the classification accuracy. The time domain EEG was tested using EEGNet and Shallow Net. To check the effect of adding the spatial dimensions, we used the Multi-branch Conv3D approach and two newly developed models based on our dense representation of input data.

2.3.1. 3D Representation of EEG Signals

Given that the relative positions of the electrodes can carry relevant information for the classifier, we used a three-dimensional representation (two spatial dimensions and one temporal dimension) of the EEG signals as the input tensor, similar to previous studies [36–38]. Our goal was to examine the impact of three-dimensional representation on classification performance. For this transformation, we used a unique arrangement of electrode placements, referred to as dense 3D transformation. In this construction, the Iz electrode is omitted, and the remaining 63 electrodes are rearranged into a 9×7-dimensional rectangle. The arrangement can be seen in Figure 2. This arrangement will be the input shape for 2D and 3D convolutional neural networks.

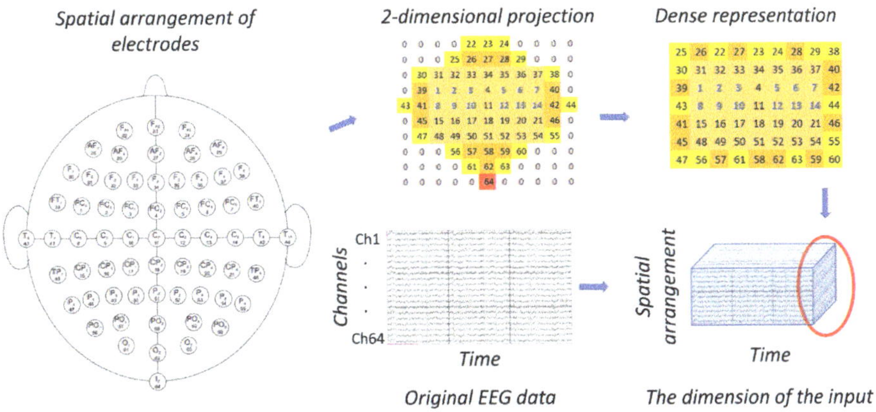

Figure 2. 3D representation of the EEG data: the channels × time initial arrangement of the EEG data is rearranged into a three-dimensional form, where the first two dimensions refer to the spatial arrangement of the electrodes, while the third dimension is time. For the spatial dimension, we used a dense electrode arrangement. While the original distribution contains placeholder zero channels, we rearranged this map to have a dense representation with no such cells. We designed this input representation to investigate the effect of explicit involvement of the spatial arrangement of electrodes on the classification performance.

2.3.2. 3D CNN (Conv2D)

The first network we developed to explore the effect of 3D representation was built of 2D-convolutional layers over the original 3D data. In this operation, kernels have a dimension of 3, where the size of the 3rd dimension is equal to the number of timesteps on the time/feature dimension. With this method, the system performs convolution on the spatial arrangement, whereas, on the last axis, all data points are summed with certain weights. As the first step, L2 normalization is performed, followed by three layers of two-dimensional convolution. After the last step, the flattened data are given as input to two layers of a fully connected network, and finally, a softmax layer is responsible for the classification. The architecture can be observed in Figure 3.

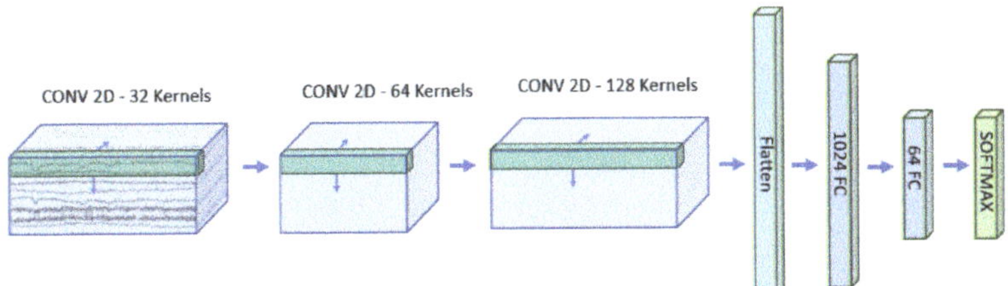

Figure 3. The structure of the designed 2D convolutional neural network: The input is the 3D representation of the EEG signal, on what 2D convolution is performed with 32 kernels of the size of [3 × 3 × number of timepoints] in the first layer. In the next two layers of convolution, firstly 64, then 128 kernels of dimension [3 × 3 × number of kernels of the previous layer] are performed. Next, the flattened representation is given to two layers of a fully connected network with 1024 and 64 neurons, and finally, a softmax layer is responsible for classification with the output of four numbers as the number of classes.

2.3.3. 3D CNN (Conv3D)

Networks with 3D convolution are widely used for video-processing tasks, and numerous articles propose this architecture for EEG classification as well [38,39]. The difference from the previous structure is that, in this case, convolution is performed even on the third dimension, resulting in a four-dimensional structure on the second layer. This requires greater memory usage but has the advantage of exploiting features scattered in time. The designed structure starts with L2 normalization. Afterward, three layers of 3D convolution are performed, with kernel dimensions [1,30] of the first layer and [2,40] for the upper layers. Between convolutions, a Batch normalization and an ELU operation are performed. As in the previous case, two layers of fully connected layers and a softmax layer are responsible for classifying the input EEG signals. Figure 4 contains the structure of this classifier.

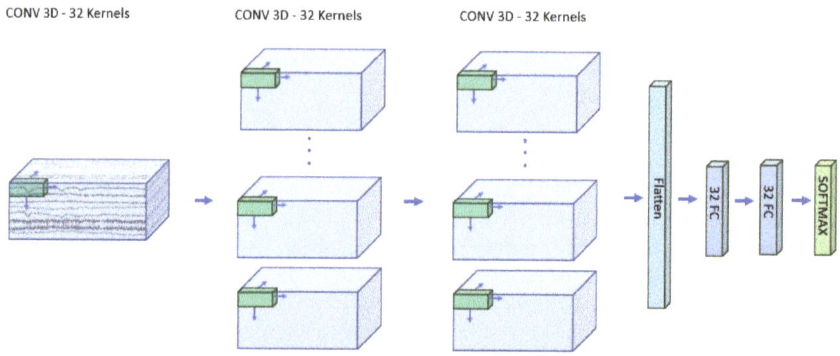

Figure 4. The structure of the 3D convolutional network. The input is identical to that of the 2D CNN. The difference is that in this structure, three-dimensional convolution is performed. There are 32 kernels in all the layers. The shape is $(1 \times 1 \times 30)$ in the first and $(2 \times 2 \times 40)$ in the second and third layers. The fully connected part consists of two layers with 32 neurons each, and finally, a softmax classification layer.

2.3.4. Multi-Branch 3D CNN

The second type of implemented 3D CNN is based on the article of Xinqiao Zhao et al. [36]. It consists of three branches, all of them with two layers of different dimensions of convolutional kernels and three layers of the fully connected network. The last layers are the size of the number of classes; thereafter, the outputs of all three branches are added up, and a softmax operation is performed for the final classification.

2.3.5. EEGNet

The fourth network we used in our study is one of the most well-known classification systems for MI-signal classification, the EEGNet, presented by V. J. Lawhern et al. [40]. We used half of the sampling frequency as the length of the kernel of the first layer without applying any further modifications. The main layers of the network are a 2D convolutional layer, a depth-wise 2D convolutional layer, a separable 2D convolutional layer, and finally, a fully connected classification layer.

2.3.6. Shallow Convolutional Neural Network

Finally, we evaluated the performance of the Shallow Convolutional Neural Network (Shallow ConvNet) architecture proposed by Schirmeister et al. [41]. This widely recognized algorithm has been extensively utilized in numerous studies due to its effectiveness [42,43]. This network consists of a temporal convolution, a spatial convolution, a mean pooling layer, and a linear classification FC part. The network was implemented using the source code provided in [40]. However, our implementation employs several modified parameters relative to the originally published article.

2.4. Transfer Learning and Fine-Tuning

A deep learning system usually requires a great amount of data to generalize features well and to have reasonable accuracy. Systems for EEG classification only acquire a limited number of samples, as recording and labeling these signals is cumbersome, requires a significant amount of time to process, and requires human intervention. The main idea of transfer learning (TL) application is pre-training a system—or a part of the system—over an independent dataset and transferring these weights as an initial state. These weights are fine-tuned during the actual training phase of the network. In the case of EEG studies, two main types of transfer learning are used [44]. One of the TL approaches is when the feature space generated from the EEG data is similar to one of those tasks for which a much larger dataset is available. An example is ImageNet, which can serve as an initial system for classifying EEG samples transformed into images [45]. The second approach, which we also used in this study, is pre-training the network on subjects that differ from the actual subject we test on. In other words, we fine-tune the globally learned network with the data from certain subjects. Multiple studies used this technique, and it yielded significant improvements in classification accuracy [2,42,44].

In our study, we divided the 105 subjects of the Physionet database into two parts: a Neural Network was pre-trained over 90% of subjects, and to address inter-subject variability, the classifier was fine-tuned in the remaining subjects individually. There are 105 subjects, and we did not repeat the process 105 times, but rather 10 times, with 10% of the subjects belonging to the test set in each iteration. During pre-training, we allocated 20% of the training data as a validation set and implemented early stopping with a patience value of 15. For fine-tuning, we trained the networks for 15 epochs without using a validation dataset, as the fine-tuning dataset contained relatively few data points, making it challenging to achieve reliable validation results. The exact method for training can be observed in Figure 5.

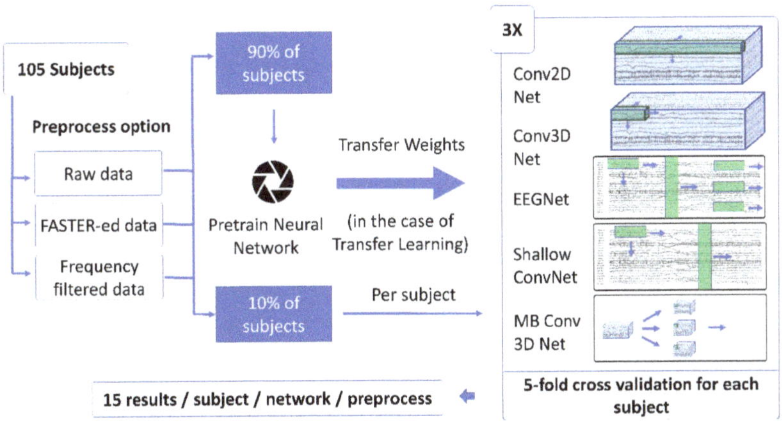

Figure 5. Transfer learning and the generation of results. To have statistically significant results, we performed a 5-fold cross-validation for each subject three times, resulting in 15 results for each subject, network, and preprocess option. 90% of the subjects' data is used for pre-training the classifier, which is fine-tuned over individual subjects' training sets. The whole process iterates 10 times for every subject to be tested.

2.5. Effect of Frequency Filtering

In the final phase of our study, we conducted experiments to find out the dependence of accuracy on the lowest frequency ranges (0.1 to 5 Hz) relative to the broad range of 5 to 75 Hz. We also examined the whole range of 0.1 to 75 Hz frequency band and the differences when no filtering was performed at all. To carry out this examination, we employed 5th-order band-pass Butterworth filters. The Butterworth filter, renowned for its characteristic of being maximally flat, ensures a uniform magnitude response within the pass band. However, a noteworthy drawback of this filter is the substantial width of its transition band.

2.6. Cropped Training

To investigate the generalizability of EEG signals with different onsets, we implemented cropped training, a method also employed by Schirrmeister et al. [41]. In this approach, the training and testing data are augmented by generating new samples through systematic time shifts of the original data. This ensures that neural networks are exposed to a broader range of temporal variations within the EEG signals, helping to simulate different onset timings and improve model generalization. In this scenario, we used 1 s long windows instead of the previous 2 s ones and performed shifts with 0.1 s long steps until 2 s. This gives 11 overlapping crops of the same epoch. Samples from a single epoch go solely to the train or solely to the test set. We repeated the previously described experiment regarding frequency dependence in this augmented dataset with all the networks, using or neglecting the FASTER method.

3. Results

3.1. The Effect of Artifact Rejection

As can be observed in Figure 6 and Table 1, without using the TL method, the average of the classification accuracy of Conv 2D, Conv 3D, Shallow ConvNet, and Multi-branch Conv3D Net models are significantly improved due to the FASTER algorithm while regarding the EEGNet there was no significant improvement. In this scenario, we used 2 s long windows, and during the FASTER algorithm, a frequency filter between 0.1 and 75 Hz was applied. For the raw data, we did not use any frequency filtering.

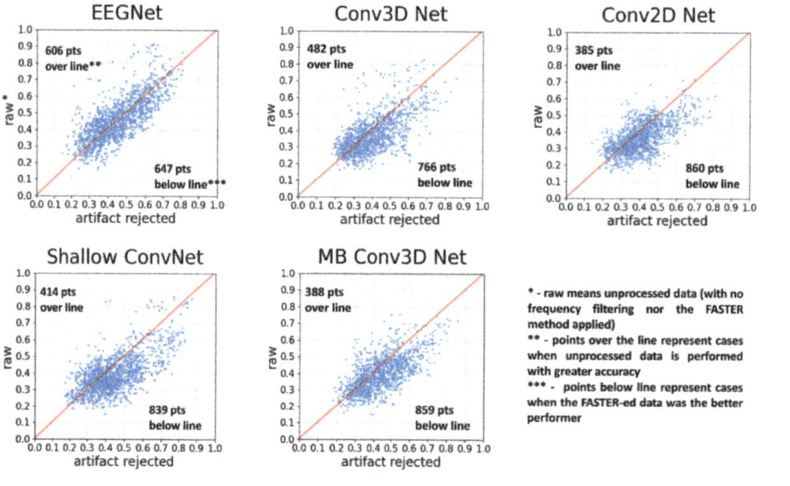

Figure 6. The comparison of neural networks with and without using the FASTER method. Each point represents two accuracies for a subject: the one obtained with AR and the other obtained without using the FASTER algorithm. The test was run 12 times to obtain significant results. The red line indicates the points with no difference between the two options, points over the line run with the raw option as the more accurate, and points below the line where the artifact-rejected version yields better results.

Upon subject-specific scrutiny, it becomes evident that the influence of artifact rejection is contingent on the particular subject under evaluation. Our analysis encompassed a comprehensive approach. Initially, we executed 5-fold cross-validation three times for all four networks, both with and without artifact rejection. To substantiate disparities, we scrutinized the distribution of 15 results obtained for an individual subject with a specific network. If the data followed a normal distribution, we executed a t-test; conversely, if non-normality was detected, a Man–Whitney U test was performed to ascertain the significance of the observed differences.

Table 1. The accuracy of each classifier with and without artifact rejection, and the p-values of the significance of differences using Wilcoxon signed ranked test. The p-values determine the significance of differences between omitting or using the FASTER method for artifact rejection.

Classifier	Original Acc.	AR Acc.	p-Value
EEGNet	0.460	0.455	0.907
Shallow ConvNet	0.394	0.439	3.46×10^{-17}
Conv2D Net	0.367	0.411	4.90×10^{-21}
Conv3D Net	0.378	0.405	4.09×10^{-09}
Multi-branch Conv3D Net	0.401	0.444	2.05×10^{-19}

We categorized subjects based on the extent of change in the corresponding classification performance across the various networks. Intriguingly, several scenarios arose in which certain networks led to a notable enhancement, while others yielded a significant decline in the performance for the same subjects. In response to a slight variance in results observed during a second examination, we iteratively conducted the calculations two times more to explore the evolving significance of the observed differences. Finally, we had 4 times (3 accuracy results for each cross-fold iteration and network), meaning four results of significance for each classifier.

In our assessment, to each subject, a numerical score was assigned, denoted as follows: a value of -1 indicated a significant decline, 0 denoted no significant difference, and +1 represented a discernible increase observed for each computational aspect across all networks. Therefore, the cumulative score per subject ranged from -20 to 20. Twenty-five subjects scored over 8, indicating substantial performance gains from the AR method, with Subject 69 achieving a remarkable 20-point increase. Conversely, some subjects, like Subject 15, experienced significant declines, with only five subjects scoring below -8. Regarding classifiers, EEGNet showed the least improvement, with only three subjects scoring at least 3, while six scored below -3. In contrast, the other four networks had at least 20 subjects exceeding the three-point mark, with fewer than five scoring -3 or worse. The subject dependence on AR is illustrated in Figure 7.

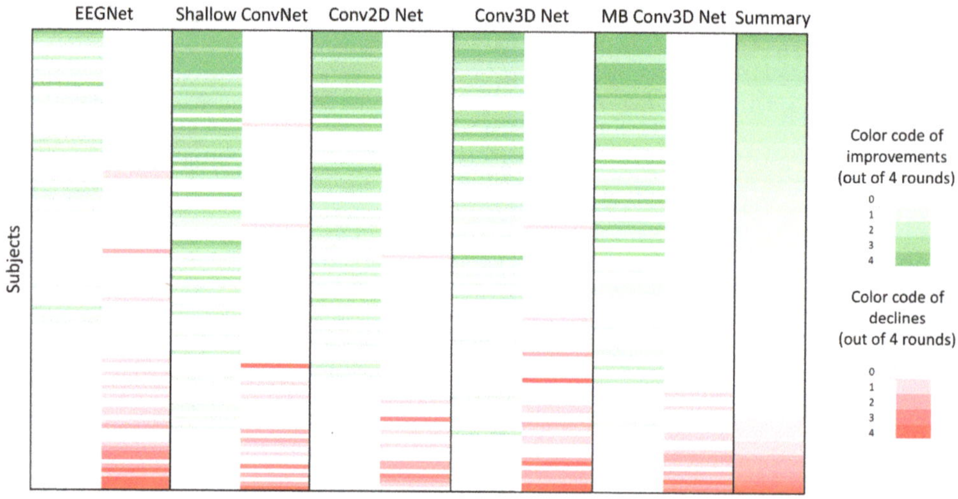

Figure 7. Subject dependence of artifact filtering. The shade of color yields on how many runs there were statistically significant improvement (green) or significant decline (red). (The deeper the shade, the more runs were significant.) Significance is based on the *t*-test in the case of normality and the Man–Whitney U test in the case of non-normality. Subjects are ordered by summary number, computed as the sum of the significance of improvements minus the sum of the significance of decline.

3.2. The Effect of Transfer Learning

The classification accuracy obtained by transfer learning was significantly better in every scenario. Based on the signed-rank Wilcoxon test, the learning process performs significantly better in the case of each network, both in the case of unfiltered data (Table 2) and artifact-rejected data (Table 3). In Figure 8, it can be observed that transfer learning does not improve as much in cases of artifact-rejected data as in the case of raw data, resulting in higher classification accuracies in the latter case. (This difference is significant in all the cases except the Multi-branch Conv3D Network.)

Table 2. The accuracy of each classifier on unfiltered data, with and without transfer learning, and the *p*-values of significance using the Wilcoxon test. The highest accuracies and the largest difference are highlighted in bold.

Classifier	Simple Acc.	TL Acc.	Difference	*p*-Value
EEGNet	**0.461**	0.587	0.126	1.50×10^{-18}
Shallow ConvNet	0.394	**0.637**	**0.243**	5.83×10^{-19}
Conv2D Net	0.366	0.528	0.162	7.78×10^{-19}
Conv3D Net	0.378	0.56	0.182	7.14×10^{-19}
Multi-branch Conv3D Net	0.401	0.561	0.16	1.30×10^{-18}

Table 3. The accuracy of each classifier on artifact-rejected data, with and without transfer learning, and the *p*-value of significance using the Wilcoxon test. The highest accuracies and the largest difference are highlighted in bold.

Classifier	Simple Acc.	TL Acc.	Difference	*p*-Value
EEGNet	**0.455**	0.538	0.083	1.67×10^{-17}
Shallow ConvNet	0.441	**0.559**	**0.118**	1.38×10^{-18}
Conv2D Net	0.41	0.491	0.081	7.20×10^{-17}
Conv3D Net	0.405	0.521	0.116	7.14×10^{-19}
Multi-branch Conv3D Net	0.444	0.557	0.113	5.52×10^{-18}

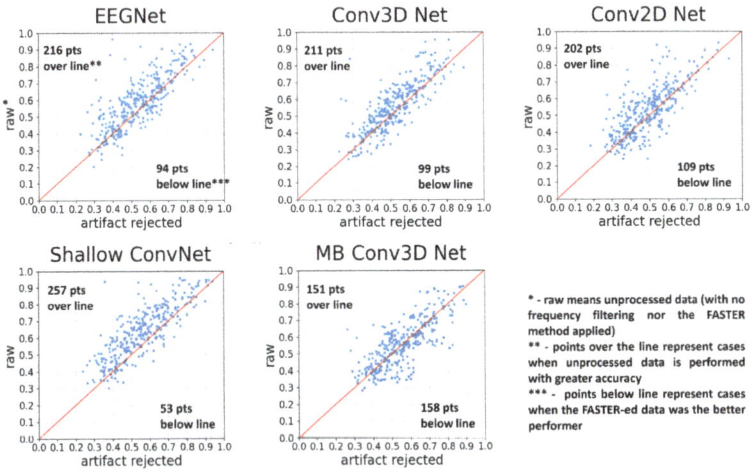

Figure 8. The comparison of neural networks with and without using the FASTER method, using transfer learning. Each point represents two accuracies for a subject: the one obtained with AR and the other obtained without using the FASTER algorithm. The test was run three times. The red line indicates the points with no difference between the two options, points over the line run with the raw option as the more accurate, and points below the line where the artifact-rejected version yields better results. Generally, the usage of transfer learning accuracies without artifact rejection tends to be higher.

3.3. Comparison of Neural Networks

In network comparisons, the primary evaluative criterion centers on classification accuracy. As illustrated in Figure 9, when transfer learning is not employed, the highest classification accuracy is achieved by the EEGNet classifier, both in raw and artifact-rejected conditions. However, in the latter case, the differences between the EEGNet, Shallow ConvNet, and MB Conv3D Net were not significant. In the raw data scenario, the Multi-branch Conv3D Net is the second-best performer, followed by the Shallow ConvNet, with our proposed networks trailing.

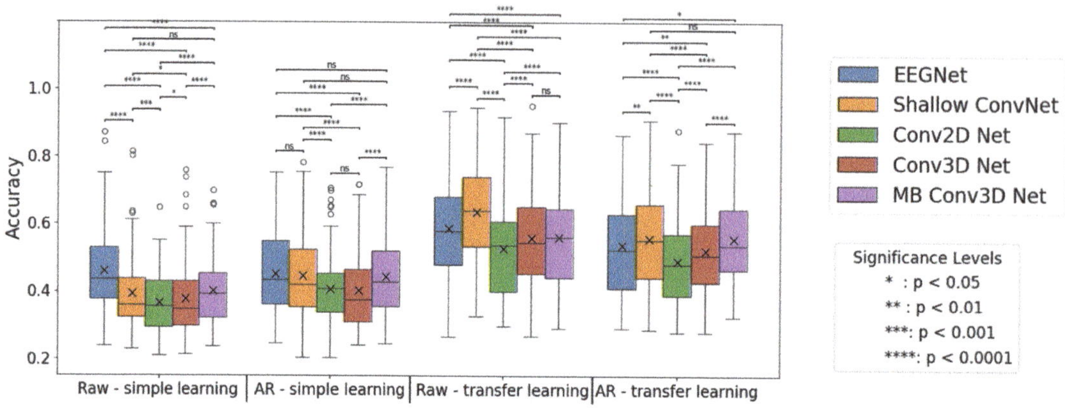

Figure 9. The accuracies obtained by neural networks, with and without transfer learning and artifact-rejection. In the box plot, x denotes the mean, while the horizontal line marks the median value. The limits of the box indicate the range of the central 50% of the data. Significance levels were determined using the Wilcoxon test. (The notation 'ns' denotes non-significant differences.) Before applying the TL method, EEG Net was the greatest performer, while after the application, Shallow ConvNet had the highest accuracy. These observations are generally true for the raw and the artifact-rejected data as well. However, for AR-ed data without TL, this difference is non-significant. The figure also presents that without transfer learning, networks generally perform better with the FASTER method, while after the application, raw data will attain a higher precision.

However, the performance landscape shifts with the introduction of TL. Shallow ConvNet emerges as the top performer, surpassing EEGNet and other classifiers. In the raw data scenario with TL, EEGNet ranks second, followed by the 3D convolutional networks. In the artifact-rejected condition with TL, the MB Conv 3D CNN outperforms EEGNet, securing the shared first position, with non-significant differences from the Shallow ConvNet. This demonstrates that TL significantly influences the performance hierarchy among different network architectures.

3.4. *The Effect of Input Representation*

Using the dense 3D representation for input did not yield significant improvements in classification performance. This indicates that incorporating spatial information did not lead to higher accuracies compared to networks without this additional spatial data. The only exception is with the artifact-rejected data using transfer learning, where the MB Conv 3D CNN slightly outperformed the EEGNet.

3.5. The Effect of Frequency Filtering

The outcomes stemming from the frequency filtering analysis have yielded unexpected findings. We compared results using the 0.1 to 5 Hz range and the 5 to 75 Hz range. As observed in Table 4, three out of the five neural networks yielded significant results only within the first frequency range (0.1–5 Hz). Shallow ConvNet and EEGNet are the only networks for which the 5 to 75 Hz range also provides results greater than the chance level. However, for EEGNet, these results were still far below the accuracy achieved in the lower frequency range.

Table 4. Results of comparing the classification accuracies of neural networks when filters of certain frequency ranges are applied. The highest classification accuracy values for each frequency range are emphasized in bold.

Frequency Range	EEGNet	Shallow ConvNet	Conv2D Net	Conv3D Net	MB Conv3D Net
0.1–5 Hz—Raw	**0.482**	0.406	0.437	0.410	0.457
5–75 Hz—Raw	0.315	**0.362**	0.262	0.258	0.271
0.1–5 Hz—AR	0.452	0.405	0.421	0.436	**0.462**
5–75 Hz—AR	0.324	**0.381**	0.261	0.261	0.289

3.6. The Effect of Simple and Cropped Training

The results derived from simple and cropped learning can be observed in Figure 10. Upon comparing results from the simple learning process, we can state that those experiments, when 5 Hz to 75 Hz frequency filtering was performed, had the lowest accuracies. Only the Shallow ConvNet attained moderately higher accuracy, as we have observed in the previous paragraph. The relation between AR data (with 0.1 to 75 Hz frequency filter) and raw data also remains the same, as we have seen in the first part of the results: classification accuracy is enhanced by the application of artifact rejection in the case of four out of five networks, with EEGNet as the only exception. When we apply the mentioned frequency filter without the FASTER algorithm, the case of Shallow ConvNet becomes similar to the EEGNet, in the sense of being the non-artifact-rejected option, the better performer. The partly unexpected finding is the superior performance of cases where frequency filtering between 0.1 and 5 Hz was applied. These signals include only the lowest part of the frequency spectrum, meaning that they do not contain the mu or beta band.

When applying cropped training, we acquired noteworthy results, depending on the network we used for classification. Contrastingly to the simple learning approach, in the specific context of the Shallow ConvNet and the EEGNet, an inverse trend was observed where the highest performances were achieved in the broader frequency range spanning from 5 to 75 Hz. On the other hand, the worst results were attained by the higher frequency range for the other three networks, where the input feature was the 3D representation. These results show that the frequency range of the mu, beta, and gamma bands are the most valuable in the case of EEGNet and Shallow ConvNet, while the other three networks are not able to extract those features effectively. It is important to emphasize that, for the two widely used networks to achieve better results at higher frequencies, as reported in the literature, cropped training was necessary.

Within the realm of neural networks, the integration of cropped training demonstrates a parallel effect akin to the observed outcomes with transfer learning. In the absence of cropped training, EEGNet emerges as the preeminent performer. However, upon application, there is a discernible transition where Shallow ConvNet surpasses EEGNet in terms of classification prowess.

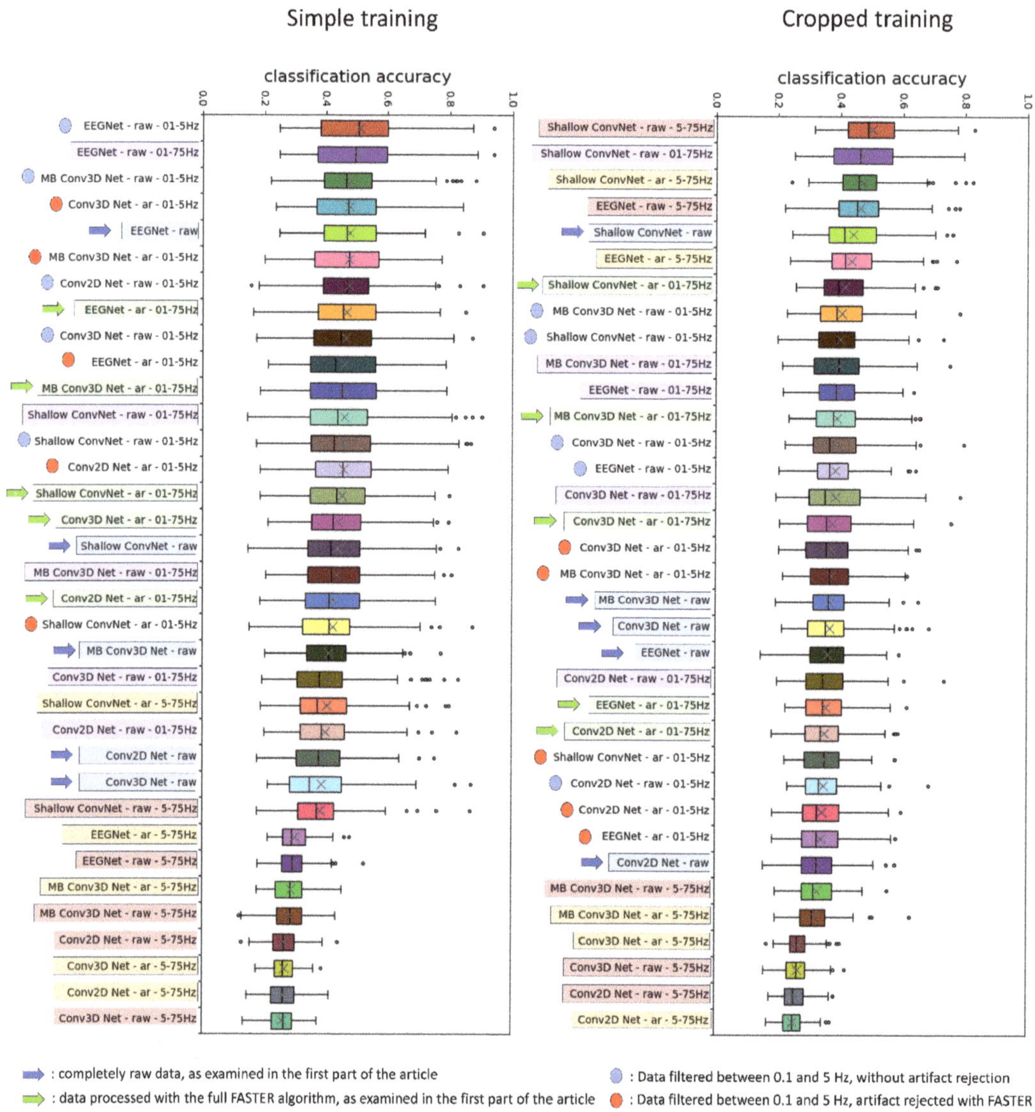

Figure 10. The results obtained with and without cropped training ordered by the classification accuracy. Arrows mark the cases examined in the first part of the article: the differences between the FASTER-ed (green arrow) and the fully raw data (blue arrow). The relation between the two datasets remains the same, as only the EEGNet yields a better raw classification accuracy. However, during cropped training, Shallow ConvNet also joined this group. Dots marked the results when frequency filtering was presented between 0.1 and 5 Hz. (Red dots for artifact rejected and blue dots for raw data). Without cropped training, these results tend to be unexpectedly high, yielding that this low-frequency range contains the most relevant information for MI classification on the Physionet database when subject-vise learning is performed without any augmentation of the training dataset. In the other cases, the color of the bars represents the corresponding preprocessing option. The most noteworthy result is that while during simple learning, almost every 5–75 Hz filtered attempt was presented on the lowest part of the list, the cases of EEGNet and Shallow ConvNet jumped to the first places of the cropped training results, corresponding to the extensive literature.

3.7. Subject Dependence in Simple Learning and Cropped Training

We extended our investigation to explore subject dependence within the context of cropped training. This involved conducting a series of experiments, comprising four sets of three trials each, on both the EEGNet and Shallow ConvNet systems. We selected these two networks because only these two classifiers achieved reasonable accuracy on the frequency-filtered data during cropped training. We meticulously assessed subject-wise performance, focusing on the frequency range of 5 Hz to 45 Hz for the FASTER method. As there were 4–4 significance values for both networks, each subject was assigned a point ranging from −8 to 8, facilitating a comparative analysis between FASTER-processed and completely raw data.

Furthermore, alongside these comparisons, we computed accuracies based on frequency-filtered raw data spanning the 5 Hz to 45 Hz range and juxtaposed these against both raw and FASTER-processed data. These findings are synthesized and presented in Figure 11, providing valuable insights into the relative efficacy of different preprocessing techniques across various subjects. As can be observed, the FASTER-applied and frequency-filtered data exhibit similar subject-wise distributions when compared to the raw data, indicating that both methods have comparable effects on subject dependence. However, when comparing these two methods, it is evident that frequency filtering alone yields significantly higher accuracy results. Another important factor to note is that the set of subjects whose classification accuracies improved during cropped training differs substantially from those who showed improvements in the simple training process.

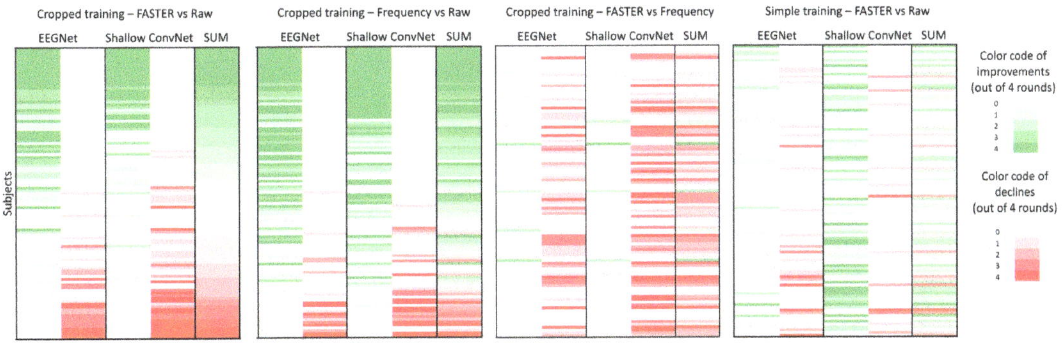

Figure 11. Subject-wise effect of the FASTER artifact rejection method and the frequency filtering on the classification performance of the EEGNet and Shallow ConvNet, related to the raw data, during cropped and simple training. The shade of the color means the number of significant increases (green) or declines (red) out of the four tests. Subjects are ordered by the score achieved during the AR method compared to the unfiltered data. The order of the subjects is almost the same in the second graph, meaning that frequency filtering has a similar subject-wise effect as the FASTER algorithm related to the raw data. However, this sequence is merely different from the result from the first part of the article, meaning that during cropped training, it changes on which subjects the AR has positive effects. The sequence is also different on the FASTER compared to the plain frequency filtering graph, where, for most of the subjects, the AR method caused a significant decline. It means that during cropped training, it is more advised to filter only the appropriate frequency ranges and not to use the more complex FASTER method to obtain better results.

4. Discussion

As can be seen from the results, classification accuracy depends on many factors. The subject dependence of the efficacy of artifact rejection is a noteworthy result. The question that emerges is, what is the reason behind these differences? As each subject has a different scale of artifact contamination, the reasoning can be the amount and quality of contained artifacts. However, it is only partially true because, for instance, subject 15 has the greatest accuracy decline after artifact rejection, but if we examine it from closer, multiple artifactual components were rejected there. The real reason can be the number of rejected components on which neural networks can base their classification. If the data from a certain subject contain artifactual yet useful components, the accuracy will decline as a result of the application of the FASTER method.

The application of transfer learning yields a noteworthy improvement for each classification task. This beneficial effect of the process is well-documented in the existing literature [45–47]. What is interesting to note is that transfer learning improves the raw data classification to the greatest extent, and results obtained this way surpass the accuracy achieved on the artifact-rejected data. A plausible underlying explanation for this phenomenon may be that, with a substantial amount of training data, neural networks can learn to disregard artifacts and focus on neuronal processes. Conversely, the FASTER artifact rejection method may filter out components that are relevant and could have been utilized by neural networks for learning.

The best accuracy is achieved by the Shallow ConvNet with raw data as input and transfer learning, replacing the EEGNet, which was the classifier with the highest accuracy when transfer learning was not applied. This confirms Shallow ConvNet's superior performance in transferring weights from the broader dataset, as was also described earlier [46].

As it was presented, adding the spatial dimension using dense 3D representation did not lead to improvements in most cases. This is likely because networks like Shallow ConvNet and EEGNet can inherently learn the spatial relationships between channels through their architectures. Shallow ConvNet achieves this via spatial filtering, while EEGNet uses DepthwiseConv2D layers. These methods allow the networks to autonomously identify and leverage the importance of spatial data and channel interactions, which proves to be more effective for classification performance than explicitly incorporating a dense 3D spatial representation.

Discussing results from the frequency filter section, the conclusion that can be drawn is when a limited number of data is given to training the classifier, such as in the case of subject-wise training, even the well-known and widely used Shallow ConvNet and EEGNet tend to concentrate on the lowest frequency ranges, which ranges can contain relevant information. This attention to lower ranges of frequency can be seen in Figure 10. An unexpected observation is that networks trained on the 0.1–5 Hz filtered dataset exhibited relatively high accuracies, often surpassing those trained on the 0.1–75 Hz filtered version. In the scenario of lowpass filtering, raw data yielded greater accuracies than the artifact-rejected data, indicating that the previously examined effect of AR does not hold under these circumstances. One possible explanation for the superior classification accuracy of the delta band is its inherently higher amplitude power compared to other frequency bands. This elevated power may provide the classifier with more prominent features, thereby enhancing performance. Additionally, the delta band is closely associated with attentional processes, which may contribute to the presence of underlying features that strongly influence classification accuracy. Furthermore, the limitations of training data without the use of cropped training or transfer learning might impair the networks' ability to effectively recognize the importance of higher-frequency bands, such as mu and beta. In this work, we demonstrated that the inclusion of cropped training significantly improves the classification of signals containing higher-frequency components, a result that aligns with prior findings in the literature.

It is important to note that when comparing raw data with the artifact-rejected dataset filtered in the 0.1–75 Hz range, we observe the same results: artifact rejection improves performance in four out of five networks, except EEGNet. However, when frequency filtering is applied to the raw data, the Shallow ConvNet without AR also outperforms the version where the FASTER method was applied. For the remaining networks, the performance relationship remains unchanged. Nevertheless, when cropped training was present, the overall picture changed. On one hand, the average classification accuracy of the networks decreases with this method. This decline can be attributed to the test set containing an extended range of data, making it more challenging for the classifiers to produce accurate results. On the other hand, via this method, we can acquire more data to train the classifier—as in the case of transfer learning. This extended version of the dataset is enough for the EEGNet and the Shallow ConvNet to learn from the higher frequency ranges—as expected from the literature. In investigating this aspect, it is shown that EEGNet and Shallow ConvNet are more capable of generalizing data, as they effectively leverage the variability introduced by cropped training. This demonstrates their ability to adapt to different temporal onsets in the EEG signals, a crucial factor in improving classification accuracy and model robustness.

Figure 11 demonstrates a notable consistency in the performance trends observed during cropped training across networks trained on frequency-filtered and FASTER-processed datasets, exhibiting a similar propensity for performance enhancement in comparison to the raw dataset. This consistency suggests that, in this case, the performance enhancement caused by the FASTER algorithm can be explained by the effect of frequency filtering. Moreover, if we compare the artifact rejection method to the frequency-filtered performance, the latter gets significantly higher accuracies in the vast majority of subjects. That means that with a simple frequency filter, we can achieve better performance during cropped training than with a complex AR method. Another important observation to note is that the subjects that had higher accuracies with the FASTER method compared to the raw data during cropped training differ from the subjects that were obtained by simple learning. This could be an effect of the phenomenon that during cropped training, different parts of the original signals are simultaneously presented, and neural networks can learn about other factors, some of which are filtered out during artifact rejection.

There are only a limited number of papers examining the subjects of the Physionet database regarding the effect of artifact rejection. Therefore, it is hard to give a thorough comparison. There are studies where filtering enhances the classification performance in various datasets [20,28]. Other articles report a decline in accuracy due to artifact rejection [21]. Subject dependence remains a significant factor, as evidenced not only in various articles exploring the BCI Competition IV Dataset 2a [25,48] but also in our study focusing on the Physionet database. Our findings highlight that the efficacy of the FASTER artifact rejection method in terms of classification accuracy is profoundly influenced by the specific subject under consideration.

For the case of frequency dependence, Hauke Dose et al. [42], who also examined the Shallow ConvNet's accuracy on the Physionet database, concluded that this architecture tends to concentrate on the lowest part of the frequency domain. In their experiment, they analyzed the squared frequency responses of the learned temporal filters, and the mean focused on the lowest frequency range (below 10 Hz). These results correspond to our findings. The frequency results obtained through cropped training and the use of Shallow ConvNet and EEGNet align with the broader literature. As imagery movements are described to be mostly classifiable in the mu and beta ranges [49,50], it is expected that signals filtered between 5 and 75 Hz exhibit higher (or at least similar) accuracies compared to unfiltered signals.

The study reveals how small changes in the preprocessing pipeline can significantly impact classification accuracy, underscoring the need for tailored solutions in EEG-based BCI systems. As discussed by Xu et al. [51], while much progress has been made in neural interface research, translating these advancements into reliable, real-world applications remains challenging. This research contributes to bridging that gap by optimizing processing techniques that can enhance the practicality of BCIs in neurorehabilitation and beyond.

5. Conclusions

In conclusion, our research indicates that the FASTER method can enhance performance in a subject and network-specific manner. There are subjects where the application of AR comes with an efficiency increase, while in other cases, it comes with the deterioration of the results. Transfer learning proved to be effective in improving the performance of all networks in both raw and artifact-rejected data. However, it was noted that the accuracy of classification for artifact-rejected data did not improve as significantly as it did for the unfiltered data, resulting in less precision. Our findings also revealed an unexpected outcome from frequency filtering, as the tested networks demonstrated strong classification performance based on the low-frequency components during learning. Notably, we observed that higher frequency ranges were more discriminative in the case of EEGNet and Shallow ConvNet when cropped training was applied. In summary, the study underscores the intricate interplay between processing techniques and neural network performance, highlighting the necessity for tailored processing approaches designed for specific subjects and network architectures.

Author Contributions: Conceptualization, A.A., C.M.K., G.M. and W.F.; methodology, A.A.; software, A.A. and C.M.K.; validation, A.A. and I.U.; formal analysis, A.A.; investigation, A.A.; resources, A.A. and C.M.K.; data curation, A.A.; writing—original draft preparation, A.A.; writing—review and editing, C.M.K., G.M., W.F. and I.U.; visualization, A.A.; supervision, I.U.; project administration, I.U.; funding acquisition, I.U. All authors have read and agreed to the published version of the manuscript.

Funding: Prepared with the professional support of the Doctoral Student Scholarship Program of the Co-operative Doctoral Program of the Ministry of Innovation and Technology, financed from the National Research, Development and Innovation Fund. Project no. FK146115 has been implemented with the support provided by the Ministry of Innovation and Technology of Hungary from the National Research, Development, and Innovation Fund, financed under the FK_23 funding scheme. The research was also funded by the Hungarian Brain Research Program Grant (NAP2022-I-2/2022) and the Pharmaceutical Research and Development Laboratory Project (PharmaLab, RRF-2.3.1-21-2022-00015).

Institutional Review Board Statement: Not applicable.

Informed Consent Statement: Not applicable.

Data Availability Statement: We used the publicly available Physionet dataset, which can be accessed by the following link: https://physionet.org/content/eegmmidb/1.0.0/ (accessed on 13 December 2023).

Acknowledgments: During the preparation of this work, the authors used ChatGPT (version GPT-4-turbo) in order to improve grammar and readability. After using this tool, the authors reviewed and edited the content as needed and took full responsibility for the content of the publication.

Conflicts of Interest: The authors declare no conflicts of interest.

References

1. Wolpaw, J.R.; Birbaumer, N.; McFarland, D.J.; Pfurtscheller, G.; Vaughan, T.M. Brain–computer interfaces for communication and control. *Clin. Neurophysiol.* **2002**, *113*, 767–791. [CrossRef] [PubMed]
2. Nicolas-Alonso, L.F.; Gomez-Gil, J. Brain Computer Interfaces, a Review. *Sensors* **2012**, *12*, 1211–1279. [CrossRef] [PubMed]
3. Fatourechi, M.; Bashashati, A.; Ward, R.K.; Birch, G.E. EMG and EOG artifacts in brain computer interface systems: A survey. *Clin. Neurophysiol.* **2007**, *118*, 480–494. [CrossRef]
4. Jiang, X.; Bian, G.-B.; Tian, Z. Removal of Artifacts from EEG Signals: A Review. *Sensors* **2019**, *19*, 987. [CrossRef] [PubMed]
5. Mannan, M.M.N.; Kamran, M.A.; Jeong, M.Y. Identification and Removal of Physiological Artifacts from Electroencephalogram Signals: A Review. *IEEE Access* **2018**, *6*, 30630–30652. [CrossRef]
6. Delorme, A.; Makeig, S.; Sejnowski, T. Automatic artifact rejection for EEG data using high-order statistics and independent component analysis. In Proceedings of the Third International ICA Conference, San Diego, CA, USA, 9–12 December 2001; Available online: https://sccn.ucsd.edu/~arno/mypapers/delormefinal01.pdf (accessed on 16 June 2023).
7. Radüntz, T.; Scouten, J.; Hochmuth, O.; Meffert, B. EEG artifact elimination by extraction of ICA-component features using image processing algorithms. *J. Neurosci. Methods* **2015**, *243*, 84–93. [CrossRef] [PubMed]
8. Winkler, I.; Brandl, S.; Horn, F.; Waldburger, E.; Allefeld, C.; Tangermann, M. Robust artifactual independent component classification for BCI practitioners. *J. Neural Eng.* **2014**, *11*, 035013. [CrossRef] [PubMed]
9. Bajaj, N.; Carrión, J.R.; Bellotti, F.; Berta, R.; De Gloria, A. Automatic and tunable algorithm for EEG artifact removal using wavelet decomposition with applications in predictive modeling during auditory tasks. *Biomed. Signal Process. Control* **2020**, *55*, 101624. [CrossRef]
10. De Clercq, W.; Vergult, A.; Vanrumste, B.; Van Paesschen, W.; Van Huffel, S. Canonical Correlation Analysis Applied to Remove Muscle Artifacts from the Electroencephalogram. *IEEE Trans. Biomed. Eng.* **2006**, *53*, 2583–2587. [CrossRef] [PubMed]

11. Grubov, V.V.; Runnova, A.E.; Efremova, T.Y.; Hramov, A.E. Artifact removal from EEG data with empirical mode decomposition. In *Dynamics and Fluctuations in Biomedical Photonics XIV*; SPIE: Bellingham, WA, USA, 2017; Volume 10063, pp. 185–191. [CrossRef]
12. Nolan, H.; Whelan, R.; Reilly, R.B. FASTER: Fully Automated Statistical Thresholding for EEG artifact Rejection. *J. Neurosci. Methods* **2010**, *192*, 152–162. [CrossRef] [PubMed]
13. Zhong, J.; Qi, F. Study on the effect of artefact rejection on BCI performance. In Proceedings of the 2018 IEEE 23rd International Conference on Digital Signal Processing (DSP), Shanghai, China, 19–21 November 2018; pp. 1–5. [CrossRef]
14. Kim, M.; Kim, S.-P. A comparsion of artifact rejection methods for a BCI using event related potentials. In Proceedings of the 2018 6th International Conference on Brain-Computer Interface (BCI), Gangwon, Republic of Korea, 15–17 January 2018; pp. 1–4. [CrossRef]
15. Mohammadi, M.; Mosavi, M.R. Comparison of two methods of removing EOG artifacts for use in a motor imagery-based brain computer interface. *Evol. Syst.* **2021**, *12*, 527–540. [CrossRef]
16. Tangermann, M.; Müller, K.-R.; Aertsen, A.; Birbaumer, N.; Braun, C.; Brunner, C.; Leeb, R.; Mehring, C.; Miller, K.J.; Müller-Putz, G.R.; et al. Review of the BCI Competition IV. *Front. Neurosci.* **2012**, *6*, 55. Available online: https://www.frontiersin.org/articles/10.3389/fnins.2012.00055 (accessed on 19 December 2023). [CrossRef] [PubMed]
17. Schreuder, M.; Rost, T.; Tangermann, M. Listen, You are Writing! Speeding up Online Spelling with a Dynamic Auditory BCI. *Front. Neurosci.* **2011**, *5*, 11793. [CrossRef]
18. Blankertz, B.; Sannelli, C.; Halder, S.; Hammer, E.M.; Kübler, A.; Müller, K.-R.; Curio, G.; Dickhaus, T. Neurophysiological predictor of SMR-based BCI performance. *NeuroImage* **2010**, *51*, 1303–1309. [CrossRef]
19. Iqbal, M.; Rahman, M.M.; Shubha, S. Effect of EOG Artifact Removal on EEG Motor-Imagery Classification. In Proceedings of the 25th International Conference on Computer and Information Technology (ICCIT), Cox's Bazar, Bangladesh, 17–19 December 2022. [CrossRef]
20. Assi, E.B.; Rihana, S.; Sawan, M. 33% Classification Accuracy Improvement in a Motor Imagery Brain Computer Interface. *J. Biomed. Sci. Eng.* **2017**, *10*, 6. [CrossRef]
21. Thompson, D.E.; Mowla, M.R.; Dhuyvetter, K.J.; Tillman, J.W.; Huggins, J.E. Automated Artifact Rejection Algorithms Harm P3 Speller Brain-Computer Interface Performance. *Brain Comput. Interfaces Abingdon Engl.* **2019**, *6*, 141–148. [CrossRef] [PubMed]
22. Frølich, L.; Winkler, I.; Müller, K.-R.; Samek, W. Investigating effects of different artefact types on motor imagery BCI. In Proceedings of the 2015 37th Annual International Conference of the IEEE Engineering in Medicine and Biology Society (EMBC), Milan, Italy, 25–29 August 2015; pp. 1942–1945. [CrossRef]
23. Mannan, M.M.N.; Kamran, M.A.; Kang, S.; Jeong, M.Y. Effect of EOG Signal Filtering on the Removal of Ocular Artifacts and EEG-Based Brain-Computer Interface: A Comprehensive Study. *Complexity* **2018**, *2018*, e4853741. [CrossRef]
24. Daly, I.; Scherer, R.; Billinger, M.; Müller-Putz, G. FORCe: Fully Online and Automated Artifact Removal for Brain-Computer Interfacing. *IEEE Trans. Neural Syst. Rehabil. Eng.* **2015**, *23*, 725–736. [CrossRef] [PubMed]
25. Merinov, P.; Belyaev, M.; Krivov, E. The comparison of automatic artifact removal methods with robust classification strategies in terms of EEG classification accuracy. In Proceedings of the 2015 International Conference on Biomedical Engineering and Computational Technologies (SIBIRCON), Novosibirsk, Russia, 28–30 October 2015; pp. 221–224. [CrossRef]
26. van Stigt, M.N.; Camps, C.R.; Coutinho, J.M.; Marquering, H.A.; Doelkahar, B.S.; Potters, W.V. The effect of artifact rejection on the performance of a convolutional neural network based algorithm for binary EEG data classification. *Biomed. Signal Process. Control* **2023**, *85*, 105032. [CrossRef]
27. Chen, Y.; Zhang, C.; Wu, X. To Assess the Influence of Artifacts on Motor Imagery Based BCI. In Proceedings of the 2019 IEEE 4th International Conference on Signal and Image Processing (ICSIP), Wuxi, China, 19–21 July 2019; pp. 925–929. [CrossRef]
28. Islam, M.K.; Ghorbanzadeh, P.; Rastegarnia, A. Probability mapping based artifact detection and removal from single-channel EEG signals for brain–computer interface applications. *J. Neurosci. Methods* **2021**, *360*, 109249. [CrossRef]
29. Anjum, M.; Sakib, N.; Islam, M.K. Effect of artifact removal on EEG based motor imagery BCI applications. In Proceedings of the Fourth International Conference on Computer Vision and Information Technology (CVIT 2023), Beijing, China, 4–6 August 2023; p. 9. [CrossRef]
30. Al-Saegh, A.; Dawwd, S.A.; Abdul-Jabbar, J.M. Deep learning for motor imagery EEG-based classification: A review. *Biomed. Signal Process. Control* **2021**, *63*, 102172. [CrossRef]
31. Salazar-Varas, R.; Vazquez, R.A. Evaluating the effect of the cutoff frequencies during the pre-processing stage of motor imagery EEG signals classification. *Biomed. Signal Process. Control* **2019**, *54*, 101592. [CrossRef]
32. Goldberger, A.L.; Amaral, L.A.N.; Glass, L.; Hausdorff, J.M.; Ivanov, P.C.; Mark, R.G.; Mietus, J.E.; Moody, G.B.; Peng, C.-K.; Stanley, H.E. PhysioBank, PhysioToolkit, and PhysioNet: Components of a New Research Resource for Complex Physiologic Signals. *Circulation* **2000**, *101*, e215–e220. [CrossRef]
33. Schalk, G.; McFarland, D.J.; Hinterberger, T.; Birbaumer, N.; Wolpaw, J.R. BCI2000: A general-purpose brain-computer interface (BCI) system. *IEEE Trans. Biomed. Eng.* **2004**, *51*, 1034–1043. [CrossRef] [PubMed]
34. Fan, C.-C.; Yang, H.; Hou, Z.-G.; Ni, Z.-L.; Chen, S.; Fang, Z. Bilinear neural network with 3-D attention for brain decoding of motor imagery movements from the human EEG. *Cogn. Neurodyn.* **2020**, *15*, 181–189. [CrossRef] [PubMed]
35. Roots, K.; Muhammad, Y.; Muhammad, N. Fusion Convolutional Neural Network for Cross-Subject EEG Motor Imagery Classification. *Computers* **2020**, *9*, 72. [CrossRef]

36. Zhao, X.; Zhang, H.; Zhu, G.; You, F.; Kuang, S.; Sun, L. A multi-branch 3D convolutional neural network for EEG-based motor imagery classification. *IEEE Trans. Neural Syst. Rehabil. Eng.* **2019**, *27*, 2164–2177. [CrossRef] [PubMed]
37. Liu, T.; Yang, D. A densely connected multi-branch 3D convolutional neural network for motor imagery EEG decoding. *Brain Sci.* **2021**, *11*, 197. [CrossRef] [PubMed]
38. Yang, L.; Song, Y.; Jia, X.; Ma, K.; Xie, L. Two-branch 3D convolutional neural network for motor imagery EEG decoding. *J. Neural Eng.* **2021**, *18*, 0460c7. [CrossRef] [PubMed]
39. Salama, E.S.; El-Khoribi, R.A.; Shoman, M.E.; Shalaby, M.A.W. EEG-based emotion recognition using 3D convolutional neural networks. *Int. J. Adv. Comput. Sci. Appl.* **2018**, *9*, 329–337. [CrossRef]
40. Lawhern, V.J.; Solon, A.J.; Waytowich, N.R.; Gordon, S.M.; Hung, C.P.; Lance, B.J. EEGNet: A Compact Convolutional Network for EEG-based Brain-Computer Interfaces. *J. Neural Eng.* **2018**, *15*, 056013. [CrossRef]
41. Schirrmeister, R.T.; Springenberg, J.T.; Fiederer, L.D.J.; Glasstetter, M.; Eggensperger, K.; Tangermann, M.; Hutter, F.; Burgard, W.; Ball, T. Deep learning with convolutional neural networks for EEG decoding and visualization. *Hum. Brain Mapp.* **2017**, *38*, 5391–5420. [CrossRef] [PubMed]
42. Dose, H.; Møller, J.S.; Iversen, H.K.; Puthusserypady, S. An end-to-end deep learning approach to MI-EEG signal classification for BCIs. *Expert Syst. Appl.* **2018**, *114*, 532–542. [CrossRef]
43. Hermosilla, D.M.; Codorniu, R.T.; Baracaldo, R.L.; Zamora, R.S.; Rodriguez, D.D.; Albuerne, Y.L.; Alvarez, J.R.N. Shallow Convolutional Network Excel for Classifying Motor Imagery EEG in BCI Applications. *IEEE Access* **2021**, *9*, 98275–98286. [CrossRef]
44. Wan, Z.; Yang, R.; Huang, M.; Zeng, N.; Liu, X. A review on transfer learning in EEG signal analysis. *Neurocomputing* **2021**, *421*, 1–14. [CrossRef]
45. Tan, C.; Sun, F.; Zhang, W. Deep Transfer Learning for EEG-Based Brain Computer Interface. In Proceedings of the 2018 IEEE International Conference on Acoustics, Speech and Signal Processing (ICASSP), Calgary, AB, Canada, 15–20 April 2018; pp. 916–920. [CrossRef]
46. Köllőd, C.M.; Adolf, A.; Iván, K.; Márton, G.; Ulbert, I. Deep Comparisons of Neural Networks from the EEGNet Family. *Electronics* **2023**, *12*, 2743. [CrossRef]
47. Mattioli, F.; Porcaro, C.; Baldassarre, G. A 1D CNN for high accuracy classification and transfer learning in motor imagery EEG-based brain-computer interface. *J. Neural Eng.* **2022**, *18*, 066053. [CrossRef]
48. Hassanpour, A.; Moradikia, M.; Adeli, H.; Khayami, S.R.; Shamsinejadbabaki, P. A novel end-to-end deep learning scheme for classifying multi-class motor imagery electroencephalography signals. *Expert Syst.* **2019**, *36*, e12494. [CrossRef]
49. McFarland, D.J.; Miner, L.A.; Vaughan, T.M.; Wolpaw, J.R. Mu and beta rhythm topographies during motor imagery and actual movements. *Brain Topogr.* **2000**, *12*, 177–186. [CrossRef] [PubMed]
50. Pfurtscheller, G.; Brunner, C.; Schlögl, A.; Da Silva, F.H.L. Mu rhythm (de)synchronization and EEG single-trial classification of different motor imagery tasks. *NeuroImage* **2006**, *31*, 153–159. [CrossRef] [PubMed]
51. Xu, S.; Liu, Y.; Lee, H.; Li, W. Neural interfaces: Bridging the brain to the world beyond healthcare. *Exploration* **2024**, *4*, 20230146. [CrossRef] [PubMed]

brain sciences

Article

Adaptive Compensatory Neurophysiological Biomarkers of Motor Recovery Post-Stroke: Electroencephalography and Transcranial Magnetic Stimulation Insights from the DEFINE Cohort Study

Guilherme J. M. Lacerda [1,2], Fernanda M. Q. Silva [1], Kevin Pacheco-Barrios [1,3], Linamara Rizzo Battistella [2,4] and Felipe Fregni [1,*]

1 Neuromodulation Center and Center for Clinical Research Learning, Spaulding Rehabilitation Hospital and Massachusetts General Hospital, Harvard Medical School, Boston, MA 02138, USA; guilherme.lacerda@hc.fm.usp.br (G.J.M.L.)
2 Instituto de Medicina Física e Reabilitação, Hospital das Clínicas, Faculdade de Medicina, Universidade de São Paulo, São Paulo 04101-300, SP, Brazil
3 Vicerrectorado de Investigación, Unidad de Investigación para la Generación y Síntesis de Evidencias en Salud, Universidad San Ignacio de Loyola, Lima 15023, Peru
4 Departamento de Medicina Legal, Bioética, Medicina do Trabalho e Medicina Física e Reabilitação, Faculdade de Medicina, Universidade de São Paulo (FMUSP), São Paulo 01246-903, SP, Brazil
* Correspondence: fregni.felipe@mgh.harvard.edu; Tel.: +1-617-952-6153; Fax: +1-617-952-6150

Abstract: Objective: This study aimed to explore longitudinal relationships between neurophysiological biomarkers and upper limb motor function recovery in stroke patients, focusing on electroencephalography (EEG) and transcranial magnetic stimulation (TMS) metrics. Methods: This longitudinal cohort study analyzed neurophysiological, clinical, and demographic data from 102 stroke patients enrolled in the DEFINE cohort. We investigated the associations between baseline and post-intervention changes in the EEG theta/alpha ratio (TAR) and TMS metrics with upper limb motor functionality, assessed using the outcomes of five tests: the Fugl-Meyer Assessment (FMA), Handgrip Strength Test (HST), Pinch Strength Test (PST), Finger Tapping Test (FTT), and Nine-Hole Peg Test (9HPT). Results: Our multivariate models identified that a higher baseline TAR in the lesioned hemisphere was consistently associated with poorer motor outcomes across all five assessments. Conversely, a higher improvement in the TAR was positively associated with improvements in FMA and 9HPT. Additionally, an increased TMS motor-evoked potential (MEP) amplitude in the non-lesioned hemisphere correlated with greater FMA-diff, while a lower TMS Short Intracortical Inhibition (SICI) in the non-lesioned hemisphere was linked to better PST improvements. These findings suggest the potential of the TAR and TMS metrics as biomarkers for predicting motor recovery in stroke patients. Conclusion: Our findings highlight the significance of the TAR in the lesioned hemisphere as a predictor of motor function recovery post-stroke and also a potential signature for compensatory oscillations. The observed relationships between the TAR and motor improvements, as well as the associations with TMS metrics, underscore the potential of these neurophysiological measures in guiding personalized rehabilitation strategies for stroke patients.

Keywords: stroke; rehabilitation; motor recovery; neurophysiological biomarkers; electroencephalography (EEG); resting-state EEG; theta/alpha ratio (TAR); transcranial magnetic stimulation (TMS)

1. Introduction

Stroke is the second leading cause of death worldwide and the primary cause of disability, impacting various functional domains such as mobility, touch, pain, cognition, and mood [1]. The incidence of stroke is increasing due to an aging global population, with approximately 16.9 million new cases each year [2]. Despite scientific advances improving rehabilitation outcomes over the past decade, many stroke survivors still experience reduced mobility and severe chronic disability, particularly in the upper limbs, after a year of inpatient rehabilitation [3]. This highlights the ongoing challenge of implementing effective rehabilitation strategies for stroke survivors, which remains in its incipient stages.

In the context of stroke recovery, which has significant heterogenicity among patients, further research is essential to understand the disability mechanisms and enhance patient-focused rehabilitation, developing predictors of outcomes for various interventions [4]. Since rehabilitation marks a complex neurobiological phase, identifying reliable biomarkers is vital [5]. Many biomarkers of stroke recovery have been discussed in the literature, such as those derived from Electroencephalography (EEG) and Transcranial Magnetic Stimulation (TMS), either in isolation or combined. For instance, recent studies have demonstrated that EEG shows potential in predicting upper limb sensorimotor recovery [6–10] and TMS shows potential for upper extremity motor function improvement, both in post-stroke patients [9,11,12]. Therefore, the EEG and TMS-metrics could serve as an effective brain biomarker for upper limb motor function recovery.

Moreover, the role of the EEG and TMS metrics in understanding and predicting stroke recovery underscores their importance in stroke rehabilitation. A 2017 systematic review identified the motor-evoked potential (MEP) derived from TMS as a brain biomarker for improved upper limb motor outcomes [7]. A 2023 systematic review further highlighted EEG's potential in predicting upper extremity sensorimotor recovery, emphasizing its ability to capture neural alterations such as interhemispheric imbalance and disruptions in beta oscillatory activity, which are closely linked to motor recovery outcomes [13]. However, despite these advances, significant gaps remain in the identification of reliable biomarkers due to methodological limitations and the complexity of interpreting interactions between biomarkers and clinical measures [5]. For instance, recent research employing multivariate models revealed a specific neural signature in which a higher EEG theta/alpha ratio (TAR) in the lesioned hemisphere combined with increased MEP amplitude in the non-lesioned hemisphere correlates with poorer upper limb motor function [14]. This finding suggests a potential neural signature of brain compensation where lower frequencies of EEG power are heightened in the lesioned hemisphere, and corticospinal excitability is reduced in the non-lesioned hemisphere. Consequently, there is a clear need for more comprehensive studies utilizing the EEG and TMS metrics to enhance our understanding of upper extremity recovery patterns post-stroke.

To advance the knowledge of upper limb recovery post-stroke, our longitudinal study employs a prospective cohort analysis exploring EEG theta/alpha oscillations as potential biomarkers for predicting upper extremity impairment and motor outcomes. We aim to determine how this measure correlates with five motor assessments: the Fugl-Meyer Assessment (FMA), Handgrip Strength Test (HST), Finger Tapping Test (FTT), Nine-Hole Peg Test (9HPT), and Pinch Strength Test (PST) [15]. Our models are designed to capture recovery patterns and biomarker changes, emphasizing the potential of the EEG-TAR in predicting motor impairment and recovery in the upper limbs of stroke patients. We hypothesize that a decreased EEG-TAR will be related to a better functional improvement and this approach could significantly enhance current rehabilitation practices by facilitating the development of personalized treatment strategies.

2. Materials and Methods

2.1. Participants, Study Design, and Sample Size

This study involves a longitudinal analysis employing baseline and post-interventional data from the "Deficit of Inhibition as a Marker of Neuroplasticity (DEFINE Study) in Rehabilitation: A Longitudinal Cohort Study Protocol" cohort study [16]. This cohort recruited patients from the conventional stroke rehabilitation program at the Instituto de Medicina Física e Reabilitação (IMREA). In the stroke arm of this cohort, 102 participants signed the informed consent, which had been previously approved by the Hospital das Clínicas da Faculdade de Medicina da Universidade de São Paulo Ethics Committee for

Research Protocol Analysis (CAAE: 86832518.7.0000.0068). Here, we analyze data from those subjects who underwent a series of clinical and neurophysiological evaluations at two key stages: before starting and after completing the IMREA rehabilitation program.

2.2. Inclusion Criteria

From 2020 to 2022, subjects of both sexes, aged 18 years and older, were included if they had a clinical and radiological diagnosis of stroke confirmed by imaging and were clinically stable as verified by a medical evaluation. They also needed to meet the eligibility criteria for the IMREA rehabilitation program. Exclusion criteria included any clinical or social conditions that could interfere with participation in the rehabilitation treatment. Pregnancy was also a criterion for exclusion. All participants provided signed informed consent forms [16].

2.3. Exclusion Criteria

Participants were excluded from the study if they had any clinical or social conditions that could interfere with their participation in the rehabilitation treatment. Additionally, pregnancy was considered an exclusion criterion.

2.4. Functional Assessments

A trained physician first evaluated a participant's eligibility, reviewed their medical history, conducted physical examinations, and collected demographic surveys. The data collected were then used to identify covariates for the regression models. Various scales were used to describe the sample and controls for confounding variables in the multivariate statistical model. To evaluate upper limb function, we selected five specific assessments, each targeting a unique aspect of motor performance: The Fugl-Meyer Assessment (FMA) measures motor recovery and coordination in the upper limb, providing a comprehensive evaluation of motor impairment. The Handgrip Strength Test (HST) assesses overall grip strength, reflecting functional hand performance and general muscle strength. The Pinch Strength Test (PST) evaluates fine motor strength required for precision tasks, such as pinching and gripping small objects. The Finger Tapping Test (FTT) measures motor speed and coordination, focusing on fine motor control and rhythm. The Nine-Hole Peg Test (9HPT) assesses dexterity and hand-eye coordination, crucial for everyday manual tasks. These assessments were selected to provide a well-rounded evaluation of upper limb function [15,17]. A more detailed description of each assessment is available in the Supplementary Materials.

2.5. Intervention

The rehabilitation interventions provided at IMREA are highly personalized, tailored to the unique needs of each patient. These interventions encompass a multidisciplinary approach, including physical therapy to enhance motor function recovery, strength, and mobility; occupational therapy to improve activities of daily living and fine motor skills; and psychological support to address emotional well-being and coping strategies post-stroke. Additional therapies, such as speech therapy or social support services, may also be incorporated based on the patient's specific condition and recovery goals. This personalized and comprehensive framework ensures that each patient receives targeted care to optimize their rehabilitation.

2.6. Neurophysiological Assessment Methods

2.6.1. Electroencephalography

A skilled clinical neurophysiologist conducted a visual analysis of the EEG data to identify artifacts and any potential clinical abnormalities. This information was subsequently exported and analyzed offline using MATLAB (R2014b, The MathWorks Inc., Natick, MA, USA) and EEGLab [18]. The EEG should will last approximately 45 min, consisting of 25 min for participant and software setup, followed by 10 min of EEG recording.

This recording will be split into a resting-EEG condition, with 5 min of open-eye and 5 min of closed-eye observation, and an 8 min task-related EEG condition. The task-related EEG part includes activities such as movement observation, imagery, and execution. The analysis focused on standard frequency bands: delta (2–4 Hz), theta (4–8 Hz), low-alpha (8–10.5 Hz), high-alpha (10.5–13 Hz), alpha (8–13 Hz), low-beta 1 (13–20 Hz), high-beta 2 (20–30 Hz), and beta (13–30 Hz). These frequency bands were measured bilaterally across the frontal, central, and parietal scalp regions and used to calculate the theta/alpha ratio (TAR), which is the simple division of theta band power (4–8 Hz) by alpha band power (8–13 Hz).

2.6.2. Transcranial Magnetic Stimulation

A Magstim Rapid® stimulator (The Magstim Company Limited, Whitland, UK) with a 70 mm figure-of-eight coil was positioned tangentially on the skull at a 45-degree angle to the sagittal line for the TMS assessment [19]. All assessments were performed bilaterally by stimulating the motor cortex, with muscular responses recorded using surface electromyography electrodes placed on the target muscle, the first dorsal interosseous of the hand. To locate the cortical area corresponding to the selected muscle, the vertex (intersection of the nasion–inion lines and zygomatic arches) was identified. Marks were then drawn 5 cm (about 1.97 in) from the vertex towards the ear tragus on both sides in (rMT) the coronal plane, identifying the "hot spot"—the area with the lowest resting motor threshold and the highest motor-evoked potential (MEP) amplitude in the target muscle.

The rMT is defined as the minimum intensity at which a single TMS pulse at the hot spot elicits an MEP with at least a 50 V peak-to-peak amplitude in 50% of attempts. The assessment also included recording ten MEPs at 130% of rMT with a 7 s interval between stimuli (the average of those measures is the MEP130); measuring the silent period (SP), which reflects the temporary suppression of electromyographic activity during a sustained MEP voluntary contraction; assessing short-interval intracortical inhibition (ICI) with two stimuli 2 milliseconds apart, with the conditioned stimulus at 80% rMT and the test stimulus at 130% rMT; and measuring intracortical facilitation (ICF) using the same stimulus values as ICI but with a 10-millisecond interval between the stimuli. Moreover, the ICI is calculated as the MEP after paired-pulse 2 ms/the MEP at baseline, and the SICF is determined by the MEP after paired-pulse 10 ms/the MEP at baseline [20].

2.7. Statistical Analysis

In this cohort study, we collected two measurements for each variable: one before and one after the rehabilitation program. We then calculated the differences by subtracting the pre-intervention values from the post-intervention values for each variable. These differences represent our outcomes and are designated as follows: FMA-diff for the Fugl-Meyer Assessment, HST-diff for the Handgrip Strength Test, PST-diff for the Pinch Strength Test, FTT-diff for the Finger Tapping Test, and 9HPT-diff for the Nine-Hole Peg Test. All other variables follow the same naming convention: variables representing the difference between post- and pre-intervention measurements have the suffix "-diff" added to their names. In contrast, variables without this suffix represent baseline (pre-intervention) data.

For the analysis, only complete cases were included. Outcomes were dichotomized into binary variables as they did not satisfy the assumptions necessary for linear regression.

A value of 0 was assigned when the difference between post- and pre-intervention measurements was zero or negative (difference ≤ 0), indicating a deterioration in the patient's condition. Conversely, a value of 1 was assigned when the difference was positive (difference > 0), indicating an improvement in the patient's condition. An exception was made for the 9HPT-diff, where a lower time reflects a better performance; hence, the categorization was reversed. In this case, a value of 1 was assigned when the difference was negative (difference < 0), indicating improvement, while a value of 0 was assigned when the difference was zero or positive (difference ≥ 0), indicating a decline in the patient's condition.

We then investigated two types of associations: first, between the outcomes (FMA-diff, HST-diff, PST-diff, FTT-diff, and 9HPT-diff) and baseline variables measured prior to the intervention; and second, between the same outcomes and the differences in neurophysiological variables, calculated as the post-intervention values minus the pre-intervention values, in the same manner as the outcomes were determined. In both cases, we first performed univariate regression analyses. Subsequently, we used the variables that had a p-value < 0.25 in the univariate associations. Finally, we constructed the final multivariate model using a combination of Forward Selection and Backward Elimination, guided by clinical reasoning in the selection of variables. During this stage, a significance level of p-value < 0.05 was adopted. It is noteworthy that we thoroughly examined potential confounders such as age, sex, educational level, and others, adjusting the model as necessary. We considered a variable to be a confounder if it changed the beta-coefficient (Log [Odds Ratio]) of another variable by more than 10%. Additionally, given the exploratory characteristic of this study, we investigated region-specific relationships between EEG metrics and motor outcomes, providing preliminary evidence for the distinct roles of the central and parietal regions. The central region, involving the primary motor cortex, was linked to gross motor functions such as strength and speed, assessed by FMA, HST, and FTT. In contrast, the parietal cortex, key for sensorimotor integration and dexterity, was associated with fine motor skills, as reflected in the 9HPT, highlighting its role in precision and coordination.

Statistical significance was determined at $p < 0.05$, and all analyses were conducted using a standard software package, R-Studio Version 2023.06.0+421 (2023.06.0+421).

3. Results

3.1. Sample Characteristics

Baseline data for 102 post-stroke patients who participated in the cohort study were collected, revealing that the cohort had a mean age of 56 years old, 58 were male (57%), the mean level of education was 11.46 years, and most participants were white, 52 (51%). Most had an ischemic stroke (81%), which was a moderate stroke (mean of National Institute of Health Stroke Scale [NIHSC] of 5) and required a mild level of assistance (Functional Independence Measure [FIM] of 96). Additionally, all five upper limb motor function metrics demonstrated a non-normal distribution. A detailed description of the sample demographic and clinical characteristics at baseline is provided in Table 1.

Table 1. Demographic and clinical features at baseline for the 102 post-stroke patients included in the study.

Patient Characteristics	Total (N = 102)
Age, mean (SD)	56.02 (13.7)
Sex, N (%)	
Male	58 (58.86)
Female	44 (43.14)
Education years, mean (SD)	11.46 (5.83)
BMI, mean (SD)	26.77 (5.17)
Smoking history, N (%)	
Yes	38 (37.2%)
No	54 (53.0%)
Missing data	10 (9.8%)
Marital status, N (%)	
Married	55 (54.06)
Not married	47 (45.94)

Table 1. *Cont.*

Patient Characteristics	Total (N = 102)
Race, N (%)	
Asian	7 (6.90)
Black	10 (9.70)
More than one race	26 (25.5)
White	52 (51)
Unknown or not reported	7 (6.90)
Stroke type, N (%)	
Ischemic	83 (81.37)
Hemorrhagic	12 (11.76)
Ischemic with hemorrhagic transformation	7 (6.87)
NIHSS, median (IQR)	5 (2–8)
Time since the injury (in months), median (IQR)	12.3 (5.29–19.38)
Patient age at the stroke), median (IQR)	56.15 (44.64–67.36)
Hand dominance aligned with the paretic side, N (%)	
Yes	91 (58.18)
No	42 (41.82)
Impairment of the thenar region on the paretic side, N (%)	
Yes	80.58
No	11 (19.42)
Presence of spasticity, N (%)	
Yes	71 (69.6%)
No	19 (18.6%)
Missing data	12 (11.8%)
CPM on the paretic side, median (IQR)	1.48 (0.65–3.44)
FIM score, median (IQR)	96 (77.75–106.0)

Abbreviations: NIHSS: National Institute of Health Stroke Scale; FIM: Functional Independence Measure; SD: Standard deviation); IQR: Interquartile range; CPM: Conditioned Pain Modulation.

3.2. Dependent Variables Characteristics

Table 2 summarizes the five upper limb motor outcomes, specifically the differences between post- and pre-intervention for each assessment. For FMA-diff, 52.7% of participants had a difference ≤0, while 47.3% had a difference >0. For HST-diff, 57.9% of participants had a difference ≤0, and 42.1% had a difference 0. For PST-diff, 66.7% of participants had a difference ≤0, and 33.3% had a difference >0. For FFT-diff, 61.5% of participants had a difference ≤0, while 38.5% had a difference >0. Lastly, for 9HPT-diff, 72.7% of participants had a difference ≤0, and 27.3% had a difference >0.

Table 2. Differences between motor function assessments after and before the rehabilitation program.

FMA-Diff	Number of Observations	%
difference ≤ 0	29	52.7
difference > 0	26	47.3
HST-diff		
difference ≤ 0	33	57.9
difference > 0	24	42.1
PST-diff		
difference ≤ 0	38	66.7
difference > 0	19	33.3
FFT-diff		
difference ≤ 0	24	61.5
difference > 0	15	38.5
9HPT-diff		
difference ≤ 0	40	72.7
difference > 0	15	27.3

3.3. *Multivariate Analysis*

When necessary, we adjusted each model for covariates such as age, sex, time of injury, education level, marital status, stroke type, BMI, smoking history, NIHSS, and SICI. This adjustment allowed us to accurately reveal the associations between the outcomes and the TAR.

3.4. *Association Between Outcomes and Baseline Neurophysiological Variables*

These models were constructed to identify associations between the five outcomes and the baseline variables, adjusted by the patients' demographic and clinical characteristics (Table 3). For FMA-diff, we found a negative association with the TAR measured in the parietal lesioned hemisphere (OR = 0.22, 95% CI: 0.04–0.82; *p*-Value = 0.04), and a positive association with MEP130 measured in the non-lesioned hemisphere, OR = 1.57 (95% CI: 1.05–2.48; *p*-Value = 0.03). For HST-diff, we found a significant negative association with the TAR measured in the central lesioned hemisphere (OR = 0.20, 95% CI: 0.04–0.75; *p*-Value = 0.0351). For PST-diff, we observed a significant negative association with the TAR measured in the central lesioned hemisphere (OR = 0.19, 95% CI: 0.03–0.78; *p*-Value = 0.0456) and a significant positive association with ICI measured in the non-lesioned hemisphere (OR = 14.62, 95% CI: 1.75–209.10; *p*-Value = 0.025). For FTT-diff, we identified a significant negative association with the TAR measured in the central lesioned hemisphere (OR = 0.08, 95% CI: 0.00–0.56; *p*-Value = 0.0393). For 9HPT-diff, we found a significant negative association with the TAR measured in the central lesioned hemisphere (OR = 0.13, 95% CI: 0.01–0.66; *p*-Value = 0.0371).

Table 3. Association between outcomes and baseline neurophysiological variables.

Multivariable Analysis	OR	95% CI	*p*-Value	N
FMA-diff [a]				48
EEG TAR (parietal lesion side)	0.22	0.04–0.82	0.040	
TMS MEP 130 (non-lesion side)	1.57	1.05–2.48	0.034	
HST-diff [b]				55
EEG TAR (central lesion side)	0.20	0.04–0.75	0.035	
PST-diff [b]				53
EEG TAR (central lesion side)	0.19	0.03–0.78	0.046	
TMS SICI (non-lesion side)	14.62	1.75–209.10	0.025	
FT-diff [b]				39
EEG TAR (central lesion side)	0.08	0.00–0.56	0.04	
9HPT-diff [b]				53
EEG TAR (central lesion side)	0.13	0.01–0.66	0.037	

[a] Adjusted by FMA baseline, marital status, education level, and smoking history. [b] Adjusted by level of education and BMI. Abbreviations: FMA: Fugl-Meyer Assessment; HST: Handgrip Strength Test; PST: Pinching Strength Test; FT: Finger Tapping; 9HPT: Nine Hole Peg Test; EEG: Electroencephalogram; TMS: Transcranial Magnetic Stimulation; TAR: theta over alpha ratio; MEP: motor evoked-potential; SICI: Short Intracortical Inhibition; OR: Odds Ratio; CI: Confidence Interval; BMI: Body Mass Index.

3.5. Association Between Outcomes and Neurophysiological Variables Differences (Post- and Pre-Intervention)

These models were constructed to identify associations between the outcomes and the differences in the variables (post-intervention minus pre-intervention), adjusted by the patients' demographic and clinical characteristics (Table 4). For FMA-diff, we found a positive association with the TAR-diff measured in the central lesioned hemisphere (OR = 1.15, 95% CI: 1.02–1.36, *p*-Value = 0.0454). For 9HPT-diff, we found a significant positive association with the TAR-diff measured in the parietal lesioned hemisphere (OR = 1.81, 95% CI: 1.15–3.62; *p*-Value = 0.029). No significant associations were found for HST-diff, PST-diff, or FT-diff.

Table 4. Association between outcomes and neurophysiological variables differences.

Multivariable Analysis	OR	95% CI	*p*-Value	N
FMA-diff [a]				41
EEG TAR (central lesion) difference	1.15	(1.02–1.36)	0.045	
9HPT-diff [b]				41
EEG TAR (parietal lesion side) difference	1.81	(1.15–3.62)	0.029	

[a] Adjusted by age, sex, NIHSS, and stroke type. [b] Adjusted by NIHSS, time since injury, EEG TAR frontal lesion side, and EEG TAR central lesion side. Abbreviations: FMA: Fugl-Meyer Assessment; 9HPT: Nine Hole Peg Test; EEG: Electroencephalogram; TMS: Transcranial Magnetic Stimulation; TAR: theta over alpha ratio; OR: Odds Ratio; CI: Confidence Interval; BMI: Body Mass Index; NIHSS: National Institute of Health Stroke Scale.

4. Discussion

4.1. Main Findings

In our current research, we explored five outcomes: FMA-diff, HST-diff, PST-diff, FT-diff, and 9HPT-diff. We identified a negative association between the outcomes and baseline TAR measured in the lesioned hemisphere, as well as a positive association between the outcomes and the TAR-diffs (post- minus pre-intervention) also measured in the lesioned hemisphere. Additionally, we observed a positive association between baseline MEP130 measured in the non-lesioned hemisphere and FMA-diff. Similarly, a positive association was found between baseline SICI, also measured in the non-lesioned hemi-sphere, and PST-diff.

4.2. Association Between Outcomes and Baseline EEG Theta/Alpha Ratio in the Lesioned Hemisphere

The models aimed to identify associations between five motor outcome measures (FMA-diff, HST-diff, PST-diff, FTT-diff, and 9HPT-diff) and baseline variables, while adjusting for patients' demographic and clinical characteristics. The TAR measured in the lesion hemisphere showed a significant negative association with the five outcomes, indicating that higher baseline TAR values were associated with less improvement across different assessments.

Building upon this, our previous study showed that a higher TAR in the frontal region of the lesioned hemisphere is associated with poorer motor outcomes, as seen in the patients' lower performance scores. This observation is consistent with prior studies that have identified an increased ratio of [slow oscillations]/[fast oscillations] as a marker of brain injury (DAR: Delta/Alpha and DTABR: (delta + theta)/(alpha + beta)). For instance, Finnigan et al. reported that the DAR and DTABR are reliable indicators for detecting acute ischemic stroke [21]. Furthermore, evidence indicates that low-frequency oscillations in the motor cortex, which relate to motor control tasks, tend to decrease shortly after a stroke and re-emerge during motor recovery [22,23]. Additionally, a transition toward alpha oscillations is associated with improvements in motor abilities [24]. In our previous cross-sectional study, we suggested that a reduced theta/alpha ratio might be associated with improved motor function [14]. In the current study, which includes two measurements (pre- and post-intervention), this hypothesis appears to be supported, as we observed consistent results across all five outcomes. In fact, the TAR has consistently been identified as a significant factor associated with motor outcomes. For example, in our models, we observed that a higher TAR in the lesion hemisphere was negatively correlated with various motor performance measures. Specifically, for FMA-diff, we found that an elevated TAR in the parietal lesion region was associated with poorer motor outcomes. Similarly, the TAR in the central lesion region was negatively associated with HST-diff, PST-diff, FTT-diff, and 9HPT-diff. These consistent findings across multiple outcome measures underscore the importance of the TAR as a potential marker for motor function, suggesting that a higher baseline TAR in the lesion hemisphere is associated with poorer improvements across all five motor outcomes. Finally, our findings solidify the role of TAR as a reliable predictor of motor outcomes in stroke patients, offering valuable insights for both research and clinical practice. Furthermore, the relationships between TAR and the different functional assessments also highlight variations in the aspects of motor function being measured. For instance, the 9HPT and FTT assess motor speed and dexterity, while the HST and PST are more focused on strength-related performance. This distinction suggests that the influence of TAR might manifest differently depending on the motor domain being assessed. Exploring these nuanced relationships further could provide deeper insights into how TAR relates to specific functional abilities and recovery trajectories.

4.3. Association Between Outcomes and EEG Theta/Alpha Ratio (Differences Post- Minus Pre-Intervention) in the Lesioned Hemisphere

The models were constructed to identify associations between the motor outcomes and the TAR-diff in the lesioned hemisphere, while adjusting for patients' demographic and clinical characteristics. For FMA-diff, a positive association was found with TAR-diff, indicating that an increase in this ratio was associated with improved motor function, as reflected by higher FMA-diff scores. Similarly, a significant positive association was observed between 9HPT-diff and TAR-diff, suggesting that increases in the TAR in the parietal lesion region were linked to a better motor performance. These findings imply that changes in the TAR post-intervention may serve as indicators of motor recovery, particularly in the central and parietal regions of the brain. Based on the previous topic, studies have identified an increased ratio of [slow oscillations]/[fast oscillations] as a marker of brain injury. Additionally, evidence indicates that low-frequency oscillations in the motor cortex, which are related to motor control tasks, tend to decrease shortly after a stroke and re-emerge during motor recovery [22,23]. Moreover, a transition toward alpha oscillations is associated with improvements in motor abilities. This aligns with our results, where we found that higher differences in the TAR were associated with greater improvements in both FMA-diff and 9HPT-diff, suggesting that these changes in the TAR could reflect a more robust recovery process in motor function. This finding suggests that the transition from delta oscillations to alpha oscillations, as represented by an increase in the theta/alpha ratio (TAR), may serve as a potential biomarker for predicting improvements in motor outcomes. The observed increase in the TAR could reflect underlying neurophysiological processes associated with recovery, such as the reorganization of neural networks or the restoration of motor control pathways. As such, monitoring changes in the TAR during rehabilitation could provide valuable insights into the effectiveness of interventions and the trajectory of motor function recovery.

It is worth mentioning that we did not find statistically significant associations between TAR-diff in the lesioned hemisphere and HST-diff, PST-diff, and FT-diff. This lack of association may be due to these three outcomes being more closely related to gross motor movements, which might not be as sensitive to changes in the theta/alpha ratio as fine motor skills. Furthermore, the differential associations observed between TAR-diff and motor outcomes highlight the varying sensitivity of these assessments to neurophysiological changes. For example, FMA-diff and 9HPT-diff primarily assess fine motor control and coordination, which may be more directly influenced by changes in the TAR. In contrast, outcomes like HST-diff and PST-diff, which focus on strength-related measures, may rely less on the neural oscillatory dynamics captured by the TAR. These differences underscore the importance of tailoring the selection of functional assessments to the specific neurophysiological mechanisms under investigation and suggest that the TAR may serve as a more robust biomarker for fine motor recovery compared to gross motor outcomes.

4.4. Association Between Outcomes and Baseline TMS Metrics in the Non-Lesioned Hemisphere

Additionally, we observed a positive association between baseline MEP130 measured in the non-lesioned hemisphere and FMA-diff. This suggests that higher baseline MEP values are associated with greater improvements in motor function. Similarly, a positive association was found between baseline SICI, also measured in the non-lesioned hemisphere, and PST-diff, indicating that a lower inhibition (as reflected by higher SICI amplitude values) may be linked to better motor recovery outcomes.

Our previous study demonstrated that increased MEP amplitude in the non-lesioned hemisphere correlates with improved motor function [14]. These relationships were consistent across all five motor assessments, reinforcing the idea that neurophysiological measures, such as MEP amplitude, may serve as reliable biomarkers for recovery. Indeed, in the current study, a higher baseline MEP130 of the motor threshold is associated with a greater score for FMA-diff, suggesting that increased activity in the non-lesioned hemisphere at baseline is linked to higher odds of improving motor skills, as measured by the Fugl-Meyer Assessment. Moreover, the finding of a positive association between SICI and PST-diff is consistent with this idea, as higher SICI values indicate less inhibition. Therefore, we could state that the lower the inhibition in the non-lesioned hemisphere, the higher the odds of improving motor skills, as measured by the Pinch Strength Test.

4.5. Strengths and Limitations

One limitation would be that the relatively small sample size may have negatively impacted our results' statistical power and precision, underlining the need for further larger longitudinal and interventional studies to establish more accurate causal associations. Additionally, it is important to acknowledge that the DEFINE cohort study was conducted in a Latin American center, which may influence the external validity of our findings. Furthermore, the dichotomization of continuous outcomes, while pragmatic for addressing data distribution challenges, may have reduced granularity and masked nuanced relationships, warranting a cautious interpretation of the results. It should be mentioned that another limitation of this study is the lack of stroke location data, which may influence EEG and recovery outcomes.

On the other hand, a key strength of our research lies in its comparison of neurophysiological measures with the differences in pre- and post-intervention values across five motor assessments within a stroke population. This approach not only enriches our understanding of the clinical validity of EEG and TMS measures, as neurophysiological biomarkers of upper limb motor recovery, but also provides a foundation for evaluating the impact of changes in EEG and TMS metrics as potential surrogate outcomes for further studies. Additionally, the longitudinal nature of our analysis offers a more dynamic representation of post-stroke recovery.

4.6. Future Perspectives: EEG Biomarkers and Emerging Parameters for Analysis

In our study, the TAR has been identified as a potential biomarker for motor recovery, reflecting cortical reorganization processes during rehabilitation. These findings align with research on event-related desynchronization (ERD) during action observation, which has been highlighted as an early predictor of motor recovery in subcortical stroke [25]. Specifically, alpha ERD during AO in the affected parietal and central regions has been correlated with better functional outcomes, such as improvements in FMA scores at 12 weeks. Similarly, the TAR captures broader oscillatory dynamics that reflect the brain's capacity for functional reorganization. While our study focuses on the TAR's predictive value, integrating ERD into future investigations alongside the TAR may enhance our understanding of recovery processes, particularly in characterizing the engagement of motor-related regions even in patients with severe impairments.

Additionally, as emphasized in [26], combining kinematic measures with EEG biomarkers provides complementary insights into post-stroke recovery. Kinematic parameters, such as movement duration and smoothness, have been strongly correlated with FMA scores, indicating their relevance as measures of motor recovery. Integrating kinematic data into studies focusing on the TAR could deepen our understanding of the neurophysiological and biomechanical correlations with motor recovery. This multimodal approach has the potential to optimize personalized rehabilitation strategies by leveraging insights from both neurophysiological and movement-related parameters.

5. Conclusions

Our study demonstrated that the TAR in the lesioned hemisphere consistently correlated with upper motor function recovery post-stroke across various prediction models. Specifically, a lower baseline TAR was associated with better outcomes in motor assessments, including the FMA-diff, HST-diff, PST-diff, FT-diff, and 9HPT-diff. These findings underscore the potential of the TAR as a reliable biomarker for predicting upper limb motor recovery post-stroke. Patients with a lower baseline TAR may possess a greater capacity for motor function improvement during rehabilitation. Additionally, our results revealed that an increase in the TAR-diff was positively associated with greater improvements in motor outcomes. This suggests that as the TAR increases after the rehabilitation process, better motor function outcomes are observed, further supporting the role of TAR as a dynamic marker of recovery. Our findings highlight the potential for the EEG theta/alpha ratio (TAR) and TMS metrics as neurophysiological biomarkers to predict motor recovery in stroke patients. Clinically, these biomarkers can provide valuable insights into individual recovery trajectories, enabling more personalized rehabilitation strategies. For example, the association between TAR and motor outcomes suggests its utility as a marker for identifying patients who may benefit from targeted interventions to enhance compensatory oscillations in the lesioned hemisphere. Moreover, metrics such as MEP amplitude and SICI in the non-lesioned hemisphere could help guide the use of neuromodulatory techniques like repetitive TMS (rTMS) or paired associative stimulation (PAS) to optimize motor recovery.

These biomarkers could also aid in stratifying patients for clinical trials, ensuring that interventions are tailored to specific neurophysiological profiles.

Supplementary Materials: The following supporting information can be downloaded at: https://www.mdpi.com/article/10.3390/brainsci14121257/s1. References [27–31] are cited in Supplementary Materials, which contains information about FAM, HST, PST, FTT, and 9HPT.

Author Contributions: Conceptualization, L.R.B. and F.F.; methodology, F.F.; software, G.J.M.L. and K.P.-B.; validation, L.R.B. and F.F.; formal analysis, G.J.M.L.; investigation, G.J.M.L. and F.F.; resources, L.R.B. and F.F.; data curation, G.J.M.L., L.R.B. and F.F.; writing—original draft preparation, G.J.M.L., F.M.Q.S. and K.P.-B.; writing—review and editing, G.J.M.L. and F.M.Q.S.; visualization, F.F.; supervision, L.R.B. and F.F.; project administration, L.R.B. and F.F.; funding acquisition, L.R.B. and F.F. All authors have read and agreed to the published version of the manuscript.

Funding: This study is supported by FAPESP (SPEC project, fund number 2017/12943-8).

Institutional Review Board Statement: The project was previously approved by the Hospital das Clínicas da Faculdade de Medicina da Universidade de São Paulo Ethics Committee for Research Protocol Analysis (CAAE: 86832518.7.0000.0068) date 6 June 2018.

Informed Consent Statement: Informed consent was obtained from all subjects involved in the study.

Data Availability Statement: The original contributions presented in the study are included in the article and Supplementary Materials, further inquiries can be directed to the corresponding author.

Conflicts of Interest: The authors declare no known competing financial or personal interests that could influence their work in this paper.

References

1. Gorelick, P.B. The Global Burden of Stroke: Persistent and Disabling. *Lancet Neurol.* **2019**, *18*, 417–418. [CrossRef] [PubMed]
2. Béjot, Y.; Daubail, B.; Giroud, M. Epidemiology of Stroke and Transient Ischemic Attacks: Current Knowledge and Perspectives. *Rev. Neurol.* **2016**, *172*, 59–68. [CrossRef] [PubMed]
3. Luengo-Fernandez, R.; Paul, N.L.M.; Gray, A.M.; Pendlebury, S.T.; Bull, L.M.; Welch, S.J.V.; Cuthbertson, F.C.; Rothwell, P.M. Population-Based Study of Disability and Institutionalization after Transient Ischemic Attack and Stroke: 10-Year Results of the Oxford Vascular Study. *Stroke* **2013**, *44*, 2854–2861. [CrossRef] [PubMed]
4. Muir, K.W. Heterogeneity of Stroke Pathophysiology and Neuroprotective Clinical Trial Design. *Stroke* **2002**, *33*, 1545–1550. [CrossRef] [PubMed]
5. Wingfield, M.; Fini, N.A.; Brodtmann, A.; Williams, G.; Churilov, L.; Hayward, K.S. Upper-Limb Motor Intervention Elements That Drive Improvement in Biomarkers and Clinical Measures Post-Stroke: A Systematic Review in a Systems Paradigm. *Neurorehabilit. Neural Repair* **2022**, *36*, 726–739. [CrossRef] [PubMed]
6. Simis, M.; Thibaut, A.; Imamura, M.; Battistella, L.R.; Fregni, F. Neurophysiological Biomarkers of Motor Improvement from Constraint-Induced Movement Therapy and Robot-Assisted Therapy in Participants with Stroke. *Front. Hum. Neurosci.* **2023**, *17*, 1188806. [CrossRef] [PubMed]
7. Hayward, K.S.; Schmidt, J.; Lohse, K.R.; Peters, S.; Bernhardt, J.; Lannin, N.A.; Boyd, L.A. Are We Armed with the Right Data? Pooled Individual Data Review of Biomarkers in People with Severe Upper Limb Impairment after Stroke. *Neuroimage Clin.* **2017**, *13*, 310–319. [CrossRef]
8. Tedesco Triccas, L.; Meyer, S.; Mantini, D.; Camilleri, K.; Falzon, O.; Camilleri, T.; Verheyden, G. A Systematic Review Investigating the Relationship of Electroencephalography and Magnetoencephalography Measurements with Sensorimotor Upper Limb Impairments after Stroke. *J. Neurosci. Methods* **2019**, *311*, 318–330. [CrossRef]
9. Thibaut, A.; Simis, M.; Battistella, L.R.; Fanciullacci, C.; Bertolucci, F.; Huerta-Gutierrez, R.; Chisari, C.; Fregni, F. Using Brain Oscillations and Corticospinal Excitability to Understand and Predict Post-Stroke Motor Function. *Front. Neurol.* **2017**, *8*, 187. [CrossRef]
10. Berger, A.; Horst, F.; Müller, S.; Steinberg, F.; Doppelmayr, M. Current State and Future Prospects of EEG and fNIRS in Robot-Assisted Gait Rehabilitation: A Brief Review. *Front. Hum. Neurosci.* **2019**, *13*, 172. [CrossRef]
11. Simis, M.; Di Lazzaro, V.; Kirton, A.; Pennisi, G.; Bella, R.; Kim, Y.-H.; Takeuchi, N.; Khedr, E.M.; Rogers, L.M.; Harvey, R.; et al. Neurophysiological Measurements of Affected and Unaffected Motor Cortex from a Cross-Sectional, Multi-Center Individual Stroke Patient Data Analysis Study. *Neurophysiol. Clin.* **2016**, *46*, 53–61. [CrossRef] [PubMed]
12. Simis, M.; Doruk, D.; Imamura, M.; Anghinah, R.; Morales-Quezada, L.; Fregni, F.; Battistella, L.R. Neurophysiologic Predictors of Motor Function in Stroke. *Restor. Neurol. Neurosci.* **2016**, *34*, 45–54. [CrossRef] [PubMed]
13. Kancheva, I.; van der Salm, S.M.A.; Ramsey, N.F.; Vansteensel, M.J. Association between Lesion Location and Sensorimotor Rhythms in Stroke—A Systematic Review with Narrative Synthesis. *Neurol. Sci.* **2023**, *44*, 4263–4289. [CrossRef] [PubMed]

14. Lacerda, G.J.; Pacheco-Barrios, K.; Barbosa, S.P.; Marques, L.M.; Battistella, L.; Fregni, F. A Neural Signature for Brain Compensation in Stroke with EEG and TMS: Insights from the DEFINE Cohort Study. *Neurophysiol. Clin.* **2024**, *54*, 102985. [CrossRef] [PubMed]
15. Gladstone, D.J.; Danells, C.J.; Black, S.E. The Fugl-Meyer Assessment of Motor Recovery after Stroke: A Critical Review of Its Measurement Properties. *Neurorehabilit. Neural Repair* **2002**, *16*, 232–240. [CrossRef] [PubMed]
16. Simis, M.; Imamura, M.; Sampaio de Melo, P.; Marduy, A.; Battistella, L.; Fregni, F. Deficit of Inhibition as a Marker of Neuroplasticity (DEFINE Study) in Rehabilitation: A Longitudinal Cohort Study Protocol. *Front. Neurol.* **2021**, *12*, 695406. [CrossRef] [PubMed]
17. Xu, J.; Ejaz, N.; Hertler, B.; Branscheidt, M.; Widmer, M.; Faria, A.V.; Harran, M.D.; Cortes, J.C.; Kim, N.; Celnik, P.A.; et al. Separable Systems for Recovery of Finger Strength and Control after Stroke. *J. Neurophysiol.* **2017**, *118*, 1151–1163. [CrossRef]
18. Delorme, A.; Makeig, S. EEGLAB: An Open Source Toolbox for Analysis of Single-Trial EEG Dynamics Including Independent Component Analysis. *J. Neurosci. Methods* **2004**, *134*, 9–21. [CrossRef]
19. Groppa, S.; Oliviero, A.; Eisen, A.; Quartarone, A.; Cohen, L.G.; Mall, V.; Kaelin-Lang, A.; Mima, T.; Rossi, S.; Thickbroom, G.W.; et al. A Practical Guide to Diagnostic Transcranial Magnetic Stimulation: Report of an IFCN Committee. *Clin. Neurophysiol.* **2012**, *123*, 858–882. [CrossRef]
20. Pacheco-Barrios, K.; Pimenta, D.C.; Pessotto, A.V.; Fregni, F. Motor Cortex Inhibition and Facilitation Correlates with Fibromyalgia Compensatory Mechanisms and Pain: A Cross-Sectional Study. *Biomedicines* **2023**, *11*, 1543. [CrossRef]
21. Finnigan, S.; Wong, A.; Read, S. Defining Abnormal Slow EEG Activity in Acute Ischaemic Stroke: Delta/Alpha Ratio as an Optimal QEEG Index. *Clin. Neurophysiol.* **2016**, *127*, 1452–1459. [CrossRef] [PubMed]
22. Cassidy, J.M.; Wodeyar, A.; Wu, J.; Kaur, K.; Masuda, A.K.; Srinivasan, R.; Cramer, S.C. Low-Frequency Oscillations Are a Biomarker of Injury and Recovery After Stroke. *Stroke* **2020**, *51*, 1442–1450. [CrossRef]
23. Ramanathan, D.S.; Guo, L.; Gulati, T.; Davidson, G.; Hishinuma, A.K.; Won, S.-J.; Knight, R.T.; Chang, E.F.; Swanson, R.A.; Ganguly, K. Low-Frequency Cortical Activity Is a Neuromodulatory Target That Tracks Recovery after Stroke. *Nat. Med.* **2018**, *24*, 1257–1267. [CrossRef]
24. Nicolo, P.; Rizk, S.; Magnin, C.; Pietro, M.D.; Schnider, A.; Guggisberg, A.G. Coherent Neural Oscillations Predict Future Motor and Language Improvement after Stroke. *Brain* **2015**, *138*, 3048–3060. [CrossRef]
25. Antonioni, A.; Galluccio, M.; Baroni, A.; Fregna, G.; Pozzo, T.; Koch, G.; Manfredini, F.; Fadiga, L.; Malerba, P.; Straudi, S. Event-Related Desynchronization during Action Observation Is an Early Predictor of Recovery in Subcortical Stroke: An EEG Study. *Ann. Phys. Rehabil. Med.* **2024**, *67*, 101817. [CrossRef] [PubMed]
26. Antonioni, A.; Galluccio, M.; Toselli, R.; Baroni, A.; Fregna, G.; Schincaglia, N.; Milani, G.; Cosma, M.; Ferraresi, G.; Morelli, M.; et al. A Multimodal Analysis to Explore Upper Limb Motor Recovery at 4 Weeks After Stroke: Insights from EEG and Kinematics Measures. *Clin. EEG Neurosci.* **2024**, *55*, 465–476. [CrossRef] [PubMed]
27. Boissy, P.; Bourbonnais, D.; Carlotti, M.M.; Gravel, D.; Arsenault, B.A. Maximal grip force in chronic stroke subjects and its relationship to global upper extremity function. *Clin. Rehabil.* **1999**, *13*, 354–362. [CrossRef]
28. Jobbágy, Á.; Marik, A.R.; Fazekas, G. Quantification of the Upper Extremity Motor Functions of Stroke Patients Using a Smart Nine-Hole Peg Tester. *J. Healthc. Eng.* **2018**, *2018*, 7425858. [CrossRef]
29. Mathiowetz, V.; Kashman, N.; Volland, G.; Weber, K.; Dowe, M.; Rogers, S. Grip and pinch strength: Normative data for adults. *Arch. Phys. Med. Rehabil.* **1985**, *66*, 69–74.
30. Schmidt, R.T.; Toews, J.V. Grip strength as measured by the Jamar dynamometer. *Arch. Phys. Med. Rehabil.* **1970**, *51*, 321–327.
31. Shimoyama, I.; Ninchoji, T.; Uemura, K. The finger-tapping test. A quantitative analysis. *Arch. Neurol.* **1990**, *47*, 681–684. [CrossRef] [PubMed]

brain sciences

MDPI

Review

Neurophysiological Markers of Reward Processing Can Inform Preclinical Neurorehabilitation Approaches for Cognitive Impairments Following Brain Injury

Miranda Francoeur Koloski [1,2,3,*], Reyana Menon [1,†] and Victoria Krasnyanskiy [1,†]

1 Mental Health Service, VA San Diego Healthcare System, La Jolla, CA 92161, USA
2 Department of Psychiatry, UC San Diego, La Jolla, CA 92093, USA
3 Center of Excellence for Stress and Mental Health, VA San Diego Healthcare System, La Jolla, CA 92161, USA
* Correspondence: mfrancoeur@health.ucsd.edu; Tel.: +1-(603)-988-7516
† These authors contributed equally to this work.

Abstract: Brain stimulation therapies may be used to correct motor, social, emotional, and cognitive consequences of traumatic brain injury (TBI). Neuromodulation applied with anatomical specificity can ameliorate desired symptoms while leaving functional circuits intact. Before applying precision medicine approaches, preclinical animal studies are needed to explore potential neurophysiological signatures that could be modulated with neurostimulation. This review discusses potential neural signatures of cognition, particularly reward processing, which is chronically impaired after brain injury. Electrophysiology, compared to other types of biomarkers, can detect deficits missed by structural measures, holds translational potential between humans and animals, and directly informs neuromodulatory treatments. Disturbances in oscillatory activity underscore structural, molecular, and behavioral impairments seen following TBI. For instance, cortico-striatal beta frequency activity (15–30 Hz) during reward processing represents subjective value and is chronically disturbed after frontal TBI in rodents. We use the example of evoked beta oscillations in the cortico-striatal network as a putative marker of reward processing that could be targeted with electrical stimulation to improve decision making after TBI. This review highlights the necessity of collecting electrophysiological data in preclinical models to understand the underlying mechanisms of cognitive behavioral deficits after TBI and to develop targeted stimulation treatments in humans.

Keywords: neurophysiology; traumatic brain injury; reward processing; beta oscillations

1. Introduction

Biomarkers serve as indicators of biological processes that are objective and reproducible [1]. Categories of biomarkers include neuroimaging (magnetic resonance imaging (MRI)), neurophysiological (Electroencephalography (EEG)), biofluid (blood), and digital biomarkers (wearables), all of which are being implored to study traumatic brain injury (TBI). A number of TBI biomarkers have been cataloged, including serum-based protein, exosomal microRNAs, and metabolic indicators [2,3]. However, these molecular markers are limited in temporal precision and ability to reflect ongoing changes in brain activity, particularly in relation to cognition. Structural imaging (MRI) often fails to detect mild injuries. Although, new imaging advancements like diffusion tensor imaging have improved sensitivity to detect microstructural damage (axonal injury) [4,5]. To address these limitations, the impact of injury on neurophysiological activity (neural oscillations) can be

https://doi.org/10.3390/brainsci15050471

examined. The search for biomarkers in TBI is extremely difficult given the heterogeneous nature of brain injury and the symptoms it produces. Even a single injury propagates multiscale disruptions to local and distributed brain areas. Since behavioral outcomes are hard to predict, establishing a reliable biomarker of injury that can estimate injury progression and response to treatment will be useful.

Brain injury triggers a cascade of insults to the central nervous system including a combination of inflammatory, cellular, and molecular changes [6–8]. Due to its superficial position in the brain, prefrontal cortex is particularly susceptible to injury [9–11]. Cortical damage leads to changes in reward-guided behavior, including impaired decision making, decreased motivation, impulsivity, and interrupted reinforcement-learning [11–13]. Most symptoms are resolved; however, in 5–15% of cases, cognitive deficits endure for years after injury [10,11]. Chronic symptoms are a result of cortical neuron loss, disconnection with networks, and neurotransmitter dysregulation [14–16].

The mechanism of injury determines if damage is focal (hematoma, hemorrhage, contusion), diffuse (axonal injury), or a combination. The most common type of injury is diffuse damage to the white matter tracts caused by shearing [17–19]. Diffuse axonal injury often fractures cortical-subcortical connectivity that can afflict cognitive networks and perpetuate neuropsychiatric symptoms [18,20]. The symptom profile is dependent on the neural pathways impacted. For example, thalamo-cortical tract damage accounts for significant alteration in executive dysfunction [21].

The goal of TBI treatment is to develop a precision medicine approach that targets a specific brain area or brain function (neurophysiological biomarker) that can be individualized based on each unique trauma pathology. Most treatment strategies center around addressing motor functions or physical impairments through rest and rehabilitation [22], leaving long-term cognitive deficits unaddressed. TBI-induced motor deficits, including gait, coordination, fine motor skills, seizures, and muscle tremors, can be alleviated with a combination of rehabilitation, brain stimulation, and pharmacologic strategies (muscle relaxation/motor inhibition drugs) [23–25]. There are no pharmacological treatments that specifically regulate TBI-related changes in mood and cognition or prevent the onset of neuropsychiatric disorders (which have a 2–3 times greater risk of developing) [11,16,26].

In line with a precision medicine approach, using neuromodulation to provide electrical stimulation is thought to activate specific neural circuits, which can strengthen its functional network to improve desired behaviors. Identifying a neurophysiological biomarker associated with specific behaviors is needed to improve the precision and reliability of neuromodulation. As opposed to pharmacological interventions, neuromodulation has the potential to influence a multitude of brain areas while maintaining specificity for functional circuits. This review aims to synthesize findings across preclinical and clinical studies related to neurophysiological biomarkers in TBI, specifically oscillatory associated with reward-guided decision-making impairments seen chronically after injury. We provide a theoretical framework by which preclinical studies can support large-scale electrophysiological recordings in healthy animals to identify brain networks and potential "biomarkers" that can be targeted with electrical stimulation in injured animals (Figure 1). While not a formal meta-analysis, our narrative review offers converging evidence to support the translational potential of using electrophysiological signatures to identify new therapeutic targets for TBI.

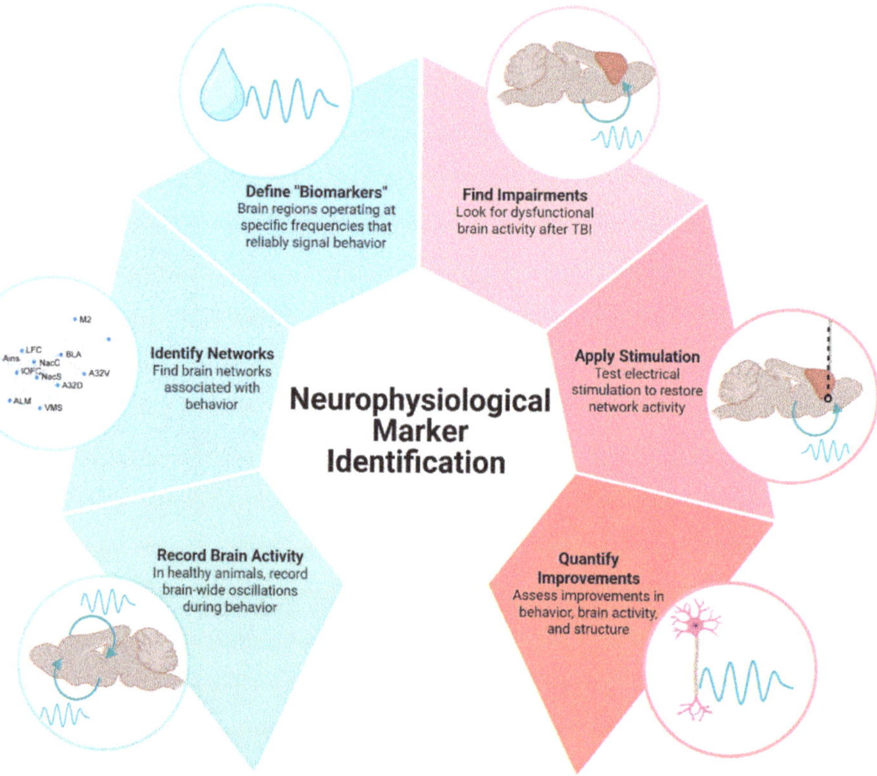

Figure 1. Neurophysiological marker identification. An illustrative overview of how identifying neural activity associated with distinct behaviors can be used to identify and test therapeutic targets in TBI. In this example of a preclinical study, brain activity recorded in healthy animals is used to identify brain networks associated with behavior. Neural signals with region and frequency specificity can be used to apply electrical stimulation to injured brains. In this review, we use beta frequency oscillations in the cortico-striatal network as a potential neurophysiological marker of reward processing that can be used to improve cognition in animals with TBI. Created in BioRender. Koloski, M. (2025) https://BioRender.com/7xr6lr3 (accessed on 21 April 2025).

2. Neurophysiological Markers Can Identify Potential Therapeutic Targets

2.1. Brain Oscillations and Their Disruption After Injury

Electrophysiological measures, including EEG, can detect neural changes after even a mild TBI that structural imaging may fail to detect [4,22]. For example, EEG can detect reduced amplitude and slowing of brain waves in the acute and subacute phases of mild TBI of patients without visible structural damage on MRI/CT scans [27]. The benefits of electrophysiological measures include good temporal resolution, translatability between humans and animals, affordability, usefulness during recovery, and the capacity to be paired with cognitive tasks [1,12]. Oscillatory activity in an EEG reflects coordinated neural activity that facilitates communication within and between brain areas. Oscillatory bands, which are operationally defined based on functional brain states, range from 0.5 to 500 Hz [28]. The average ionic movements (including synaptic activity, calcium fluctuations, spike after-hyperpolarization, gap junctions, glial, etc.) recorded from an electrode are known as the local field potential (LFP) [28]. LFP activity patterns are associated with distinct behaviors and can predict disease states or response to treatment [28,29].

Numerous studies have identified persistent abnormalities in oscillatory dynamics following TBI, including theta suppression during working memory tasks [30], delta elevations during wakefulness [31], and reductions in beta/gamma frequencies during cognition [32]. Generally, lower frequencies (8–30 Hz) are thought to reflect top–down information processing driven by thalamo-cortical connections and higher frequencies (40–100 Hz) to drive bottom–up processing [33,34]. Importantly, connectivity data, reflecting widespread changes across networks, better predict symptoms severity than local changes in power, suggesting brain injury should be treated as a network disorder (opposed to focal injury) [5]. Changes in oscillations are identified after even mild TBI and therefore may be sensitive to the underlying neural alterations associated with injury [5]. Thus, neurophysiological signals have the potential to aid in TBI diagnosis, prognosis, and treatment response monitoring [1,22,35].

In both humans and animals, TBI disrupts brain oscillations. More recently, studies have shown that electrical stimulation can restore oscillatory rhythms to improve cognitive outcomes following TBI. Generally, after injury, there is a decrease in power across frequency bands which returns to baseline after the acute injury response [1,36]. However, deficits in brain activity can persist chronically (>10 weeks) [36] and, thus, may represent a therapeutic window for clinical intervention. In patients with TBI, altered EEG activity can be seen years after injury [12,32,37,38]. Specifically, alterations of thalamo-cortical circuits are associated with persistent cognitive symptoms seen with TBI [5,21]. The dorsolateral prefrontal cortex, which is normally engaged during cognition, shows reduced power and EEG abnormalities after TBI [8]. Electrophysiological abnormalities occur across brain states (rest, sleep, etc.) and are consistent enough to distinguish between injured and non-injured patients and predict patient outcomes [32,38].

2.2. Neuromodulation to Restore Brain Activity

Modulating brain activity through electrical stimulation can restore rhythmic patterns of the brain disrupted by TBI [39]. To address the widespread neural impairments, stimulation would need to restore activity beyond the focal injury site. Functional MRI in humans and electrophysiological recordings in rodents show that brain stimulation emulates changes across networks [40].

Methods of neuromodulation vary in their level of invasiveness, continuous or transient application, and pairing with functional behavior or rest. Stimulation protocols can be open-loop (applied at/for a certain amount of time, not tied to a behavior or brain function), "on-demand" (triggered by a behavioral event), or closed-loop (triggered by brain feedback) [41,42]. High-frequency stimulation (>5 Hz) is thought to induce excitability and increase cerebral blood flow, whereas low-frequency stimulation (<1 Hz) typically induces inhibitory effects [43]. Factors such as the type of neuromodulation, duration of stimulation, and regularity required will determine the best target for neuromodulation.

Neurostimulation in the form of transcranial magnetic stimulation (TMS), transcranial direct current stimulation (tDCS), and deep brain stimulation (DBS), applied in preclinical models have successfully ameliorated behavioral symptoms of TBI, including motor, attention, memory, mood, and impulsivity impairments [35,44]. In preclinical studies, stimulation applied immediately to several weeks after trauma improved behavioral outcomes, whereas clinical applications of neuromodulation are not usually introduced until later stages of recovery [44]. Repeated TMS applied to the left dorsolateral prefrontal cortex has FDA approval to treat depression and therefore may also be useful in TBI patients exhibiting mood dysregulation, central pain, and blunted cognition or affect [43,45,46].

Importantly, compared to other interventions, growing evidence supports that TMS can have a long-lasting impact on neural circuits by inducing plasticity [47].

TMS has been used in rodent models with non-invasive coils, but the translational potential is obscured by stimulation site specificity and the use of restraints or anesthesia [48,49]. Instead, "TMS-like" approaches that use implanted electrodes opposed to topical coils, may be optimal when designing preclinical studies [35]. "TMS-like" protocols of 20 Hz stimulation trains applied repeatedly for 10 days to prelimbic cortex normalize depressive-like behaviors and reduce brain-derived neurotrophic factor levels in reward-related regions [50].

2.3. *How Does Stimulation Restore Brain Function Following TBI?*

Axonal shearing, swelling, chronic inflammation, and microglia activation perpetuate neural loss and tissue degeneration after trauma [22,51–53]. Stimulation provides neuroprotection, decreases apoptosis, reduces inflammation, directs cerebral blood flow, and induces neuroplasticity to improve cell health and restore neural circuits [35,43,44]. White matter regeneration (which may be particularly important for cognition) is achieved through optogenetic, pharmacogenetic, or indirect brain stimulation by encouraging glial cells to myelinate active axons [54,55]. In preclinical models of TBI, reversed cortical tissue loss, white matter regeneration, and increased intracellular signaling are directly related to behavioral improvements [56]. TDCS and TMS increase c-fos expression (a marker of neural activity) [57]. TDCS also increases brain-derived neurotrophic factor (BDNF) in stimulated cortical regions of injured brains, indicating a focal neuroplasticity response that is associated with cognitive improvements in spatial memory [58]. Similarly, TMS reduces glial fibrillary acidic protein (GFAP) expression, associated with astroglia cells and neural regeneration [2,59]. Improved neural health and signal transduction can re-establish physiological patterns. DBS reduces spontaneous neural firing and encourages activation of efferent pathways by increasing the release of GABA from interneurons to restore the excitatory/inhibitory balance [45,60].

3. A Potential Biomarker of Reward-Processing Deficits in TBI

The lack of mechanistic information regarding failure of cognitive networks precludes the utility of neuromodulation to treat chronic TBI deficits. Precision medicine approaches to treat TBI depend on finding neurophysiological biomarkers that reliably mark discrete cognitive functions like decision-making, planning, and memory.

3.1. *Preclinical Models of Frontal TBI*

Rodent models provide information about neurobiological mechanisms which are necessary to inform clinical interventions. Aside from notable differences in brain size, relative volume of brain regions, and presence/absence of cortical folding, the architecture and functional networks are largely preserved between humans and rodents [61]. Animal models of TBI offer control over injury specifications, ability to monitor injury progression, and invasive techniques that are not feasible in humans [62]. Limitations in these models include lack of continuity between protocols, challenge replicating the heterogeneous nature of TBI, broad timeframe of injury (acute, subacute, and chronic), and inability to model exact biomechanical parameters of injury [37,48]. Despite these limitations, rodent models of frontal TBI produce robust and predictable cognitive deficits [62]. Bilateral frontal controlled cortical impact (CCI) injury reliably produces impairments in impulse control and decision-making [63].

Animal models also enable large-scale recordings of in vivo brain activity. Multi-site electrodes simultaneously capture field potentials at different sites across, or between, networks. "Brain-wide" recordings measuring activity from up to 32 brain areas simultaneously, can be used to characterize networks operating at distinct oscillatory frequencies to support unique behaviors [64–66]. Although these electrodes record from within the brain, the information they provide (large-scale measure of brain oscillations) is like human EEG. The following section provides an example of how this technique can identify neurophysiological biomarkers and potential targets for neuromodulation.

3.2. Deficient Cortical Beta Oscillations Indicate Reward-Processing Issues After TBI

As shown in humans and animals, an injury to prefrontal cortex results in chronic executive function deficits, including attention, memory, and reward-guided decision making [13,67,68]. Damage causes lasting disruption to brain circuits through inflammation, gliosis, cell death, and alterations in microstructure, ultimately effecting neural communication. Changes in brain oscillations perpetuate disruptions throughout functional networks after cortical damage. Related to executive function deficits, fronto-parietal (attention) and fronto-striatal (decision-making) networks show reduced electrophysiological responses and connectivity consistent with slower reaction times and poor decision making [69,70].

Oscillatory activity, particularly at beta frequencies, deriving from the prefrontal cortex is important for top–down attention [71], executive control [72], sensorimotor integration [73], motor planning [74], and decision making [75]. Beta oscillations are correlated with the cortical microstructure (i.e., myelin density/integrity) and therefore susceptible to damage through TBI [76]. Even mild TBI leads to localized and widespread disruptions in beta oscillatory activity (measured by magnetoencephalography) [5]. Beta activity in cortico-striatal regions (prefrontal cortex, orbitofrontal cortex, anterior insula, ventral striatum, and basolateral amygdala) marks positive valence (i.e., rewarded outcomes) [77–80]. In humans, reward-evoked beta oscillations correlate with activation of (and coupling between) ventral striatum and medial prefrontal cortex, suggesting that beta frequency oscillations may coordinate the neural circuits involved in reward processing [80]. In rats, across multiple tasks of reward-guided decision-making, we consistently find increases in beta-frequency (15–30 Hz) oscillations during reward processing that reflects reward magnitude, reward probability, and subjective value [77].

Due to its superficial position in the brain, cortex is particularly susceptible to injury [10], giving rise to socially inappropriate behavior, poor impulse control, and trouble decision-making [21]. When cortical areas are damaged, their participating networks are also dysfunctional. As a network, cortico-striatal brain areas mediate adaptive reward-guided decision-making by creating action-outcome associations, controlling impulsive choices, and responding flexibly to changing conditions [81]. Bifrontal TBI caused by CCI decreases the ability of rats to detect reinforced outcomes, impairs behavioral flexibility, and increases impulsivity [63,82].

Consistent with these behavioral changes, we find that CCI-TBI also blunts reward-locked beta oscillations and cellular activity in lateral orbitofrontal cortex (measured by c-Fos staining) [82]. Beta activity in the orbitofrontal cortex does not discriminate between reward outcomes as efficiently following injury (effects observed up to 12 weeks after injury). Beta oscillations are correlated with structural deficits observed after TBI, including myelin density and morphology [76]. Local changes in beta power and deficits in functional connectivity following TBI has also been characterized in thalamo-cortical circuits [5,30,83]. Interestingly, we found that reductions in orbitofrontal activity were rescued by a behavioral intervention in which reinforced outcomes were cued, thus suggesting reward signals in the brain are malleable and can be altered to improve cognitive behavior [82]. Brain-based interventions may work similarly to increase beta oscillations during positive reward outcomes thereby reversing reward-related deficits from TBI. For these reasons, reward-evoked beta oscillations in the cortico-striatal network may represent a neurophysiological marker that can be targeted with electrical stimulation to improve cognitive symptoms of TBI. In accordance with the theoretical framework for neurophysiological marker identification (Figure 1), we have shown that large-scale recordings of brain activity can identify behaviorally relevant networks to define putative "biomarkers" with region and frequency specificity (Figure 2). We have found impairments of reward-evoked beta oscillations following TBI that may be remediated by electrical stimulation (Figure 2).

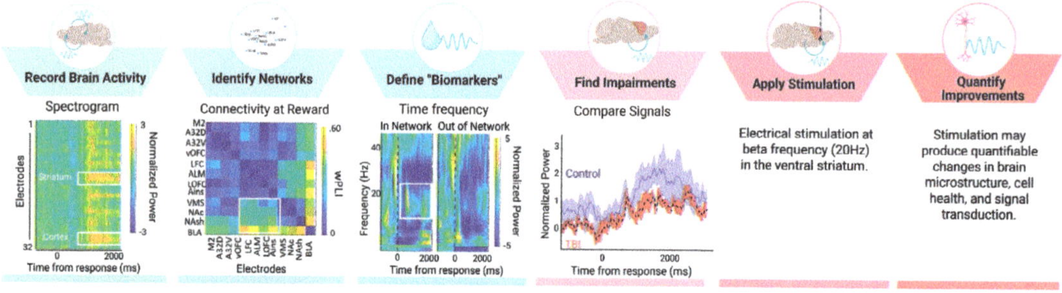

Figure 2. Reward-evoked beta oscillations as a putative neurophysiological marker. We apply large-scale beta oscillations to our theoretical framework to provide one example of how to identify neural activity associated with distinct behaviors to test neuromodulation targets in TBI. Neural signals were recorded across 32 CH in rodents performing behavioral tasks. We find areas of the cortico-striatal network that have heightened activity (spectrogram shows increased power at cortical and striatal electrodes) and connectivity (weighted phase lag index (wPLI) connectivity between pairs of electrodes) during reward. Evoked activity during reward occurs at beta frequencies (time frequency plot shows increased beta frequency activity in "within network" electrodes), thereby providing region (striatum) and frequency (beta) specificity needed for "biomarker" identification. In animals with TBI (red), beta oscillations during reward are blunted compared to controls (blue) (mean and SEM of evoked signal), and therefore may represent a logical target for neuromodulation. In the future, beta frequency (20 Hz) stimulation applied to the striatum should be tested in TBI animals and improvements in cell health, microstructure, and signal integrity should be quantified. Importantly, beta oscillations represent just one example of a neurophysiological signal that can be targeted to improve TBI symptoms. Our discussion includes examples of other neurophysiological signals. Created in BioRender. Koloski, M. (2025) https://BioRender.com/7xr6lr3 (accessed on 21 April 2025).

Although, we have not tested electrical stimulation on this putative beta signal in TBI animals, we have stimulated beta oscillations to influence decision making in healthy animals. Beta oscillations were modified in healthy rats performing a delayed discounting task, in which they chose between a small, immediate reward or a large, delayed reward [77]. In total, 20 Hz electrical stimulation was applied "on-demand", triggered by an animal selecting the large, delayed reward outcome. Stimulation biased behavior toward the large reward choice, despite the temporal delay (2–10 s). Multiple cortico-striatal targets (including OFC) had this effect [77]. Although these results are limited due to lack of control/replication, they warrant testing beta-frequency modulation in an injured cohort.

While much of the current evidence regarding beta oscillations observed after TBI stems from our research, other studies have identified beta activity for reward processing and cognitive control across both humans and animal models [71–73,78,79]. Beta-evoked oscillations are seen during decision-making, particularly in situations with high cognitive demand [73]. In humans with mild TBI, deficits in frontal beta power (measured with magnetoencephalography) reflect thalamo-cortical network damage [5]. Beta-functional connectivity, compared with other frequencies, best predicted symptoms severity and mild TBI classification [5].

Contrary to these findings, some researchers have found evoked beta power following the omission of expected rewards [84]. Thus, although beta likely plays a role in reward processing, under what conditions it signals expectancy and reward outcome remain unclear. However, the role of beta oscillations following TBI warrants further investigation.

3.3. Other Potential Neurophysiological Biomarkers

Beta oscillations in the cortico-striatal network may represent a promising neurophysiological biomarker tied to a specific cognitive domain (reward valuation). Other neurophysiological markers pertaining to different cognitive domains may also be used to improve TBI outcomes with neuromodulation. For example, theta oscillations are suppressed after TBI and return over the course of recovery [30,83]. Theta oscillations are an appealing target to consider, as they are believed to modulate long-term potentiation underlying learning and memory [85,86]. Stimulating the medial septum (increasing theta oscillations) restored cognitive performance in rats with TBI [83,86,87]. Likewise, in rodent TBI models, gamma activity is decreased near the focal injury site and signal (and associated behavior consequences) is rescued with 40 Hz modulation (Blue LED photobiomodulation therapy) [88].

Another avenue that has been explored is modulating delta oscillations to change sleep dynamics (and thereby wakeful neural function) that are interrupted by TBI. Delta oscillations typically present in deep sleep are elevated during wakefulness after TBI [31]. Hypothalamic stimulation recovers delta oscillations to restore organized sleep after TBI [31].

4. Limitations and Challenges (In Translation and Beyond)

Despite promising preliminary findings, research on beta oscillations as a biomarker in TBI is still in its infancy. Generally, a role for beta oscillations in facilitating cognition (and reward evaluation) is accepted, but the disruption of beta signals in the case of TBI is less known. Beta oscillations are believed to facilitate top–down control of cognitive behaviors [89], and although these coincide with TBI sequalae, the involvement of beta oscillations explicitly still needs to be explored. Studies examining neurophysiological markers pertaining to cognitive deficits in TBI are sparse and often limited by methodological variability (injury models, severity, timeline, region of interest) across studies. Compared with deficits in motor function, cognitive networks are harder to precisely target and are often confounded by emotional/neuropsychiatric symptoms [90]. More preclinical research is needed to delineate the neurobiological substrates of cognitive behavior. Neurophysiological measures sometimes show weak or no associations with behavioral or functional outcomes. Barone et al., 2024, find that EEG alterations in mild TBI did not significantly correlate with clinical outcomes measures, underscoring the need to further refine which markers can best predict patient outcomes [27]. Although EEG is used clinically in an intensive-care setting to monitor seizure activity or drug effects, it is not common in outpatient settings [90]. Clinical assessment of EEG activity over a more chronic time course will help identify electrophysiological signatures related to individual pathologies.

Moreover, the success of neuromodulation to ameliorate cognitive symptoms of TBI is occluded by inconsistent protocols and translational barriers. Animal stimulation protocols are inconsistent in how they apply neuromodulation to TBI. Methodological factors like the type of stimulation (deep brain v. external), location, duration, and onset of stimulation relevant to TBI must be considered [44]. Further, factors like the use of anesthesia complicate methodological similarities and obscure translatability. While rodent models offer a high degree of experimental control, their translational value is often obscured by differences in anatomical division, cytoarchitecture, network structure, and behavior complexity. Moreover, clinical safety profiles and ideal stimulation parameters may be hard to gleam from rodent studies. The risk of seizures and adverse side effects with TMS or tDCS in TBI patients must be considered when generalizing across species [39,44,45]. Since patients with TBI present a high-risk population, the risk profile and safety guidelines of clinical neuromodulation must be carefully considered. Future work will need to identify homologous neural circuits, validate neurophysiological markers across species, and refine neuromodulation parameters for clinical application.

Finally, due to the heterogenous nature of TBI, multimodal treatments will likely be effective. Guided by biomarkers, stimulation may be paired with therapy or pharmacological treatment to address multiple symptoms. In humans, neuromodulation (tDCS) paired with cognitive training can enhance its success [90]. In some cases, enhancements in cognition were small and non-significant, further supporting the need to identify the networks to apply neuromodulation that will have maximal benefit when paired with cognitive rehabilitation [91]. Multimodal treatments create a challenge in ascertaining the exact mechanism, pathological target, and duration/frequency of treatment needed [44,90]. Without knowledge about which networks to target, it is possible that each treatment modality would target opposing networks and produce a null effect [90].

5. Conclusions

Even after identifying a reliable biomarker for intervention, important questions, like when in the disease progression to intervein and how long benefits persist, will need to be researched. Electrophysiology allows us to measure impairments which are undetectable by other methods, but using multi-modal diagnostic biomarkers (blood, structural, physiological, genetic) offers the most precision to treat individual pathologies by capturing the heterogeneous nature of TBI. The current lack of objective biomarkers for TBI poses a significant challenge for translational research. Future work should employ biomarkers to confirm that injuries induced in preclinical models are comparable to humans and that a proposed treatment would be clinically effective.

Author Contributions: Conceptualization, M.F.K.; Writing—original draft, M.F.K.; Writing—review and editing, R.M. and V.K.; Visualization, M.F.K., R.M. and V.K.; Supervision, M.F.K.; Funding Acquisition, M.F.K. All authors have read and agreed to the published version of the manuscript.

Funding: This research was funded by VA Office of Research and Development (Career Development Award IK2BX0006125 to M.F.K.).

Institutional Review Board Statement: Not Applicable.

Informed Consent Statement: Not Applicable.

Acknowledgments: We would like to acknowledge Christopher M. O'Hearn, Alyssa Terry, Brianna Newton, Morteza Salimi, and Jonathan Mishler for their contributions to data collection and analysis in projects discussed here. We also acknowledge Cole Vonder Haar and Dhakshin Ramanathan for project oversight and discussions pertaining to this review.

Conflicts of Interest: The authors declare no conflicts of interest.

References

1. Wilde, E.A.; Wanner, I.-B.; Kenney, K.; Gill, J.; Stone, J.R.; Disner, S.; Schnakers, C.; Meyer, R.; Prager, E.M.; Haas, M.; et al. A Framework to Advance Biomarker Development in the Diagnosis, Outcome Prediction, and Treatment of Traumatic Brain Injury. *J. Neurotrauma* **2022**, *39*, 436–457. [CrossRef] [PubMed]

2. Ghaith, H.S.; Nawar, A.A.; Gabra, M.D.; Abdelrahman, M.E.; Nafady, M.H.; Bahbah, E.I.; Ebada, M.A.; Ashraf, G.M.; Negida, A.; Barreto, G.E. A Literature Review of Traumatic Brain Injury Biomarkers. *Mol. Neurobiol.* **2022**, *59*, 4141–4158. [CrossRef] [PubMed]
3. Gutierre, M.U.; Telles, J.P.M.; Welling, L.C.; Rabelo, N.N.; Teixeira, M.J.; Figueiredo, E.G. Biomarkers for Traumatic Brain Injury: A Short Review. *Neurosurg. Rev.* **2021**, *44*, 2091–2097. [CrossRef]
4. Bigler, E.D. Neuroimaging Biomarkers in Mild Traumatic Brain Injury (mTBI). *Neuropsychol. Rev.* **2013**, *23*, 169–209. [CrossRef] [PubMed]
5. Zhang, J.; Safar, K.; Emami, Z.; Ibrahim, G.M.; Scratch, S.E.; Da Costa, L.; Dunkley, B.T. Local and Large-Scale Beta Oscillatory Dysfunction in Males with Mild Traumatic Brain Injury. *J. Neurophysiol.* **2020**, *124*, 1948–1958. [CrossRef]
6. Freire, M.A.M.; Rocha, G.S.; Bittencourt, L.O.; Falcao, D.; Lima, R.R.; Cavalcanti, J.R.L.P. Cellular and Molecular Pathophysiology of Traumatic Brain Injury: What Have We Learned So Far? *Biology* **2023**, *12*, 1139. [CrossRef]
7. Ladak, A.A.; Enam, S.A.; Ibrahim, M.T. A Review of the Molecular Mechanisms of Traumatic Brain Injury. *World Neurosurg.* **2019**, *131*, 126–132. [CrossRef]
8. Risbrough, V.B.; Vaughn, M.N.; Friend, S.F. Role of Inflammation in Traumatic Brain Injury–Associated Risk for Neuropsychiatric Disorders: State of the Evidence and Where Do We Go From Here. *Biol. Psychiatry* **2022**, *91*, 438–448. [CrossRef]
9. Chen, Y.-H.; Huang, E.Y.-K.; Kuo, T.-T.; Ma, H.-I.; Hoffer, B.J.; Tsui, P.-F.; Tsai, J.-J.; Chou, Y.-C.; Chiang, Y.-H. Dopamine Release Impairment in Striatum after Different Levels of Cerebral Cortical Fluid Percussion Injury. *Cell Transplant.* **2015**, *24*, 2113–2128. [CrossRef]
10. Chou, A.; Morganti, J.M.; Rosi, S. Frontal Lobe Contusion in Mice Chronically Impairs Prefrontal-Dependent Behavior. *PLoS ONE* **2016**, *11*, e0151418. [CrossRef]
11. Polinder, S.; Cnossen, M.C.; Real, R.G.L.; Covic, A.; Gorbunova, A.; Voormolen, D.C.; Master, C.L.; Haagsma, J.A.; Diaz-Arrastia, R.; Von Steinbuechel, N. A Multidimensional Approach to Post-Concussion Symptoms in Mild Traumatic Brain Injury. *Front. Neurol.* **2018**, *9*, 1113. [CrossRef]
12. Gosselin, N.; Bottari, C.; Chen, J.-K.; Huntgeburth, S.C.; De Beaumont, L.; Petrides, M.; Cheung, B.; Ptito, A. Evaluating the Cognitive Consequences of Mild Traumatic Brain Injury and Concussion by Using Electrophysiology. *Neurosurg. Focus* **2012**, *33*, E7. [CrossRef] [PubMed]
13. Shaver, T.K.; Ozga, J.E.; Zhu, B.; Anderson, K.G.; Martens, K.M.; Vonder Haar, C. Long-Term Deficits in Risky Decision-Making after Traumatic Brain Injury on a Rat Analog of the Iowa Gambling Task. *Brain Res.* **2019**, *1704*, 103–113. [CrossRef] [PubMed]
14. Rabinowitz, A.R.; Levin, H.S. Cognitive Sequelae of Traumatic Brain Injury. *Psychiatr. Clin. N. Am.* **2014**, *37*, 1–11. [CrossRef]
15. Rutherford, W.H.; Merrett, J.D.; McDonald, J.R. Symptoms at One Year Following Concussion from Minor Head Injuries. *Injury* **1979**, *10*, 225–230. [CrossRef] [PubMed]
16. Kundu, B.; Brock, A.A.; Englot, D.J.; Butson, C.R.; Rolston, J.D. Deep Brain Stimulation for the Treatment of Disorders of Consciousness and Cognition in Traumatic Brain Injury Patients: A Review. *Neurosurg. Focus* **2018**, *45*, E14. [CrossRef]
17. Kraus, M.F.; Susmaras, T.; Caughlin, B.P.; Walker, C.J.; Sweeney, J.A.; Little, D.M. White Matter Integrity and Cognition in Chronic Traumatic Brain Injury: A Diffusion Tensor Imaging Study. *Brain* **2007**, *130*, 2508–2519. [CrossRef]
18. Smith, D.H.; Meaney, D.F. Axonal Damage in Traumatic Brain Injury. *Neuroscientist* **2000**, *6*, 483–495. [CrossRef]
19. Stephens, J.A.; Williamson, K.-N.C.; Berryhill, M.E. Cognitive Rehabilitation After Traumatic Brain Injury: A Reference for Occupational Therapists. *OTJR Occup. Ther. J. Res.* **2015**, *35*, 5–22. [CrossRef]
20. Yin, B.; Li, D.-D.; Huang, H.; Gu, C.-H.; Bai, G.-H.; Hu, L.-X.; Zhuang, J.-F.; Zhang, M. Longitudinal Changes in Diffusion Tensor Imaging Following Mild Traumatic Brain Injury and Correlation With Outcome. *Front. Neural Circuits* **2019**, *13*, 28. [CrossRef]
21. Dockree, P.M.; Robertson, I.H. Electrophysiological Markers of Cognitive Deficits in Traumatic Brain Injury: A Review. *Int. J. Psychophysiol.* **2011**, *82*, 53–60. [CrossRef] [PubMed]
22. McNerney, M.W.; Gurkoff, G.G.; Beard, C.; Berryhill, M.E. The Rehabilitation Potential of Neurostimulation for Mild Traumatic Brain Injury in Animal and Human Studies. *Brain Sci.* **2023**, *13*, 1402. [CrossRef] [PubMed]
23. Corrigan, F.; Wee, I.C.; Collins-Praino, L.E. Chronic Motor Performance Following Different Traumatic Brain Injury Severity-A Systematic Review. *Front. Neurol.* **2023**, *14*, 1180353. [CrossRef]
24. Hornby, T.G.; Reisman, D.S.; Ward, I.G.; Scheets, P.L.; Miller, A.; Haddad, D.; Fox, E.J.; Fritz, N.E.; Hawkins, K.; Henderson, C.E.; et al. Clinical Practice Guideline to Improve Locomotor Function Following Chronic Stroke, Incomplete Spinal Cord Injury, and Brain Injury. *J. Neurol. Phys. Ther. JNPT* **2020**, *44*, 49–100. [CrossRef] [PubMed]
25. Tani, J.; Wen, Y.-T.; Hu, C.-J.; Sung, J.-Y. Current and Potential Pharmacologic Therapies for Traumatic Brain Injury. *Pharmaceuticals* **2022**, *15*, 838. [CrossRef] [PubMed]
26. Stein, M.B.; Jain, S.; Giacino, J.T.; Levin, H.; Dikmen, S.; Nelson, L.D.; Vassar, M.J.; Okonkwo, D.O.; Diaz-Arrastia, R.; Robertson, C.S.; et al. Risk of Posttraumatic Stress Disorder and Major Depression in Civilian Patients After Mild Traumatic Brain Injury: A TRACK-TBI Study. *JAMA Psychiatry* **2019**, *76*, 249–258. [CrossRef]

27. Barone, V.; De Koning, M.E.; Van Der Horn, H.J.; Van Der Naalt, J.; Eertman-Meyer, C.J.; Van Putten, M.J.A.M. Neurophysiological Signatures of Mild Traumatic Brain Injury in the Acute and Subacute Phase. *Neurol. Sci.* **2024**, *45*, 3313–3323. [CrossRef]
28. Buzsáki, G.; Anastassiou, C.A.; Koch, C. The Origin of Extracellular Fields and Currents--EEG, ECoG, LFP and Spikes. *Nat. Rev. Neurosci.* **2012**, *13*, 407–420. [CrossRef]
29. Masimore, B.; Kakalios, J.; Redish, A.D. Measuring Fundamental Frequencies in Local Field Potentials. *J. Neurosci. Methods* **2004**, *138*, 97–105. [CrossRef]
30. Cavanagh, J.F.; Rieger, R.E.; Wilson, J.K.; Gill, D.; Fullerton, L.; Brandt, E.; Mayer, A.R. Joint Analysis of Frontal Theta Synchrony and White Matter Following Mild Traumatic Brain Injury. *Brain Imaging Behav.* **2020**, *14*, 2210–2223. [CrossRef]
31. Dong, X.; Ye, W.; Tang, Y.; Wang, J.; Zhong, L.; Xiong, J.; Liu, H.; Lu, G.; Feng, Z. Wakefulness-Promoting Effects of Lateral Hypothalamic Area-Deep Brain Stimulation in Traumatic Brain Injury-Induced Comatose Rats: Upregulation of A1-Adrenoceptor Subtypes and Downregulation of Gamma-Aminobutyric Acid β Receptor Expression Via the Orexins Pathway. *World Neurosurg.* **2021**, *152*, e321–e331. [CrossRef] [PubMed]
32. Antón Álvarez, X.; Sampedro, C.; Pérez, P.; Laredo, M.; Couceiro, V.; Hernández, Á.; Figueroa, J.; Varela, M.; Arias, D.; Corzo, L.; et al. Positive Effects of Cerebrolysin on Electroencephalogram Slowing, Cognition and Clinical Outcome in Patients with Postacute Traumatic Brain Injury: An Exploratory Study. *Int. Clin. Psychopharmacol.* **2003**, *18*, 271–278. [CrossRef]
33. Sherman, M.T.; Kanai, R.; Seth, A.K.; VanRullen, R. Rhythmic Influence of Top–Down Perceptual Priors in the Phase of Prestimulus Occipital Alpha Oscillations. *J. Cogn. Neurosci.* **2016**, *28*, 1318–1330. [CrossRef]
34. Xiong, Y.; Fries, P.; Bastos, A.M. Which Rhythms Reflect Bottom-Up and Top-Down Processing? In *Intracranial EEG*; Axmacher, N., Ed.; Studies in Neuroscience, Psychology and Behavioral Economics; Springer International Publishing: Cham, Switzerland, 2023; pp. 389–414, ISBN 978-3-031-20909-3.
35. Surendrakumar, S.; Rabelo, T.K.; Campos, A.C.P.; Mollica, A.; Abrahao, A.; Lipsman, N.; Burke, M.J.; Hamani, C. Neuromodulation Therapies in Pre-Clinical Models of Traumatic Brain Injury: Systematic Review and Translational Applications. *J. Neurotrauma* **2023**, *40*, 435–448. [CrossRef] [PubMed]
36. Pevzner, A.; Izadi, A.; Lee, D.J.; Shahlaie, K.; Gurkoff, G.G. Making Waves in the Brain: What Are Oscillations, and Why Modulating Them Makes Sense for Brain Injury. *Front. Syst. Neurosci.* **2016**, *10*, 30. [CrossRef]
37. Slobounov, S.; Sebastianelli, W.; Hallett, M. Residual Brain Dysfunction Observed One Year Post-Mild Traumatic Brain Injury: Combined EEG and Balance Study. *Clin. Neurophysiol.* **2012**, *123*, 1755–1761. [CrossRef] [PubMed]
38. Thatcher, R.W.; Walker, R.A.; Gerson, I.; Geisler, F.H. EEG Discriminant Analyses of Mild Head Trauma. *Electroencephalogr. Clin. Neurophysiol.* **1989**, *73*, 94–106. [CrossRef]
39. Demirtas-Tatlidede, A.; Vahabzadeh-Hagh, A.M.; Bernabeu, M.; Tormos, J.M.; Pascual-Leone, A. Noninvasive Brain Stimulation in Traumatic Brain Injury. *J. Head Trauma Rehabil.* **2012**, *27*, 274–292. [CrossRef]
40. Tanaka, Y.; Matsuwaki, T.; Yamanouchi, K.; Nishihara, M. Exacerbated Inflammatory Responses Related to Activated Microglia after Traumatic Brain Injury in Progranulin-Deficient Mice. *Neuroscience* **2013**, *231*, 49–60. [CrossRef]
41. Bundy, D.T.; Nudo, R.J. Preclinical Studies of Neuroplasticity Following Experimental Brain Injury: An Update. *Stroke* **2019**, *50*, 2626–2633. [CrossRef]
42. Ghasemi, P.; Sahraee, T.; Mohammadi, A. Closed- and Open-Loop Deep Brain Stimulation: Methods, Challenges, Current and Future Aspects. *J. Biomed. Phys. Eng.* **2018**, *8*, 209–216. [CrossRef] [PubMed]
43. Calderone, A.; Cardile, D.; Gangemi, A.; De Luca, R.; Quartarone, A.; Corallo, F.; Calabrò, R.S. Traumatic Brain Injury and Neuromodulation Techniques in Rehabilitation: A Scoping Review. *Biomedicines* **2024**, *12*, 438. [CrossRef] [PubMed]
44. Ziesel, D.; Nowakowska, M.; Scheruebel, S.; Kornmueller, K.; Schäfer, U.; Schindl, R.; Baumgartner, C.; Üçal, M.; Rienmüller, T. Electrical Stimulation Methods and Protocols for the Treatment of Traumatic Brain Injury: A Critical Review of Preclinical Research. *J. Neuroeng. Rehabil.* **2023**, *20*, 51. [CrossRef] [PubMed]
45. Perera, T.; George, M.S.; Grammer, G.; Janicak, P.G.; Pascual-Leone, A.; Wirecki, T.S. The Clinical TMS Society Consensus Review and Treatment Recommendations for TMS Therapy for Major Depressive Disorder. *Brain Stimul.* **2016**, *9*, 336–346. [CrossRef]
46. Pink, A.E.; Williams, C.; Alderman, N.; Stoffels, M. The Use of Repetitive Transcranial Magnetic Stimulation (rTMS) Following Traumatic Brain Injury (TBI): A Scoping Review. *Neuropsychol. Rehabil.* **2021**, *31*, 479–505. [CrossRef]
47. Klomjai, W.; Katz, R.; Lackmy-Vallée, A. Basic Principles of Transcranial Magnetic Stimulation (TMS) and Repetitive TMS (rTMS). *Ann. Phys. Rehabil. Med.* **2015**, *58*, 208–213. [CrossRef]
48. Vahabzadeh-Hagh, A.M.; Muller, P.A.; Gersner, R.; Zangen, A.; Rotenberg, A. Translational Neuromodulation: Approximating Human Transcranial Magnetic Stimulation Protocols in Rats. *Neuromodulation Technol. Neural Interface* **2012**, *15*, 296–305. [CrossRef]
49. Boonzaier, J.; Petrov, P.I.; Otte, W.M.; Smirnov, N.; Neggers, S.F.W.; Dijkhuizen, R.M. Design and Evaluation of a Rodent-Specific Transcranial Magnetic Stimulation Coil: An In Silico and In Vivo Validation Study. *Neuromodulation* **2020**, *23*, 324–334. [CrossRef]
50. Moshe, H.; Gal, R.; Barnea-Ygael, N.; Gulevsky, T.; Alyagon, U.; Zangen, A. Prelimbic Stimulation Ameliorates Depressive-Like Behaviors and Increases Regional BDNF Expression in a Novel Drug-Resistant Animal Model of Depression. *Brain Stimul.* **2016**, *9*, 243–250. [CrossRef]

51. Collins-Praino, L.E. Traumatic Axonal Injury as a Key Driver of the Relationship between Traumatic Brain Injury, Cognitive Dysfunction, and Dementia. In *Cellular, Molecular, Physiological, and Behavioral Aspects of Traumatic Brain Injury*; Elsevier: Amsterdam, The Netherlands, 2022; pp. 475–486, ISBN 978-0-12-823036-7.
52. Smith, D.H.; Kochanek, P.M.; Rosi, S.; Meyer, R.; Ferland-Beckham, C.; Prager, E.M.; Ahlers, S.T.; Crawford, F. Roadmap for Advancing Pre-Clinical Science in Traumatic Brain Injury. *J. Neurotrauma* **2021**, *38*, 3204–3221. [CrossRef]
53. Xu, H.; Wang, Z.; Li, J.; Wu, H.; Peng, Y.; Fan, L.; Chen, J.; Gu, C.; Yan, F.; Wang, L.; et al. The Polarization States of Microglia in TBI: A New Paradigm for Pharmacological Intervention. *Neural Plast.* **2017**, *2017*, 5405104. [CrossRef] [PubMed]
54. Johnson, V.E.; Stewart, J.E.; Begbie, F.D.; Trojanowski, J.Q.; Smith, D.H.; Stewart, W. Inflammation and White Matter Degeneration Persist for Years after a Single Traumatic Brain Injury. *Brain* **2013**, *136*, 28–42. [CrossRef]
55. Maas, A.I.R.; Menon, D.K.; Manley, G.T.; Abrams, M.; Åkerlund, C.; Andelic, N.; Aries, M.; Bashford, T.; Bell, M.J.; Bodien, Y.G.; et al. Traumatic Brain Injury: Progress and Challenges in Prevention, Clinical Care, and Research. *Lancet Neurol.* **2022**, *21*, 1004–1060. [CrossRef] [PubMed]
56. Carballosa Gonzalez, M.M.; Blaya, M.O.; Alonso, O.F.; Bramlett, H.M.; Hentall, I.D. Midbrain Raphe Stimulation Improves Behavioral and Anatomical Recovery from Fluid-Percussion Brain Injury. *J. Neurotrauma* **2013**, *30*, 119–130. [CrossRef]
57. Yoon, Y.-S.; Cho, K.H.; Kim, E.-S.; Lee, M.-S.; Lee, K.J. Effect of Epidural Electrical Stimulation and Repetitive Transcranial Magnetic Stimulation in Rats With Diffuse Traumatic Brain Injury. *Ann. Rehabil. Med.* **2015**, *39*, 416–424. [CrossRef]
58. Yu, K.P.; Yoon, Y.-S.; Lee, J.G.; Oh, J.S.; Lee, J.-S.; Seog, T.; Lee, H.-Y. Effects of Electric Cortical Stimulation (ECS) and Transcranial Direct Current Stimulation (tDCS) on Rats With a Traumatic Brain Injury. *Ann. Rehabil. Med.* **2018**, *42*, 502–513. [CrossRef] [PubMed]
59. Sekar, S.; Zhang, Y.; Miranzadeh Mahabadi, H.; Parvizi, A.; Taghibiglou, C. Low-Field Magnetic Stimulation Restores Cognitive and Motor Functions in the Mouse Model of Repeated Traumatic Brain Injury: Role of Cellular Prion Protein. *J. Neurotrauma* **2019**, *36*, 3103–3114. [CrossRef]
60. Chiken, S.; Nambu, A. Mechanism of Deep Brain Stimulation: Inhibition, Excitation, or Disruption? *Neuroscientist* **2016**, *22*, 313–322. [CrossRef]
61. Xu, N.; LaGrow, T.J.; Anumba, N.; Lee, A.; Zhang, X.; Yousefi, B.; Bassil, Y.; Clavijo, G.P.; Khalilzad Sharghi, V.; Maltbie, E.; et al. Functional Connectivity of the Brain Across Rodents and Humans. *Front. Neurosci.* **2022**, *16*, 816331. [CrossRef]
62. Ma, X.; Aravind, A.; Pfister, B.J.; Chandra, N.; Haorah, J. Animal Models of Traumatic Brain Injury and Assessment of Injury Severity. *Mol. Neurobiol.* **2019**, *56*, 5332–5345. [CrossRef]
63. Vonder Haar, C.; Lam, F.C.W.; Adams, W.K.; Riparip, L.-K.; Kaur, S.; Muthukrishna, M.; Rosi, S.; Winstanley, C.A. Frontal Traumatic Brain Injury in Rats Causes Long-Lasting Impairments in Impulse Control That Are Differentially Sensitive to Pharmacotherapeutics and Associated with Chronic Neuroinflammation. *ACS Chem. Neurosci.* **2016**, *7*, 1531–1542. [CrossRef] [PubMed]
64. Francoeur, M.J.; Tang, T.; Fakhraei, L.; Wu, X.; Hulyalkar, S.; Cramer, J.; Buscher, N.; Ramanathan, D.R. Chronic, Multi-Site Recordings Supported by Two Low-Cost, Stationary Probe Designs Optimized to Capture Either Single Unit or Local Field Potential Activity in Behaving Rats. *Front. Psychiatry* **2021**, *12*, 678103. [CrossRef]
65. Fakhraei, L.; Francoeur, M.; Balasubramani, P.; Tang, T.; Hulyalkar, S.; Buscher, N.; Claros, C.; Terry, A.; Gupta, A.; Xiong, H.; et al. Mapping Large-Scale Networks Associated with Action, Behavioral Inhibition and Impulsivity. *eNeuro* **2021**, *8*, ENEURO.0406-20.2021. [CrossRef] [PubMed]
66. Fakhraei, L.; Francoeur, M.; Balasubramani, P.P.; Tang, T.; Hulyalkar, S.; Buscher, N.; Mishra, J.; Ramanathan, D.S. Electrophysiological Correlates of Rodent Default-Mode Network Suppression Revealed by Large-Scale Local Field Potential Recordings. *Cereb. Cortex Commun.* **2021**, *2*, tgab034. [CrossRef] [PubMed]
67. Paterno, R.; Folweiler, K.A.; Cohen, A.S. Pathophysiology and Treatment of Memory Dysfunction After Traumatic Brain Injury. *Curr. Neurol. Neurosci. Rep.* **2017**, *17*, 52. [CrossRef]
68. VanSolkema, M.; McCann, C.; Barker-Collo, S.; Foster, A. Attention and Communication Following TBI: Making the Connection through a Meta-Narrative Systematic Review. *Neuropsychol. Rev.* **2020**, *30*, 345–361. [CrossRef]
69. Kim, N.; Jamison, K.; Jaywant, A.; Garetti, J.; Blunt, E.; RoyChoudhury, A.; Butler, T.; Dams-O'Connor, K.; Khedr, S.; Chen, C.-C.; et al. Comparisons of Electrophysiological Markers of Impaired Executive Attention after Traumatic Brain Injury and in Healthy Aging. *NeuroImage* **2023**, *274*, 120126. [CrossRef]
70. De Simoni, S.; Jenkins, P.O.; Bourke, N.J.; Fleminger, J.J.; Hellyer, P.J.; Jolly, A.E.; Patel, M.C.; Cole, J.H.; Leech, R.; Sharp, D.J. Altered Caudate Connectivity Is Associated with Executive Dysfunction after Traumatic Brain Injury. *Brain* **2018**, *141*, 148–164. [CrossRef]
71. Schmidt, R.; Herrojo Ruiz, M.; Kilavik, B.E.; Lundqvist, M.; Starr, P.A.; Aron, A.R. Beta Oscillations in Working Memory, Executive Control of Movement and Thought, and Sensorimotor Function. *J. Neurosci.* **2019**, *39*, 8231–8238. [CrossRef]
72. Spitzer, B.; Haegens, S. Beyond the Status Quo: A Role for Beta Oscillations in Endogenous Content (Re)Activation. *eNeuro* **2017**, *4*, ENEURO.0170-17.2017. [CrossRef]

73. Dunkley, B.T.; Freeman, T.C.A.; Muthukumaraswamy, S.D.; Singh, K.D. Cortical Oscillatory Changes in Human Middle Temporal Cortex Underlying Smooth Pursuit Eye Movements. *Hum. Brain Mapp.* **2013**, *34*, 837–851. [CrossRef] [PubMed]
74. Little, S.; Bonaiuto, J.; Barnes, G.; Bestmann, S. Human Motor Cortical Beta Bursts Relate to Movement Planning and Response Errors. *PLOS Biol.* **2019**, *17*, e3000479. [CrossRef]
75. Hoy, C.W.; De Hemptinne, C.; Wang, S.S.; Harmer, C.J.; Apps, M.A.J.; Husain, M.; Starr, P.A.; Little, S. Beta and Theta Oscillations Track Effort and Previous Reward in Human Basal Ganglia and Prefrontal Cortex during Decision Making. *Proc. Natl. Acad. Sci. USA* **2024**, *121*, e2322869121. [CrossRef]
76. Hunt, A.W.; Mah, K.; Reed, N.; Engel, L.; Keightley, M. Oculomotor-Based Vision Assessment in Mild Traumatic Brain Injury: A Systematic Review. *J. Head Trauma Rehabil.* **2016**, *31*, 252–261. [CrossRef] [PubMed]
77. Koloski, M.F.; Hulyalkar, S.; Barnes, S.A.; Mishra, J.; Ramanathan, D.S. Cortico-Striatal Beta Oscillations as a Reward-Related Signal. *Cogn. Affect. Behav. Neurosci.* **2024**, *24*, 839–859. [CrossRef] [PubMed]
78. HajiHosseini, A.; Holroyd, C.B. Sensitivity of Frontal Beta Oscillations to Reward Valence but Not Probability. *Neurosci. Lett.* **2015**, *602*, 99–103. [CrossRef]
79. Marco-Pallarés, J.; Münte, T.F.; Rodríguez-Fornells, A. The Role of High-Frequency Oscillatory Activity in Reward Processing and Learning. *Neurosci. Biobehav. Rev.* **2015**, *49*, 1–7. [CrossRef]
80. Mas-Herrero, E.; Ripollés, P.; HajiHosseini, A.; Rodríguez-Fornells, A.; Marco-Pallarés, J. Beta Oscillations and Reward Processing: Coupling Oscillatory Activity and Hemodynamic Responses. *NeuroImage* **2015**, *119*, 13–19. [CrossRef]
81. D'Cruz, A.-M.; Ragozzino, M.E.; Mosconi, M.W.; Pavuluri, M.N.; Sweeney, J.A. Human Reversal Learning under Conditions of Certain versus Uncertain Outcomes. *NeuroImage* **2011**, *56*, 315–322. [CrossRef]
82. Koloski, M.F.; O'Hearn, C.M.; Frankot, M.; Giesler, L.P.; Ramanathan, D.S.; Vonder Haar, C. Behavioral Interventions Can Improve Brain Injury-Induced Deficits in Behavioral Flexibility and Impulsivity Linked to Impaired Reward-Feedback Beta Oscillations. *J. Neurotrauma* **2024**, *41*, e1721–e1737. [CrossRef]
83. Lee, D.J.; Gurkoff, G.G.; Izadi, A.; Seidl, S.E.; Echeverri, A.; Melnik, M.; Berman, R.F.; Ekstrom, A.D.; Muizelaar, J.P.; Lyeth, B.G.; et al. Septohippocampal Neuromodulation Improves Cognition after Traumatic Brain Injury. *J. Neurotrauma* **2015**, *32*, 1822–1832. [CrossRef] [PubMed]
84. Yaple, Z.; Martinez-Saito, M.; Novikov, N.; Altukhov, D.; Shestakova, A.; Klucharev, V. Power of Feedback-Induced Beta Oscillations Reflect Omission of Rewards: Evidence From an EEG Gambling Study. *Front. Neurosci.* **2018**, *12*, 776. [CrossRef]
85. Hyman, J.M.; Wyble, B.P.; Goyal, V.; Rossi, C.A.; Hasselmo, M.E. Stimulation in Hippocampal Region CA1 in Behaving Rats Yields Long-Term Potentiation When Delivered to the Peak of Theta and Long-Term Depression When Delivered to the Trough. *J. Neurosci.* **2003**, *23*, 11725–11731. [CrossRef] [PubMed]
86. Lee, D.J.; Gurkoff, G.G.; Izadi, A.; Berman, R.F.; Ekstrom, A.D.; Muizelaar, J.P.; Lyeth, B.G.; Shahlaie, K. Medial Septal Nucleus Theta Frequency Deep Brain Stimulation Improves Spatial Working Memory after Traumatic Brain Injury. *J. Neurotrauma* **2013**, *30*, 131–139. [CrossRef] [PubMed]
87. Lee, D.J.; Izadi, A.; Melnik, M.; Seidl, S.; Echeverri, A.; Shahlaie, K.; Gurkoff, G.G. Stimulation of the Medial Septum Improves Performance in Spatial Learning Following Pilocarpine-Induced Status Epilepticus. *Epilepsy Res.* **2017**, *130*, 53–63. [CrossRef]
88. Yang, X.; Li, X.; Yuan, Y.; Sun, T.; Yang, J.; Deng, B.; Yu, H.; Gao, A.; Guan, J. 40 Hz Blue LED Relieves the Gamma Oscillations Changes Caused by Traumatic Brain Injury in Rat. *Front. Neurol.* **2022**, *13*, 882991. [CrossRef]
89. Lundqvist, M.; Miller, E.K.; Nordmark, J.; Liljefors, J.; Herman, P. Beta: Bursts of Cognition. *Trends Cogn. Sci.* **2024**, *28*, 662–676. [CrossRef]
90. Zaninotto, A.L.; El-Hagrassy, M.M.; Green, J.R.; Babo, M.; Paglioni, V.M.; Benute, G.G.; Paiva, W.S. Transcranial Direct Current Stimulation (tDCS) Effects on Traumatic Brain Injury (TBI) Recovery: A Systematic Review. *Dement. Neuropsychol.* **2019**, *13*, 172–179. [CrossRef]
91. Leśniak, M.; Polanowska, K.; Seniów, J.; Członkowska, A. Effects of Repeated Anodal tDCS Coupled With Cognitive Training for Patients With Severe Traumatic Brain Injury: A Pilot Randomized Controlled Trial. *J. Head Trauma Rehabil.* **2014**, *29*, E20–E29. [CrossRef]

MDPI AG
Grosspeteranlage 5
4052 Basel
Switzerland
Tel.: +41 61 683 77 34

Brain Sciences Editorial Office
E-mail: brainsci@mdpi.com
www.mdpi.com/journal/brainsci

MDPI

www.ingramcontent.com/pod-product-compliance
Lightning Source LLC
Chambersburg PA
CBHW061556210926
51850CB00009B/2115